Radioactivity in the Environment
Volume 18

Tropical Radioecology

A companion series to the Journal of Environmental Radioactivity

Series Editor
M.S. Baxter
Ampfield House
Clachan Seil
Argyll, Scotland, UK

Volume 1: Plutonium in the Environment
(A. Kudo, Editor)
Volume 2: Interactions of Microorganisms with Radionuclides
(F.R. Livens and M. Keith-Roach, Editors)
Volume 3: Radioactive Fallout after Nuclear Explosions and Accidents
(Yu.A. Izrael, Author)
Volume 4: Modelling Radioactivity in the Environment
(E.M. Scott, Editor)
Volume 5: Sedimentary Processes: Quantification Using Radionuclides
(J. Carroll and I. Lerche, Authors)
Volume 6: Marine Radioactivity
(H.D. Livingston, Editor)
Volume 7: The Natural Radiation Environment VII
(J.P. Mclaughlin, S.E. Simopoulos and F. Steinhäusler, Editors)
Volume 8: International Conference on Isotopes and Environmental Studies
(P.P. Povinec and J.A. Sanchez-Cabeza, Editors)
Volume 9: Deep Geological Disposal of Radioactive Waste
(W.R. Alexander and L.E. Mckinley, Editors)
Volume 10: Radioactivity in the Terrestrial Environment
(G. Shaw, Editor)
Volume 11: Analysis of Environment Radionuclides
(P.P. Povinec, Editor)
Volume 12: Radioactive Aerosols
(C. Papastefanou, Author)
Volume 13: U-Th Series Nuclides in Aquatic Systems
(S. Krishnaswami and J. Kirk Cochran, Editors)
Volume 14: Remediation of Contaminated Environments
(G. Voigt and S. Fesenko, Editors)
Volume 15: Airborne Radioactive Contamination in Inhabited Areas
(K.G. Andersson, Editor)
Volume 16: Environmental Radionuclides: Tracers and Timers of Terrestrial Processes
(K. Froehlich, Editor)
Volume 17: Technologically Enhanced Natural Radiation
(A.S. Paschoa and F. Steinhäusler, Authors)

Radioactivity in the Environment
Volume 18

Tropical Radioecology

John R. Twining
Austral Radioecology, Oyster Bay, NSW, Australia

ELSEVIER

AMSTERDAM • BOSTON • HEIDELBERG • LONDON • NEW YORK • OXFORD
PARIS • SAN DIEGO • SAN FRANCISCO • SINGAPORE • SYDNEY • TOKYO

Elsevier
The Boulevard, Langford Lane, Kidlington, Oxford OX51 GB, UK
Radarweg 29, PO Box 211, 1000 AE Amsterdam, The Netherlands
30 Corporate Drive, Suite 400, Burlington, MA 01803, USA
525 B Street, Suite 1900, San Diego, CA 92101-4495, USA

First edition **2012**

British Library Cataloguing in Publication Data
A catalogue record for this book is available from the British Library

Library of Congress Cataloging-in-Publication Data
A catalog record for this book is available from the Library of Congress

For information on all **Elsevier** Publications
visit our Web site store.elsevier.com

This book has been manufactured using Print On Demand technology. Each copy is produced to order and is limited to black ink. The online version of this book will show colour figures where appropriate.

ISBN: 978-0-08-097510-8

ISSN: 1569-4860

Transferred to Digital Printing in 2016

Contents

COMPANION WEB SITE
http://www.elsevierdirect.com/v2/companion.jsp?ISBN=9780080450162

Foreword

This book is intended to help fill a significant gap in the published scientific literature, one caused by the fact that military and peaceful nuclear developments and their associated research activities began, and thereafter continued to focus, in developed countries within the temperate climatic zone of the Northern Hemisphere. The result has been that the great majority of all published knowledge on the distribution, speciation, behaviour, transfer and dosimetry of radionuclides in the environment relates to temperate conditions. Recently, however, the construction of nuclear reactors in tropical and sub-tropical countries has flourished, with many more units planned for the future. Relative to the temperate zone, extremely different climatic conditions, terrains, rates of weathering, soil types, biological productivity, biodiversity, food production systems, dietary and cooking habits, industrial activities and so forth apply in tropical regions, so there were clear needs to review and collate into one volume what relatively little is known and published on tropical radioecology and perhaps to identify the priority areas requiring further study. It was therefore a great pleasure when John Twining accepted an invitation to produce and edit this book. Not only is John one of relatively few scientists who have published extensively and authoritatively on key aspects of the subject, but he has proved to be an extremely professional editor who has subjected the chapters that follow to independent peer review and has led a fine team of chapter authors to produce what I believe is a high-quality addition to the literature in a much needed subject area.

This volume begins with an essential review of some basics, that is, on the nature of radioactivity, its interactions with matter and with biological systems, its natural and anthropogenic inputs to the environment, its methods of measurement, its health consequences, and its environmental tracer applications. Tropical radioecology is immediately revealed as a modern multidisciplinary subject, with its origins solidly based in a range of fundamental sciences. Throughout this volume, lists of the available data and data links are provided so that the reader can see exactly what is known, and perhaps establish how his/her own observations fit in, or identify gaps in the databases. The volume then guides the reader sequentially through the physical, chemical, and biological factors that influence radioecology, both with regard to natural and man-made radionuclides, and the modelling tools that are available to integrate those factors, in a range of tropical systems. Primary amongst the unique characteristics of tropical systems is the warm monsoonal climate, leading to extremes of wash-out, interception, weathering, wash-down, soil penetration, leaching and runoff. Fundamental differences (from temperate conditions) also apply to terrain and climate type as the tropical environment ranges from desert to rainforest to mountains to coral islands. Soils, in general, tend to be very low in organic matter, nutrients, and exchange capacity, while those based on carbonates show quite different radionuclide cycling and retention properties compared to temperate soils. Many tropical and subtropical countries are developing or emerging, with high and rapidly growing population densities and exponentially increasing industrial activity and associated major changes in land use, such as encroaching urbanisation, conversion of forest to pasture, and so forth. Around 40% of the world's human population currently lives in the tropical zone, increasing to 60% in the next five decades. About two-thirds of the world's agricultural land lies within the tropics. Therefore, this is a key developing zone featuring an expanding nuclear industry and a huge discrepancy in environmental conditions and radionuclide transfer parameters from those extensively published for temperate latitudes. Different food-production systems (e.g., paddy fields) apply, flooding is common, and these processes can induce relatively poorly quantified redox-related recycling of radionuclides. The main tropical food items are of course quite different from those in temperate zones and radionuclide transfer factors need to be further studied. Radiological models need to take account of all of the above differences, whilst also extending the past anthropocentric approach to risk assessment in order to include a more ecologically aware perspective. This will involve a shift toward considering not only the biotas and ecosystems on which humans depend but also the preservation of biodiversity *per se*, irrespective of any perceived economic or social benefits to humans. For this to be achieved, many more data, particularly on the effects of chronic low dose, are required. This is all the more pertinent for tropical systems given the relative lack of such data in these environments, particularly so given that the current data for temperate species are found to be only poorly capable of extrapolation to tropical and sub-tropical species and environments. In any case, where critical environmental assessments are required, site- and species-specific data will be preferred.

In conclusion, this volume is published at the end of an extensive book series on radioactivity in the environment and, as series editor, I want to thank all past contributors and readers. The series has paralleled the *Journal of Environmental Radioactivity*, which I founded with Elsevier back in the early 1980s. This has been a most enjoyable collaborative editorial experience and I do hope that the legacy will live on and be extended, updated and improved in the future. In the meantime, may this volume prove to be as informative and stimulating as I think it is—and many thanks and congratulations again to John Twining and his team of co-authors, for I believe that this book provides an excellent information source and a basis both for future scientific investigations and for appropriate environmental management of tropical and sub-tropical environments.

Murdoch Baxter
Series Editor

Contributors

Numbers in Parentheses indicate the pages on which the author's contributions begin.

Peter Airey (1), Institute for Environmental Research, Australian Nuclear Science and Technology Organisation (retired)

Jagoda Crawford (281), Institute for Environmental Research, Australian Nuclear Science and Technology Organisation, Lucas Heights, NSW, Australia

Robert Edis (93), Melbourne School of Land and Environment, University of Melbourne, Victoria, Australia

Thomas Hinton (1), French Institute for Radiological Protection and Nuclear Safety (IRSN), Bat. 159-BP3, 13115 Saint-Paul-Lez-Durance, France

Scott J. Markich (231), Aquatic Solutions International, Australia

Paul Martin (59), Australian Radiation Protection and Nuclear Safety Agency, Yallambie, Victoria, Australia

John L. McBride (59), School of Earth Sciences, University of Melbourne, Parkville, Australia

Timothy E. Payne (93), Australian Nuclear Science and Technology Organisation, Kirrawee, New South Wales, Australia

Gerhard Pröhl (281), International Atomic Energy Agency, Vienna, Austria

Ron Szymczak (121), Nuclear and Oceanographic Consultant, Cronulla, New South Wales, Australia

Keiko Tagami (155), Radiation Safety Research Center, National Institute of Radiological Sciences, Chiba, Japan

John R. Twining (1, 155, 231, 281), Austral Radioecology, Oyster Bay, New South Wales, Australia

M. Angelica V. Wasserman (155), Instituto de Engenharia Nuclear, Rio de Janeiro, Brasil

Chapter 1

The Scientific Basis

Peter Airey[1], Thomas Hinton[2], and John Twining[3]
[1] *Institute for Environmental Research, Australian Nuclear Science and Technology Organisation (retired)*
[2] *French Institute for Radiological Protection and Nuclear Safety (IRSN), Bat. 159 -BP3, 13115 Saint-Paul-Lez-Durance, France*
[3] *Austral Radioecology, Oyster Bay, New South Wales, Australia*

1.1. INTRODUCTION

There is a strong expectation that nuclear energy will play a substantial role in the social and economic development of human societies over the next few decades. That is particularly so given the current concerns regarding climate change; limitations on the availability of and access to fossil fuels; and the knowledge that well-managed nuclear power is one of the cleanest, cheapest, and most reliable means of providing base-load electricity. The anticipated developments will be more pronounced in tropical regions of the planet with the current population levels and predicted growth. Given those assumptions, it is also pertinent to ensure that the possible consequences of a nuclear future are well considered and that the affected populations are well placed to make use of the many advantages that nuclear science has to offer. Hence, knowledge of the nature of radiation in the environment, as well as of the behaviour of any radioactivity released into it, is essential to best manage such development. Tropical radioecology encompasses those needs.

At the outset, it is important to have an understanding of the basic elements of a subject before embarking on a journey through the topic. Radioecology is a science that: (1) endeavours to identify and understand the multitude of mechanisms that control the behaviour and impact of radioactive materials within the biosphere and (2) uses the unique characteristics of radionuclides to trace ecological processes. Radioecology is an integrated science formed from components of nuclear physics, chemistry, biology, ecology, physiology, toxicology, and risk assessment. Given that very few universities have courses specific to this science, many of the researchers coming into the field of radioecology are well trained in one aspect of the science, but not in all of the integrated components. As a consequence, students of radioecology

Radioactivity in the Environment, Vol. 18. http://dx.doi.org/10.1016/B978-0-08-045016-2.00001-1

inevitably need to enhance their knowledge from fields in which they are not experts. This volume will provide the bases required to address those needs.

Common to all radioecologists is the need for a sound knowledge of radioactivity and its effects. To this end, the introductory chapter will open with a discussion of the nature of radioactivity per se, relating it to atomic structure and stability. Following that, this chapter will cover basic principles of detecting and measuring radioactivity in environmental samples, including an introduction to counting statistics and radiation protection. An outline of the mechanisms and effects of radiation on biota at different levels of organisational structure will also be provided. This chapter will conclude with an outline of the sources and distribution of radionuclides found in the environment, particularly as they pertain to tropical and subtropical regions of our planet.

A number of authoritative Web sites maintained by national laboratories offer a general introduction to nuclear science (e.g., http://www.lbl.gov/abc/Basic.html). This chapter focuses on those aspects of the science underpinning the studies of tropical radioecology described in all the other chapters of this volume.

1.2. RADIOACTIVITY

In brief, the cosmos comprises approximately 4.6% atoms; 23% cold, dark matter; and 72% dark energy (http://map.gsfc.nasa.gov/news/5yr_release.html; Freedman and Kolb, 2006). Of the visible universe, approximately 73% of the mass is hydrogen, 25% is helium, and only 2% is comprised of the remaining 115 or so chemical elements. Radioactivity as we know it is a feature of the visible universe. Far less is known about dark matter and dark energy. Definitions and concepts underpinning the science of radioactivity are discussed in the following sections.

1.2.1 Introduction to Radioactivity

Radioactivity is a natural phenomenon. It occurs when atomic nuclei in higher energy states achieve greater stability by emitting energy in the form of radiation. The amount of energy and the forms of radiation emitted vary tremendously among the radioactive elements. It is because of this variation that the applications of radioactive materials range from powerful tracers of biological, physiological, and geological cycles to healing medicine to weapons of mass destruction. In this introductory chapter, you will learn about the fundamental characteristics of radiation from which such variation emerges. You will learn about the types of radiation, decay rates, half-lives, natural versus anthropogenic radionuclides, how to detect radiation, and what happens following exposure to radiation. Understanding such fundamentals is required to confidently evaluate the human and environmental risks from radiological exposures.

The famous physicist, Ernest Rutherford, received a Nobel Prize for investigations into the disintegration of the elements and the chemistry of radioactive substances. In 1910, Rutherford bombarded a sheet of gold foil with ionized helium atoms (i.e., alpha particles; see Section 1.2.2.2). To his total astonishment, a small percentage of the alpha particles bounced back. Rutherford is quoted as saying that it was as if he had fired a fifteen inch shell at a sheet of paper and it rebounded into his lap (Bryson, 2003). After considerable contemplation, Rutherford came to the conclusion that atoms were largely empty space, with something small and incredibly dense in the centre. Indeed, an atom consists of a positively charged nucleus surrounded by a cloud of negatively charged electrons. The nucleus is typically less than one part in 10,000 of the diameter of the atom but accounts for over 99.9% of the mass. Thus, the solidity that we perceive as encompassing our world is really an illusion. All things in our world are really dominated by empty space. The nucleus comprises an assemblage of nucleons (protons and neutrons) bound together by nucleon–nucleon, or residual, strong nuclear forces. Properties of some elementary particles are listed in Table 1.1.

1.2.1.1 Isotopes

The masses of protons and neutrons are both close to 1 atomic mass unit (u). Hence, to a first approximation, the mass of an atom is the sum of the number of protons (the atomic number, Z) and the number of neutrons, N (i.e., the mass number, $A = Z + N$). Although the chemical behaviour of each atom is uniquely identified by its atomic number (Z), the neutron number (N), and hence the mass number, can vary within limits. Atoms with the same atomic number, but with different mass numbers, are known as isotopes. The term isotope is derived from Greek meaning the same place (in the Periodic Table; see Section 1.2.1.2) and is often (but erroneously) used interchangeably with the word nuclide. For example, the element carbon has an atomic number (Z) of 6 and three well-known isotopes, ^{12}C, ^{13}C, and ^{14}C, that vary in mass due to the different number of neutrons in its nucleus (neutron numbers, N, are 6, 7, and 8, respectively).

Isotopes that are unstable and emit radiation (such as ^{14}C) are said to be *radioactive* and are often referred to as *radioisotopes*. Comprehensive catalogues comprise over 3200 nuclides from about 117 elements, of which 292 are stable (Firestone and Shirley, 1998; http://ie.lbl.gov/toi/). Their essential properties may be presented as a Chart of Nuclides (http://www.nndc.bnl.gov/chart/) in which the neutron number is presented on the x-axis and the atomic number on the y-axis.

1.2.1.2 The Periodic Table

The periodic table is an organisation of elements arranged in increasing atomic number into a series of columns (groups) and rows (periods). The

TABLE 1.1 Selected Subatomic Particles and Radioactive Decay Products

Entity	Symbol	Mass (MeV)[a]	Charge ($1.6\ 10^{-19}$ C)	Comments
Neutrino	ν	0	0	Neutrino emission accompanies beta (negatron) decay; anti-neutrino emission accompanies both electron capture and positron emission processes.
Neutron	n	938.6	0	The free neutron is unstable and decays with a half-life of 613 s emitting a proton, an electron, and an (electron) antineutrino. n $0 \rightarrow p + e^{-} + \nu_e$
X-rays and γ rays	X,γ	0	0	Electromagnetic radiation emitted in many radioactive decay processes has vanishingly small inertial mass.[b]
Electron	e^{-}	0.51	1 unit −ve	Using conventional terminology, beta particles (negatrons) and positrons are emitted from the nucleus of the atom, while electrons are derived from the outer shell of the atom.
Beta particle or negatron	β^{-}	0.51	1 unit −ve	
Positron	β^{+}	0.51	1 unit +ve	
Proton	p	938.3	1 unit +ve	
Alpha particle	α	3730	2 units +ve	Alpha particles are the nuclei of the helium atom, each comprised of two protons and two neutrons.

By definition, an atomic mass unit is 1/12 of the mass of ^{12}C at rest and in its ground state (i.e., $1.6605\ 10^{-27}$ kg or 931.5 MeV).
[a]*The Einstein mass energy conversion equation $m = E/c^2$ is the basis for converting mass (kg) to other measures of energy in consistent SI units. The SI unit of energy is the joule, which is equivalent to the work required to move an electric charge of 1 coulomb through a potential difference of 1 volt. In atomic or nuclear science, energies are frequently expressed as electron volts (eV), which is the work required to move an electron through a potential difference of 1 volt. Since the electron charge is $1.602\ 10^{-19}$ C and the speed of light c is $3.00\ 10^{8}\ m\ s^{-1}$, 1 eV is equivalent to $1.602\ 10^{-19}/(3.00\ 10^{8})^2$ or $1.783\ 10^{-36}$ kg.*
[b]*Photon energy may be expressed as a mass through the Einstein equation $E = h\nu = mc^2$, where E is the photon energy, h is Plank's constant, ν is the photon frequency, and c is the velocity of light.*

classification was chosen to emphasize the periodic nature of the chemical properties (http://www.webelements.com/; Figure 1.1). The observed periodicity is a direct consequence of atomic structure and is understood in terms of the arrangement of electrons in the outer valence shells of the atoms. The invention of the table is generally attributed to Dimitri Mendeleev in 1869 and is further discussed in Chapters 4 and 6.

Group → ↓ Period	1	2		3	4	5	6	7	8	9	10	11	12	13	14	15	16	17	18
1	1 H																		2 He
2	3 Li	4 Be												5 B	6 C	7 N	8 O	9 F	10 Ne
3	11 Na	12 Mg												13 Aj	14 Si	15 P	16 S	17 Cl	18 Ar
4	19 K	20 Ca		21 Sc	22 Ti	23 V	24 Cr	25 Mn	26 Fe	27 Co	28 Ni	29 Cu	30 Zn	31 Ga	32 Ge	33 As	34 Se	35 Br	36 Kr
5	37 Rb	38 Sr		39 Y	40 Zr	41 Nb	42 Mo	43 Tc	44 Ru	45 Rh	46 Pd	47 Ag	48 Cd	49 In	50 Sn	51 Sb	52 Te	53 I	54 Xe
6	55 Cs	56 Ba	*	71 Lu	72 Hf	73 Ta	74 W	75 Re	76 Os	77 Ir	78 Pt	79 Au	80 Hg	81 Tl	82 Pb	83 Bi	84 Po	85 Al	86 Rn
7	87 Fr	88 Ra	**	103 Lr	104 Rf	105 Db	106 Sg	107 Bh	108 Hs	109 Mt	110 Ds	111 Rg	112 Uub	113 Uut	114 Uuq	115 Uup	116 Uuh	117 Uus	118 Uuo
* Lanthanides			*	57 La	58 Ce	59 Pr	60 Nd	61 Pm	62 Sm	63 Eu	64 Gd	65 Tb	66 Dy	67 Ho	68 Er	69 Tm	70 Yb		
** Actinides			**	89 Ac	90 Th	91 Pa	92 U	93 Np	94 Pu	95 Am	96 Cm	97 Bk	98 Cf	99 Es	100 Fm	101 Md	102 No		

FIGURE 1.1 Periodic table of the elements. *Design adapted from http://www.webelements.com.*

1.2.1.3 Radioactive Decay

Radioactive decay involves the transition of a nucleus from a less to a more stable energy state. The protons and neutrons within the nucleus are held together by the so-called residual strong nuclear force. The magnitude of this force is indicated by the *binding energy*, which is the energy required to disassemble the nucleus into its constituent protons and neutrons.

The binding energy for a single nucleus is given by:

$$\Delta E = \Delta m.c^2 = (m_s - m_b)c^2 = (Z.m_p + N.m_n - m_b)c^2 \quad (1.1)$$

where c is the speed of light, m_s is the mass of the separated nucleons, m_b is the mass of the bound nucleus, Z is the atomic number of the bound nucleus, m_p is the mass of one proton, N is the number of neutrons, and m_n is the mass of one neutron.

Generally speaking, the average binding energy per nucleon increases with mass number to a maximum for isotopes of iron and nickel, and then gradually decreases (Figure 1.2). Values of binding energies are included in the comprehensive table of atomic masses compiled by Audi and Wapstra (1993) and in a searchable database (http://ie.lbl.gov/toi2003/MassSearch.asp).

1.2.1.4 Units of Energy

Einstein's famous equation shows us that energy can be expressed in units of mass and vice versa. In nuclear and radiation science, energy is normally expressed as changes in atomic mass units (u) or as electron volts (eV).

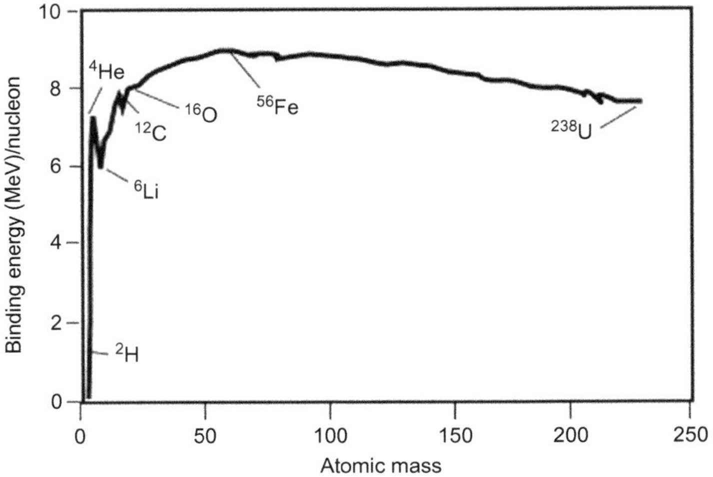

FIGURE 1.2 Binding energies.The variation with atomic mass of the binding energy per nucleon. *Plotted from http://ie.lbl.gov/toi2003/MassSearch.asp.*

1 eV is equivalent to 1.783 10^{-36} kg. Since an atomic mass unit is 1.6605 10^{-27} kg, then[1]:

$$1\,\text{atomic mass unit (u)} = 1.6605\,10^{-27}/1.783\,10^{-36}\text{eV or } 931\,\text{MeV}$$

The energy released during radioactive decay is measurable and can reach several million electron volts (MeV). Radiation in the form of alpha particles is often in the MeV range (e.g., plutonium-239 emits an alpha particle with an energy of 5.2 MeV); whereas gamma emissions are often less energetic (e.g., ^{137}Cs emits gamma radiation with energy of 662 keV). This is in stark contrast to chemical energies that are of the order of eV.

1.2.2 Radioactive Decay

Radioactive decay is accompanied by the emission of high-energy radiation (Table 1.1) and is associated with the transition of the nucleus from a higher to a lower energy state. It is constrained by the conservation of mass/energy, of charge, and of other parameters associated with nuclear structure and occurs at a rate described by a decay constant (λ; Section 1.2.3.1). The decay constant is a property of the nucleus and is independent of its surroundings and of cosmological time.

1. International System of Units (National Institute of Standards and Technology; NIST) quotes 931.494 10^6 eV. Readers are referred to the interactive Web site maintained by NIST: http://physics.nist.gov/cuu/Constants/energy.html.

The process of radioactive decay transforms one element into another. The resulting product loses all the characteristics of the parent element and acquires the characteristics of the newly formed daughter element. Characteristics such as colour, melting point, hardness, and even physical state change with each transformation. For example, within the U-decay series, radium, a chemically reactive solid with a half-life of more than 1600 y, is transformed by radioactive decay into radon, an inert gas with a half-life of just under 4 days. Radioactive decay is nature's alchemist (Hinton, 1998).

1.2.2.1 Radioactive Decay Processes

Decay emissions are normally classified as primary or secondary radiation. The emission of primary radiation leads to the formation of one or more daughter products that are usually also in an excited state. Subsequent deactivation of the daughter is accompanied by the formation of secondary gamma rays. For example, the nuclide $^{137}{}_{55}Cs$ (half-life 30.01 y) emits a β^- particle and decays through the isomer $^{137m}{}_{56}Ba$ (half-life 2.55 m) to the ground state with the emission of a 662 keV gamma. A Web-based table of isotopes has decay schemes for all known nuclides linked to a periodic chart of the elements at http://ie.lbl.gov/toi/perchart.htm.

Three classes of radioactive processes will be discussed: (1) Beta (β^-) decay and electron capture (of which there are 17,015 examples reported in the Chart of Nuclides), (2) alpha (α) decay (2076 examples), and (3) gamma ray (γ) transitions (223,791 cases). In each case there is a conservation of mass/energy, of charge, and of other properties of the subatomic particles.

1.2.2.1.1 Beta Particle Emission and Electron Capture

1.2.2.1.1.1 Beta Particle Emission In general, the emission of a β^- particle (and an electron antineutrino) leads to the conversion of a neutron to a proton within the nucleus. A free neutron can also decay by that process with a half-life of 615 s:

$$n \rightarrow p + \beta^- + \nu_e \,(\text{electron antineutrino}) \tag{1.2}$$

Hence, in the general case, the atomic number (Z) is increased and the neutron number (N) decreased by one, with no change in the mass number, ($A = N + Z$). Beta emission is characteristic of neutron-rich isotopes (i.e., nuclear stability is increased by decreasing the neutron to proton ratio in the nucleus). A well-known example is the beta decay of ^{14}C to ^{14}N (half-life 5730 y):

$$^{14}C \rightarrow {}^{14}N + \beta^- + \nu_e E_{\beta,\max} = 156\,keV \tag{1.3}$$

The rest masses of the ^{14}C and the ^{14}N nuclei are 14.003242 and 14.003074 u, respectively. The decrease in the rest masses is 0.0000168 u (or 156 keV). This is the total energy released during the radioactive decay process, and is known as the Q_β value. The energy is distributed between the beta particle

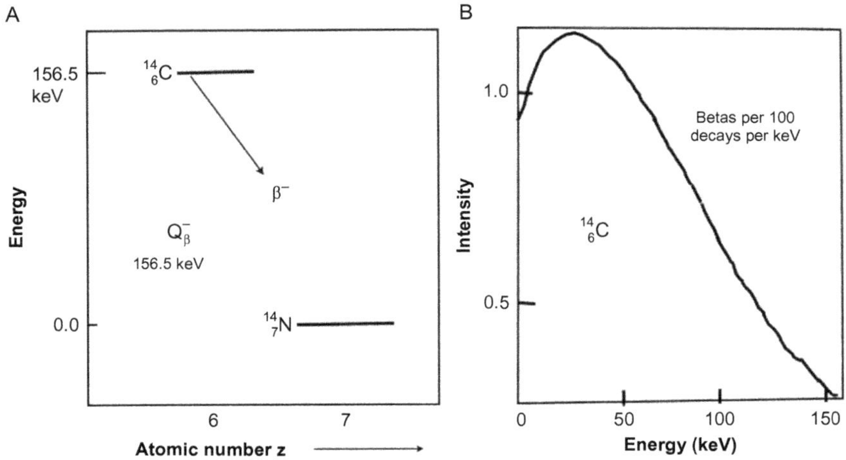

FIGURE 1.3 ^{14}C decay. (A) The ^{14}C decay scheme and energy spectrum. Q_β is the energy available for decay by the emission of the beta particle (156 keV). (B) The energy spectrum of the emitted beta particles (in units of number of betas per 100 decays per keV. *From http://ie.lbl.gov/decay/betas/BM60014.htm, with permission.*

and the electron antineutrino. The maximum in the β^- energy $E_{\beta,\ \max}$ corresponds to zero neutrino energy.

This information is encapsulated in radioactive decay schemes. The ^{14}C decay is a particularly simple example as shown in Figure 1.3A. The energy distribution of the emitted beta is depicted in Figure 1.3B.

1.2.2.1.1.2 Positron Emission The positron (β^+) is the antiparticle of the electron (β^- particle). Positron emission leads to the conversion of a proton within the nucleus to a neutron (i.e., a unit decrease in the Z value with no change in the mass number, A). Positron emission generally competes with the electron capture process (ε) in which an orbital electron from the atom is captured by the nucleus. The two processes have the same effect on the nucleus and are frequently written $\varepsilon + \beta^+$. An example is the decay of ^{22}Na in which 90% of the decay is by positron emission:

$$^{22}\text{Na} \rightarrow {}^{22}\text{Ne} + \beta^+ + \nu_e(\text{electron neutrino}) \tag{1.4}$$

$$^{22}\text{Na} + \varepsilon \rightarrow {}^{22}\text{Ne} + \nu_e \tag{1.5}$$

The conversion of the proton (mass 1.007825 u) to the neutron (mass 1.008665 u) leads to an increase in the mass of the nucleons in the daughter ^{22}Ne nucleus, and this must be more than compensated for in the change in the binding energies. The electron capture decay energy Q (EC) is the mass difference $m(^{22}\text{Na}) - m(^{22}\text{Ne})$ or 2.842 MeV.

1.2.2.2 *Alpha Emission*

An alpha particle is the helium nucleus $^{4}He^{++}$. The emission of an alpha particle reduces the atomic number of the parent nucleus by 2, the neutron number by 2, and, consequently, the mass number by 4. An example is the decay of ^{238}U to ^{234}Th:

$$^{238}U \rightarrow {}^{234}Th + {}^{4}He^{++}(\alpha) \tag{1.6}$$

Unlike beta particles, emitted alpha particles have well-defined energies. In the case of ^{238}U, they are 4.198 MeV (79.0%) and 4.151 MeV (20.9%). The energies of alpha particles from most nuclides of interest in radioecological studies fall in the range of 4–6 MeV. Further information on the uranium series decay chain is presented in Section 1.2.3.2 and Figure 1.4.

1.2.2.3 *Gamma Emission*

1.2.2.3.1 Excited States

Most nuclei are associated with a range of excited states with energies above the ground state. The primary radioactive decay process normally accesses one or more of the excited states of the daughter product. Subsequent de-excitation is accompanied by the emission of gamma rays. These processes are normally

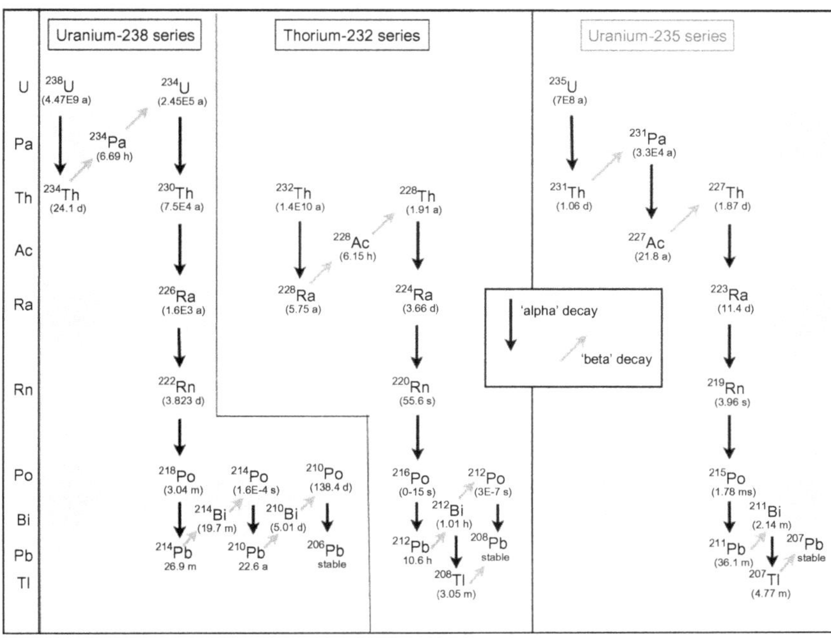

FIGURE 1.4 Uranium and Thorium series decay. The decay chain series for ^{238}U, ^{232}Th, and ^{235}U. The decay modes and the half-lives of the principal elements in the series are shown (Swarzenski, 2009).

very rapid with the lifetimes of the excited states measured in picoseconds. However, quantum mechanical (spin) effects may inhibit the de-excitation leading to excited states with a lifetime of the order of a nanosecond or greater. These are referred to as nuclear isomers or meta-stable states and are designated with an 'm.' Perhaps the best known example is technetium 99 m, the meta-stable state of ^{99}Tc, which is widely used in nuclear medicine. ^{99m}Tc has a half-life of 6.0 hours and decays with the emission of a 140 keV gamma ray to ^{99}Tc.

X-rays, like gamma rays, are photons, but are emitted following the excitation of the outer electrons of atoms. A radioactive decay process is normally accompanied by the emission of a range of X-rays, for example, ignoring intensities below 0.001%, ^{137}Cs emits 2 beta particles, 1 gamma ray, and 19 barium X-rays. The X-rays accompany the de-excitation of the outer electron shells of the daughter nuclide barium-137.

1.2.3 Rate of Radioactive Decay

1.2.3.1 Half-lives

The rate of radioactive decay is determined by the radionuclide's half-life, $t_{1/2}$, defined as the time required for 50% of the atoms of a radionuclide to decay. Half-lives range over about 40 orders of magnitude (Figure 1.5;

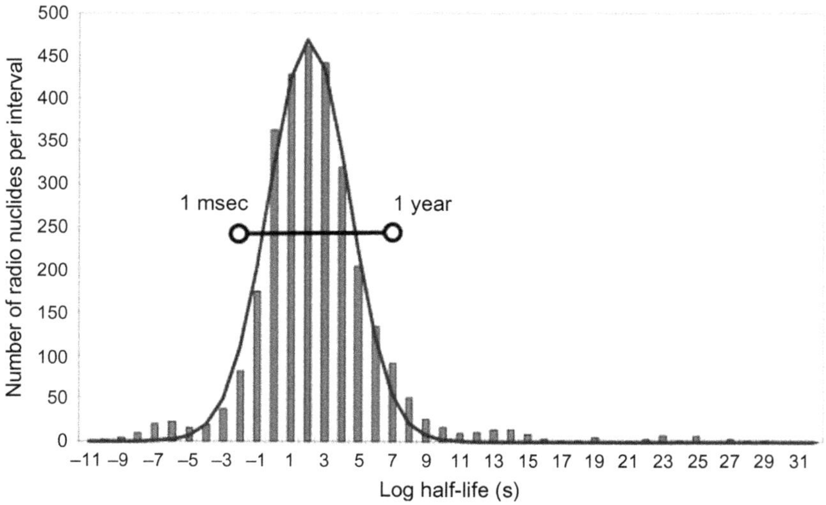

FIGURE 1.5 Half-life distribution. Frequency distribution of the half-lives of 3015 nuclides. The nuclides are listed in the Table of Isotopes (http:ie.lbl.gov/toi/nucSearch.asp) and include meta-stable states listed in this database. The time scales are expressed in a logarithmic scale and range over 42 orders of magnitude. The majority of half-lives are within the range of 1 ms to 100 y. The best-fit normal distribution is shown.

Airey and Howard, 2003). They are independent of the environment and have not varied over cosmological time.

1.2.3.1.1 Decay Rates

The rate of decay (λ) is proportional to the number of radioactive atoms (N):

$$-dN/dt = \lambda N \tag{1.7}$$

From Equation 1.8:

$$N = N_0 \exp^{-\lambda t} \tag{1.8}$$

where N_0 is the number of radioactive atoms at time t_0. From Equation 1.9:

$$N/N_0 = 0.5 \times \exp^{-\lambda t1/2} \text{and the half-life}, t_{1/2} = 0.693/\lambda \tag{1.9}$$

A characteristic half-life exists for each radionuclide, and this provides an excellent tool for radioecologists to evaluate environmental processes. A well-known example of radioactive decay is the use of ^{14}C to determine the age of artefacts and geohydrological samples over the past 40,000 years.

1.2.3.2 Decay Chains

Many radionuclides undergo a simple decay process to a stable daughter product. Examples include the decay of ^{14}C to ^{14}N (Eq. 1.3) and of tritium ^{3}H to ^{3}He as shown in the following:

$$^{3}H \rightarrow {}^{3}He + \beta^{-} + \nu_e(\text{electron anti neutrino}) \quad E_{\beta,\max} = 18.6\,\text{keV} \tag{1.10}$$

The rate of accumulation of the stable product may be used as a measure of the cumulative decay of the parent over a defined time period. This has a number of practical applications including:

- Measuring environmental levels of tritium from the precise measurement of the accumulation of ^{3}He using mass spectrometry (Eq. 1.10). For very low levels of tritium, mass spectrometry is more accurate than radioactive counting.
- Dating geological samples based on the potassium/argon and uranium/lead methods. A comprehensive survey of radioactive dating methods is presented in Geyh and Sleicher (1990).

In the previous examples, the radioactive parent decays to a stable daughter product. More commonly, the daughter product is also radioactive and decays through one or more steps to a stable end product. The resulting decay chains vary enormously in complexity.

An important example of a simple decay chain is the decay of ^{99}Mo ($t_{1/2}$ 66 hours) through ^{99m}Tc ($t_{1/2}$ 6.0 hours) and ^{99}Tc ($t_{1/2}$ 2.1 10^{5} y) to the stable ^{99}Ru. It is the basis of the molybdenum/technetium generator widely used in nuclear medicine and in environmental tracer applications as a source of the gamma-emitting ^{99m}Tc. The ^{99}Mo is strongly adsorbed onto alumina. With the passage of time (e.g., 5 half-lives or 30 hours), the daughter product ^{99m}Tc builds up. The technetium is not adsorbed strongly, and may be eluted from the alumina with a saline solution uncontaminated with ^{99}Mo.

The decay chains of uranium and thorium are more complex.

1.2.3.2.1 Uranium and Thorium Series Decay

Reference was made in Section 1.2.2.2 to the decay of ^{238}U to ^{234}Th by alpha decay. ^{234}Th is also radioactive and decays by β^- emission to ^{234}Np and then, through a chain of β^- or alpha emissions, to stable ^{206}Pb. This is the so-called ^{238}U decay chain and, leaving aside minor branches, involves 14 decay steps (Figure 1.4). The ^{235}U and the ^{232}Th chains are also illustrated in this figure. These chains are important in many radioecological studies and dose assessments (Chapter 7). However, it is vitally important to understand that the physical process of radioactive decay takes time to occur and happens in a complex environment. Combining the well-understood temporal process of decay with dynamic chemical kinetics can help radioecologists and others relying on radioactive decay to interpret a wide range of environmental processes.

1.2.3.2.2 Secular Equilibrium

Many daughter products are radioactive and their observed concentrations are a balance between their rate of formation from their parent radionuclide and their own rate of decay. Secular equilibrium is achieved in a closed system when the rate of formation matches the rate of decay. An example is the buildup of ^{222}Rn from the decay of ^{226}Ra (Figure 1.6), which is the basis for the measurement of radium in environmental waters. ^{226}Ra (half-life 1601 y) is a member of the ^{238}U chain and decays through alpha emission to the gas ^{222}Rn (half-life 3.82 days). After the lapse of sufficient time (e.g., 10 half-lives of ^{222}Rn, or 38 days), the radon reaches a steady concentration where its decay rate matches that of the parent ^{226}Ra. The parent and daughter are said to be in secular equilibrium.

To use the radon isotope to measure the level of radium in natural waters, it is necessary to isolate the sample for a sufficient time to establish secular equilibrium (e.g., 38 days) under conditions that prevent escape of the radon. ^{222}Rn, and hence the parent radium, is measured from the gamma emissions of the short-lived progeny (including ^{214}Pb). Secondary standards are used to convert the observed count rates to absolute concentrations.

General equations describing the buildup of radioactive daughters in more complex chains have been developed, but are beyond the scope of this discussion.

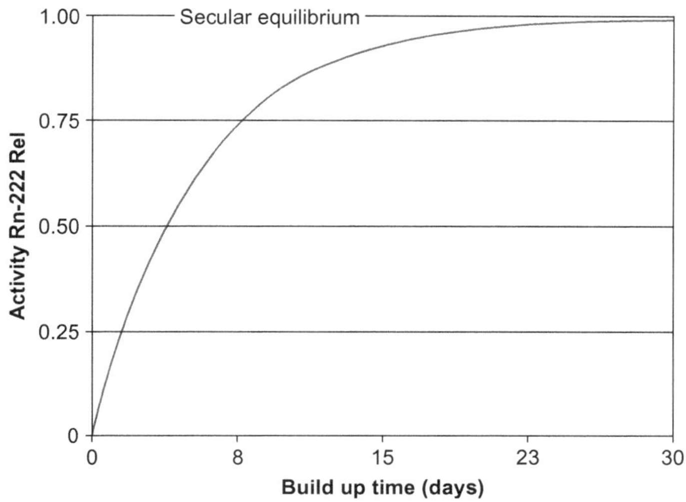

FIGURE 1.6 Buildup of ^{222}Rn from the decay of ^{226}Ra. (See Figure 1.4.)

1.2.3.2.3 Uranium Series Disequilibria and their Applications

Radioactive decay processes lead to the sequential appearance of a range of different elements within any decay chain. The chemical and physical differences between these nuclides, together with their different behaviours in biotic systems, means that secular equilibrium is very hard to attain in most environmental situations where the half-life of the daughter is long compared to the environmental processes that it may be experiencing.

In some cases the stability of the system can be quite good and long-term processes can be evaluated using the relative abundance of sequential members of the decay chain with long half-lives. Examples include sedimentary rocks and other geological archives, such as stalagmites. However, in many other instances, this is not the case and disequilibrium can be established within the chain. For example, inert gaseous ^{222}Rn with a half-life of 3.8 days is often able to be transported in air away from its parent ^{226}Ra. Given that this is more likely over land than ocean, Rn is a powerful tracer of air masses crossing continents. In another example, the different chemical behaviours of U and Th often lead to separation of isotopes of those elements when river waters mix into the ocean (as discussed in more detail in Chapter 4), which in turn can be used to quantify many of the sedimentary and bioaccumulation processes in the coastal zone.

Despite this apparent disruption in the ideal process of sequential radioactive decay, the combination of physical and chemical differences between members of a decay chain can be an extremely powerful aid in environmental studies, and the reader is directed toward other books (e.g., Ivanovich and Harmon, 1982 and, in this series, Carroll and Lerche, 2003; McLaughlin et al., 2005; Krishnaswami and Cochran, 2008; Froehlich, 2010) for a wider discussion of the potential applications of decay chain disequilibria. Also see Section 1.4.4 for more examples.

1.2.4 Detection and Measurement of Radioactivity

1.2.4.1 Introduction

From the previous section, it is clear that each radionuclide has its own characteristic mode of decay and half-life. Accordingly, radionuclides are detected, quantified, and in many cases, uniquely identified by the radiation they emit. This variability provides both opportunities and challenges to practitioners in the field of radioecology. Many readers of this volume will need to apply a range of radioactivity measurement techniques to their investigations. This section will be helpful as an introduction to this field. However, others may find the discussion on the science underpinning the measurements too detailed for their needs and are invited to skip Sections 1.2.4.2–1.3.1.4.

This outline has been organised into four topics:

1. Interaction of high-energy radiation with matter (Section 1.3.1)[2]
2. Detection of radioactivity (Section 1.2.4.2)
3. Measurement of the specific activity of an environmental sample (Section 1.2.4.3)
4. Precision of radioactivity measurements (Section 1.2.4.4)

Two characteristics of radioactivity make analyses of contaminated samples very different from similar samples containing nonradioactive contamination. The first characteristic is that most gamma emissions are sufficiently energetic that they can be detected external to the sample. The analysis of such samples involves identifying the nuclide using its characteristic gamma energy and estimating the concentration of the nuclide through the rate of emission of the gamma rays from the sample (Section 1.2.4.3). Such analyses can often be done without destructive sample preparation steps. Thus, repetitive whole-body analyses of living specimens are possible for some radioactively contaminated samples. This powerful analytical method is not possible for stable contaminants. For example, ^{137}Cs can be quantified in a living fish by determining the gamma emissions from its body. In contrast, the determination of stable Cs would require sacrificing the fish, ashing the body, and performing an analysis on the remaining residues. The ability to determine radioactive contaminant levels in a living organism provides a very powerful tool for determining the kinetics of the contaminant over time (e.g., the rate of contaminant uptake from ingesting contaminated food can

2. The discussion is included within the general topic of radiation science.

be estimated and/or the rate of elimination once the animal resumes a contaminant-free diet).

The second characteristic of radioactivity analyses is that the ability to detect the radioactivity can often be enhanced by analysing the sample for a longer period of time. The longer the analysis, the more emissions detected, and the greater the probability of positively discerning the sample emissions from interfering background emissions. For example, a plant leaf contaminated with ^{137}Cs can be analysed by quantifying the gamma emissions from the leaf. Each time a ^{137}Cs atom within the plant leaf undergoes radioactive decay the emitted gamma energy can be detected with a prescribed efficiency and quantified. The total number of emissions is accumulated and quantified by the analytical equipment as a function of time that the sample is analysed. Thus, a hypothetical sample containing very low levels of ^{137}Cs activity might emit 5 gamma rays over a 10 minute period and not be discernible above background emissions. However, by analysing the same sample for 1000 minutes, some 500 gamma rays would be quantified by the analytical equipment and would now be clearly discernible. The reader is referred to Section 1.2.4.3 and Equation 1.16 for further information. In contrast, nonradioactive samples are prepared for analyses and their concentrations in the sample are determined once. A second analysis, or a longer analytical time, does not improve the detection limits of nonradioactive samples as it does for radioactive samples. This attribute of radiation detection often permits the quantification of extremely low levels of radioactivity.

1.2.4.2 Detection of Radioactivity

A radioactive counting system is comprised of a detector (to convert incoming radiation to charge or light pulses), sensitive electronics to amplify and process the output pulses from the detector, and computing capability to present the data in an optimum form. Radiation detectors are broadly classified into three groups:

1. Gas phase detectors
2. Scintillation detectors
3. Semiconductor detectors

Further details, including applications, are listed in Table 1.2.

1.2.4.2.1 Gas Phase Detectors

Gas phase detectors depend on the fact that exposure to radiation leads to the ionization of a specialised gas mixture, R. The characteristics of the gas are such that one ion pair (R^+ e^-) is formed with every 34 eV of energy captured from the radiation emitted. Application of an electric field within the detector causes a separation of the positive and negative charges to the anode and

TABLE 1.2 Radiation Detector Systems

Class	Subclass	Comment (Technical)	Comment (Application)
Gas phase	Ionization chambers	Gas phase counters generally comprise a central anode and surrounding cathode, as shown in Figure 1.7.	Calibrates isotope yields in production processes and dose rate measurements in industrial plants, nuclear power plants, reprocessing plants, etc.
	Xe high-pressure ionization chambers	Detectors built with operating pressures up to 50 atm. Energy resolutions in the range of 2% to 3% at 662 keV achieved.	Detectors are rugged, thermally stable, and operate at room temperature.
	Proportional	Filled with a gas (e.g., Ar, 90% and CO_2, 10%) or P10 and operates at a higher voltage field than an ionization chamber.	Measures ionizing radiation and neutrons (when BF_3 is the counting gas).
	G-M counter	Operates at a still higher voltage field than a proportional counter.	Detects α and β^- particles and (with less efficiency) γ radiation.
Scintillation (solid)	NaI(Tl)	Sodium iodide crystals have a density of 3.67 and the highest light conversion efficiency of common scintillator materials.	Detectors are rugged, highly efficient, and are commonly used in field detection applications.
	BGO	Bismuth germinate (BGO) has a density of 7.13, but a light conversion efficiency only 15% to 20% that of NaI(Tl).	BGO detectors operate at room temperature, but have a higher energy resolution than sodium iodide.
	Plastic scintillators	Organic crystals, e.g., anthracene, dissolved in polystyrene or polyvinyl-toluene. Typically they have a light conversion efficiency 25% to 50% that of NaI(Tl).	Although the density of the materials is close to unity, they may be machined into large dimensions and complex shapes.
	$LaCl_3$(Ce), $LaBr_3$(Ce)	Cerium-doped lanthanum halides have higher light output, higher density, comparable energy resolution to NaI(Tl), and good temperature stability (Pirard et al., 2005).	Modern detector material under ongoing development.

continued

TABLE 1.2 Radiation Detector Systems—Cont'd

Class	Subclass	Comment (Technical)	Comment (Application)
Scintillation (liquid)	Organic	Formerly, benzene was widely used for ^{14}C dating as it has good optical properties and C/CO_2 can be converted to benzene in high yield. Now largely replaced by accelerator mass spectrometry.	Very widely used in biochemical and related tracer studies, particularly of tritium or ^{14}C-labelled organic compounds.
	Aq. emulsions	A range of scintillant 'cocktails' comprising solutes, scintillation agents, and blending agents form emulsions with the aqueous test sample.	Routinely used for the measurement of tritiated water and the radioactivity levels of β^- and α emitting aqueous solutes.
Semi-conductor	HPGe	High-resolution γ spectrometry over an energy range of 40 keV to >10 MeV.	Used to identify and measure complex mixtures of gamma emitting radionuclides.
	Silicon	Si(Li) X-ray detectors (a few hundred eV to 50 keV or so). Modern passivated implantation Si detectors are used for α particle detection.	Used, e.g., X-ray fluorescence, X-ray microanalysis, PIXE, and X-ray diffraction.

G-M, Geiger-Müller counter; PIXE, particle-induced X-ray emission.

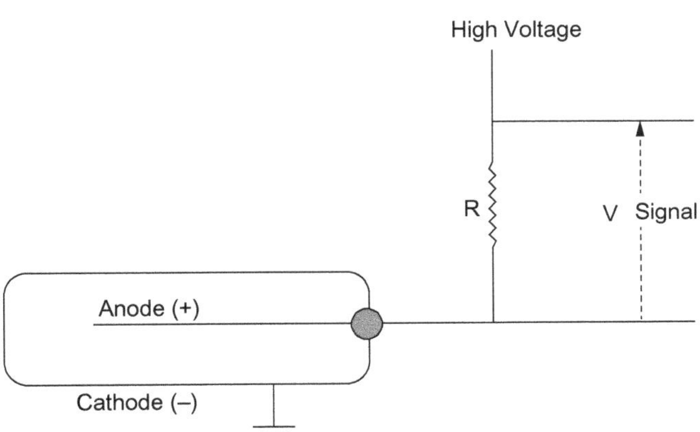

FIGURE 1.7 Schematic sketch of a gas phase detector.

cathode (Figure 1.7) and the generation of a small but measurable current. Within *ionization chambers*, the voltage applied to the instrument is such that the electrons drift to the anode without multiple ionization events. Ionization chambers are widely used in assessing the activity of artificial radionuclides during processing. The efficiency of an ionization detector for the detection of gamma rays increases with the pressure of the gas. High-pressure xenon ionization chambers have been constructed with an energy resolution of 2% to 3% for the 662 keV ^{137}Cs peak. The detector is rugged, operates at room temperature, and is temperature stable. *Gas phase proportional counters* operate at a higher voltage than ionization chambers. The ion pairs form with sufficient energy to cause subsequent ionization while drifting to the electrodes. This is known as a Townsend avalanche. The voltage is chosen so that the avalanche events are independent of one another. Hence, the accumulated charge is proportional to the number of pairs in the original event (i.e., to the energy deposited in the counter gas). Gas phase proportional counters filled with $^{10}BF_3$ at a pressure of 0.5–1 atmosphere are used to measure neutrons. The *Geiger-Müller counter* operates at a still higher voltage leading to the amplification of the initial ion pair production in a cascade effect. Ionization within the chamber results in a substantial drop in the electrode voltage during which the detector remains unresponsive (dead time).

1.2.4.2.2 Scintillation Detectors

In a scintillation detector, the interaction of radiation with the material used to construct the detector results in the production of small light pulses that can be captured by a photocathode, converted into electrons, and amplified to give a measurable electrical signal. A wide range of scintillation materials is available and each is selected on the basis of its gamma ray-stopping power (density), light output, response time, and cost.

1.2.4.2.3 Semiconductor Detectors

A semiconductor detector uses silicon, germanium, or other crystalline matrices to detect charged particles or gamma rays from the electron-hole pairs formed by the deposition of the energy. Electrons are promoted to the conduction band and holes to the valence band. The most successful semiconductor detectors for gamma or X-rays are based on high-purity germanium (usually termed HPGe). HPGe detectors can identify the energy of the absorbed gamma radiation with high resolution and thus facilitate the identification of radionuclides from their energy spectra. Such detectors are found in all industrial and research laboratories undertaking serious work in gamma ray spectroscopy. Alpha particle detectors are manufactured from thin silicon wafers because of their low penetrating power (Section 1.3.1.1).

1.2.4.3 Measurement of the Specific Activity of an Environmental Sample

The unit of radioactivity is the Becquerel (Bq), which is defined as one disintegration per second (s^{-1}). For environmental studies, the International Commission on Radiation Units and Measurements (ICRU, 2001) also refers to the activity density (or activity concentration) in units of Bq/kg, Bq/L, or Bq/m^2, which is a measure of the radioactivity per unit mass, volume, or area of the sample.

The measurement involves converting the response of the counting system C, (counts per second) to the disintegration rate A, (Bq) of the nuclide of interest. The conversion factor F, is the product of the counter efficiency ε, and the nuclear efficiency η:

$$\text{Conversion factor} \quad F = \varepsilon\eta \tag{1.11}$$

$$\text{Count rate} \quad C = \varepsilon\eta A\ (\mathrm{s}^{-1}) \tag{1.12}$$

The counter efficiency depends on the distribution of the sample around the counter (the geometric effect) and the efficiency with which the detector absorbs the radiation (or radiations) from the sample and converts the resulting signal to a recorded count (intrinsic efficiency). The nuclear efficiency is the probability that the radiation or package of radiations detected by the counter will be emitted with each disintegration of the parent nucleus.

Since it is normally very difficult to measure or calculate these parameters, the specific activity of a sample A_s is measured relative to a standard or reference material A_R:

$$A_s = (R_s/R_{\mathrm{ref}}) \times A_{\mathrm{ref}} \tag{1.13}$$

where R_s and R_{ref} are the count rates of the sample and the reference material under identical counting conditions. Equation 1.13 strictly applies only when the standard and the unknown comprise the same radionuclide dispersed on matrices of similar properties.

To assist laboratories worldwide, the International Atomic Energy Agency (IAEA) provides reference materials to facilitate and improve the quality of the results they obtain by nuclear analytical techniques (http://nucleus.iaea.org/rpst/ReferenceProducts/About/index.htm). They include a range of radionuclides in plants and animal materials, water, soils, and minerals.

1.2.4.4 Precision of Radioactivity Measurements

Radioactive decay is a random process. Although the rate of decay for a specific radionuclide can be calculated from knowledge of the number of radioactive atoms and the half-life, there is no way of knowing which specific radioactive atom will decay in which time interval. Furthermore, in all practical circumstances, the probability of a given radioactive atom decaying in a particular time interval is extremely small.[3] This situation is best described by the Poisson statistics.

The Poisson distribution may be generated from the following expression:

$$P(n) = m^n \times e^{-m}/n! \qquad (1.14)$$

where $P(n)$ is the probability of n (= 1,2,3,. . .etc.) events occurring in, for example, a one second interval and m is the mean of the distribution.

Suppose a radioactive source is analysed for a time interval $T(s)$ and that the mean number of counts accumulated in this interval is N counts. When Poisson statistics apply, the variance (v) is numerically equal to N and the standard deviation (σ_p) is $N^{1/2}$:

$$N = v = \sigma_p{}^2 \qquad (1.15)$$

Hence, the number of accumulated counts is the following:

$$N \pm N^{1/2} \qquad (1.16)$$

The proportional uncertainty is given by the ratio of the standard deviation to the number of accumulated counts, in other words:

$$N^{1/2}/N = N^{-1/2} \qquad (1.17)$$

Hence, the larger the number of accumulated counts, the lower is the proportional uncertainty. If the data are collected as count rates (r_m, counts s^{-1}) over a time interval (T), the standard deviation is $(r_mT)^{1/2}$ (compare Eq. 1.16) and the proportional uncertainty is $(r_m/T)^{1/2}/r_mT$ or $(r_mT)^{-1/2}$ (compare Eq. 1.17). The higher the count rate, the lower the proportional uncertainty for a given counting time.

1.2.4.5 Uncertainties in Estimating the Activity of Environmental Samples

The Normal, or Gaussian, distribution is a good approximation to the Poisson distribution except at low disintegration rates (for example, less than five counts in each counting interval). As a consequence, it is possible to use simple methods based on the assumption of normality to:

- Separate the total estimated uncertainties into two components: (1) those due to counting statistics (Eq. 1.14) and (2) those due to random experimental errors
- Estimate counting errors in a low-level sample when corrected for background
- Estimate counting errors in an unknown sample measured relative to a standard (Eq. 1.13)

---

3. Consider, for instance, 1 Bq Ra-226 (1 disintegration per second). The number of atoms of Ra-226 in this sample is approximately 7.2×10^{10} atoms. Of these, only one will disintegrate every one second interval on the average.

1.2.4.5.1 Propagation of Errors Associated with Background Subtraction

The levels of radioactive contaminants in many environmental samples are close to the limit of detection and reliable background corrections are essential. Background subtraction affects not only the net count rate, but also the precision of the estimate.

When count rates are added or subtracted, the cumulative standard deviation (σ_s) is calculated by a process known as quadrature summing:

$$\sigma_s{}^2 = \sigma_a{}^2 + \sigma_b{}^2 + \ldots\ldots \quad (1.18)$$

This rule is widely used when calculating the counting 'error' or standard deviation following the subtraction of the background counts from the total. For instance, if T, S, and B are the 'total,' 'sample,' and 'background' counts accumulated in a given time interval (e.g., 1 second), then:

$$S = T - B \qquad \text{and} \quad (1.19)$$

$$\sigma_s{}^2 = \sigma_t{}^2 + \sigma_b{}^2 \quad (1.20)$$

The uncertainty associated with the counting statistics for N counts is $\pm\, N^{1/2}$ (Eq. 1.16) and is increased if background counts are subtracted.

In practice, uncertainties associated with other aspects of the experimental design must also be considered. If these noncounting errors are significant, a large number of repeat measurements taken over a fixed time interval will yield a standard deviation of the distribution that significantly exceeds the estimate $N^{1/2}$ from the mean number of counts.

1.2.4.5.2 Propagation of Random Errors Associated with the Multiplication and Division of Data

This is important, as the specific activity of an unknown, A_s, is frequently calculated by direct comparison with a standard reference material A_{ref}, as noted in Equation 1.13:

$$A_s = (R_s/R_{\text{ref}}) \times A_{\text{ref}}$$

where R_s and R_{ref} are the count rates of the sample and the reference material under identical counting conditions. The uncertainty in the estimate of the activity of the sample may be calculated from uncertainties in A_{ref} specified by the supplier or by a reference laboratory and the calculated uncertainties in the count rates R_s and R_{ref}.

If count rates are combined with other measurements by a process of multiplication or division, the standard deviation of the product P ($= R_1 \times R_2 \times \ldots$) or the quotient ($R_1/R_2$) is given by:

$$(\sigma_p/P)^2 = (\sigma_1/R_1)^2 + (\sigma_2/R_2)^2 + \ldots. \quad (1.21)$$

In applying the quadrature summing rule (Eq. 1.21), measurements must be taken under identical conditions including counting time. If the counting times vary, a scaling factor F may be used to normalise the counting times too, for instance, counts in one second. Suppose N_1 counts are accumulated in T seconds, the number of counts accumulated in 1 second is N_1/T. Applying Eq. 1.21 to this nuclide, $P = N_1/T$ and:

$$\sigma_p/(N_1/T) = \sigma_1/N_1 \quad or \quad \sigma_p = \sigma_1/T \tag{1.22}$$

In other words, the standard deviation is scaled by the counting time.

These descriptions provide a basic set of tools for the propagation of errors within radioactivity measurement. However, the reader should be aware that there are many other sources of advice in relation to error reporting, including standard approaches within the scientific community in the field of radiation science. The readers are referred to IAEA (2004) TECDOC-1401 on "Quantifying uncertainty in nuclear analytical measurements" for more detailed information on errors relevant to a range of nuclear analytical procedures.

1.3. RADIATION SCIENCE

1.3.1 Interaction of Radiation with Matter

This brief discussion of the interaction of radiation with matter serves as an introduction to radiation shielding and the biological effects of radiation. It is also relevant to the science of radiation detection (Section 1.2.4.2). This section will be particularly useful to those readers who have an interest on how radiation acts on the world in general. However, it is not essential to understand this in order to appreciate that radiation has environmental impacts. Radioecologists more interested in radiation effects than mechanisms are invited to skip to Section 1.3.2 on biological effects.

1.3.1.1 Alpha Particles and Protons

Properties of alpha particles and protons are listed in Table 1.1. Strong coulombic interactions lead to the ionization and excitation of the atoms in the target material along the tracks made by the alpha and proton particles. In physics, the target material is usually some inanimate material, whereas in radioecology, the target often refers to plant or animal tissue. The rate at which energy is dissipated in the target material (MeV/cm) is known as the stopping power in nuclear physics and the linear energy transfer (LET) in radiobiology. Alpha particles have a high LET, meaning that all of their energy is absorbed by the biological tissue over a very short distance. The particles do not move in straight lines, but exhibit a circuitous path influenced by their interactions with the target material. LET data for different classes of 1 MeV radiation in water are listed in Table 1.3. The projected range for 1 and 5 MeV alphas in water are 0.006 and 0.037 mm, respectively.

TABLE 1.3 Interaction of Radiation (1 MeV) with Water

Radiation	Projected Range (α,β particles, protons); half thicknesses (gamma/neutron) in Water (g/cm^{-2})	Average LET (KeV/μ)	Comment
Alpha	$5.93\ 10^{-4}$ cm[a]	169	The projected range for a 5 MeV α (the approx. energy of many actinide α's) is $37.3\ 10^{-4}$ cm.
Protons	$2.44\ 10^{-3}$ cm	41	
Beta	0.437 cm	0.23	
Gamma	9.9 cm	0.1	At 1 MeV, gamma attenuation is dominated by the Compton scattering process.
Neutron	0.13 cm	—	D_2O is a less efficient moderator of neutrons. The half thickness for 1 MeV neutron in D_2O is 5.1 cm.

[a]*Assuming a density of 1 g cm*$^{-3}$.

The Bragg curve is the variation of the stopping power with distance along the initial direction of the radiation. Examples for 5.5 MeV alpha particles and protons in water are depicted in Figure 1.8 (constructed from data published by NIST, http://physics.nist.gov/PhysRefData/Star/Text/ASTAR.html). The rate of energy dissipation reaches a maximum at the Bragg Peak, shortly before the end of the particle track. The location of the Bragg Peak varies with the energy of the radiation and is important in the management of dose distribution to patients receiving proton therapy.

1.3.1.1.1 Alpha Particle Shielding

Alpha particles are positively charged, huge in size (relative to electrons), readily interact with the media they intersect, and dissipate their energy rapidly. For example, 5.5 MeV alphas (typical of the energy emitted by many actinides) are stopped, and thus shielded, by a few centimetres of air or about 0.04 mm of human tissue. The alpha particle is absorbed in the epidermis (i.e., the outer layer of skin). Alpha particles do not penetrate the outer layer of dead skin and consequently pose little radiological danger. However, if inhaled or ingested, alpha emitters can be very damaging to internal tissues.

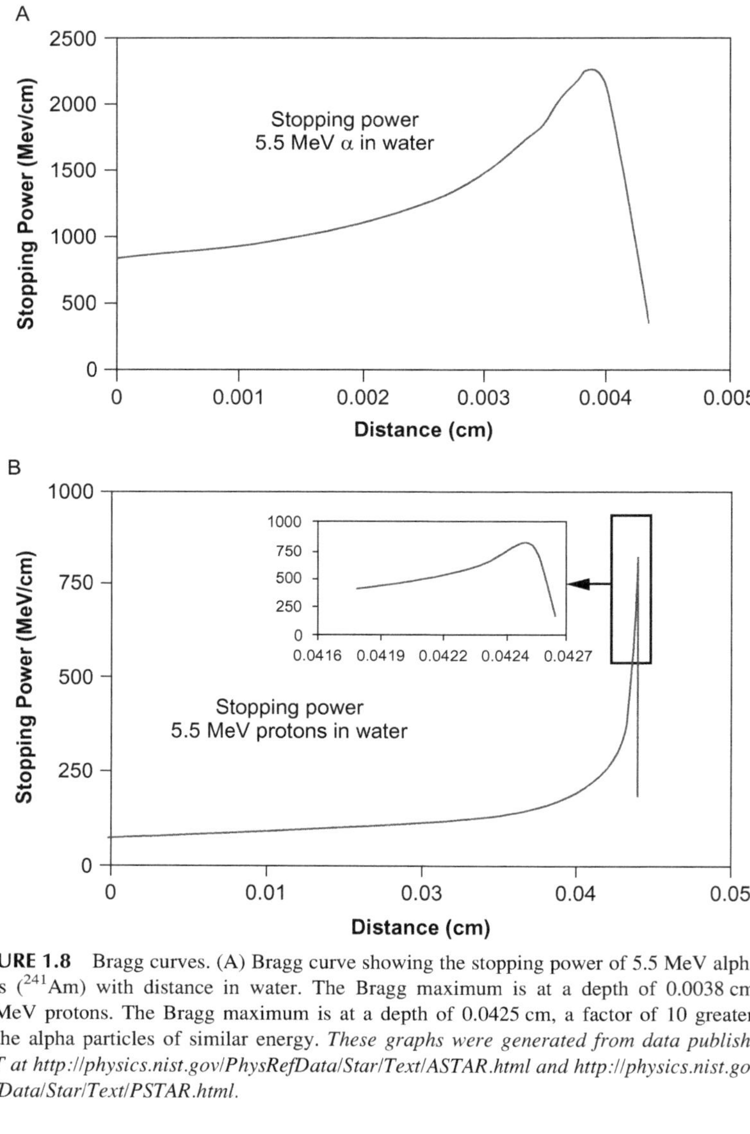

FIGURE 1.8 Bragg curves. (A) Bragg curve showing the stopping power of 5.5 MeV alpha particles (^{241}Am) with distance in water. The Bragg maximum is at a depth of 0.0038 cm. (B) 5.5 MeV protons. The Bragg maximum is at a depth of 0.0425 cm, a factor of 10 greater than for the alpha particles of similar energy. *These graphs were generated from data published by NIST at http://physics.nist.gov/PhysRefData/Star/Text/ASTAR.html and http://physics.nist.gov/PhysRefData/Star/Text/PSTAR.html.*

1.3.1.2 Beta Particles

Beta particles are negatively charged electrons and exhibit coulombic interaction with atoms leading to ionization, excitation, and the generation of X-rays known as bremsstrahlung radiation. Beta particles interact less strongly with matter than alpha particles and have a lower LET, thus traversing further through tissues. For example, a 1 MeV beta particle would travel some

1000 times further in water (~4.4 mm) than a similarly energetic alpha particle in the same medium (~5.9 μm; Table 1.3).

1.3.1.2.1 Bremsstrahlung

During the attenuation process, the beta trajectories are changed and the particles are decelerated. To conserve momentum, X-rays known as bremsstrahlung, or braking radiation, are emitted. The energy spectrum forms a continuum that increases with intensity and moves to shorter wavelengths as the bombarding energy increases. Very little bremsstrahlung is formed from the interaction of alpha particles with matter.

1.3.1.2.2 Shielding

The range of beta particles in Plexiglas, a common shielding material, increases from 3.8 mm for 1 MeV particles to 24 mm for 5 MeV. Light materials, such as Plexiglas, are used to minimise the yield of bremsstrahlung, which can contribute significantly to the dose.

1.3.1.3 Gamma Rays and X-rays

Gamma photons and X-rays are packets of energy and have no electric charge and zero rest mass. In the absence of coulombic effects, the interaction with matter is much reduced and the penetration distances are larger (~9.9 cm for a 1 MeV photon in water; Table 1.3). There are three principal modes of interaction: (1) the photoelectric effect, (2) Compton scattering, and (3) pair production.

1.3.1.3.1 Photoelectric Effect

The incident gamma ray or X-ray is absorbed in the target material by ejecting an electron from the surrounding electron cloud leaving a vacancy, normally in the inner shell. The energy required to eject an electron increases with the atomic number, ranging from about 50 eV for lithium to about 114 keV for uranium. Subsequent deactivation involves the refilling of the electron vacancy and the release of the energy as Auger electrons or X-rays. The photoelectric effect is the dominant process for energies below 150 keV.

1.3.1.3.2 Compton (Incoherent) Scattering

Gamma rays with energies in the range of 250 keV–3 MeV (mass equivalent 2.68 10^{-4}–3.22 10^{-3} u) tend to interact with an electron (mass equivalent 5.48 10^{-4} u) leading to an increase in electron energy and the emission of a new gamma ray of reduced energy. The scattered electron and gamma radiation can interact with matter, as described previously.

1.3.1.3.3 Pair Production

When high-energy gamma rays (>3 MeV) are absorbed, an electron and positron are formed that share the energy of the gamma ray. Positrons and electrons lose energy through ionization and excitation. A positron at rest interacts with an electron to create two gamma rays with energies of 0.511 keV (annihilation radiation). The two gamma rays can escape or interact with matter through Compton scattering or the photo-electric effect.

1.3.1.3.4 Gamma Shielding

The gamma dose rate in air, at any distance from a point source, is given by the specific gamma ray constant (Γ). Values for some commonly used nuclides are listed in Table 1.4. The purpose of gamma shielding is to reduce the unshielded doses to levels consistent with the application of modern radiation protection principles (Section 1.3.2.3).

The intensity $I(t)$ of a narrow collimated beam of mono-energetic photons after passing through t kg m^{-2} of matter is given by the following:

$$I(t) = I(0)\exp(-\mu_t t) \text{ or, } \quad I(0)\exp(-(\mu_m/\rho)t) \qquad (1.23)$$

where μ_t is the linear attenuation coefficient (m^{-1}), ρ is the bulk density (kg m^{-3}), and μ_m/ρ (kg m^{-2}) the mass absorption coefficient.

The attenuation equation (Eq. 1.23) underpins shielding calculations, although application in practice is normally complex and ultimately relies on measurement. NIST has published extensive tables of mass attenuation

TABLE 1.4 Specific Gamma Ray Constants (Γ) and Thicknesses of Lead Shielding for 95% Attenuation (T_{Pb})

Nucide[a]	Half-life	Γ (mSv/h per MBq at 1 m)	T_{Pb} (cm)
^{51}Cr	27.7 d	6.32E-6	0.78
^{60}Co	5.27 y	3.70E-4	4.41
^{82}Br	35.3 h	4.36E-4	3.31
^{99m}Tc	6.0 h	3.32E-5	0.11
^{131}I	8.02 d	7.64E-5	1.24
^{137}Cs	30.2 y	1.02E-4[b]	2.45
^{192}Ir	73.8 d	1.60E-4	1.28
^{198}Au	2.70 d	7.88E-5	1.29
^{241}Am	433 y	8.48E-5	0.011

[a]*The energies of the major gamma rays are listed in Table 1.8.*
[b]*From ^{137m}Ba.*

coefficients (Hubbell and Seltzer, 1996). The thicknesses of lead to reduce the doses from point sources to 5% of the unshielded values are listed in Table 1.4.

1.3.1.4 Neutrons

Neutrons have a rest mass similar to that of protons and are electrically neutral (Table 1.1). Therefore, they do not interact with electrons but undergo elastic or inelastic scattering with nuclei or nuclear reactions. All neutrons are emitted as fast neutrons and lose kinetic energy by collision until they are absorbed. Free neutrons are unstable and decay in a vacuum by β^- decay with a half-life of 615 s.

1.3.1.4.1 Elastic Scattering

Elastic scattering leads to the loss of kinetic energy (slowing down) of neutrons by billiard ball-type collisions. This process leads to the generation of highly energetic recoil nuclei, which lose energy by excitation and ionization of the surrounding material. The recoil process is governed by the laws of conservation of momentum and energy. Up to half of the kinetic energy of a neutron is lost on collision with a hydrogen nucleus, since the masses are almost equal. Thus, materials with high hydrogen content, such as water or paraffin wax, are effective for neutron shielding.

1.3.1.4.2 Inelastic Scatter

Inelastic scatter occurs when the neutron is absorbed by the target nucleus to form an excited state that subsequently emits a lower energy neutron and a gamma ray or particle. The decay process is rapid (of the order 10^{-12} s or less) and there is no change in the atomic number or the atomic mass.

1.3.1.4.3 Neutron Capture

In neutron capture, as in inelastic scatter, the target nucleus is elevated to an excited state by the absorption of the incident neutron. However, in this case the deactivation (i.e., decay) is relatively slow ($>10^{-9}$ s) and involves one of the processes described in Section 1.2.2. The most common process involves the emission of a beta particle to form a daughter product in an excited state with a rapid transition to the ground state.

Although the processes are complex, the attenuation of neutrons absorbed in shielding materials can be represented by a simple equation analogous to that for gamma rays (Eq. 1.23). The rate of decrease of neutron dose from a beam of fast neutrons passing through a material of thickness d, with a total cross-section σ_{nT} is approximately:

$$I/I_o = \exp[-N\sigma_{nT}d] \quad (1.24)$$

where I_o and I are the incident and transmitted neutron intensities in the collimated beam and N is the volume density of atoms.

The macroscopic cross-section Σ_{nT} ($= N\sigma_{nT}$) has the dimensions of reciprocal length. The mean free path:

$$mfp = 1/\Sigma_{nT} \quad (1.25)$$

The microscopic cross-section (σ_{nT}) is a property of the nucleus and has the dimension of area. It is conventionally presented in terms of the area that the neutron 'sees' and is expressed in barns where 1 barn $= 10^{-24}$ cm^2 or 10^{-28} m^2.

1.3.2 Biological Effects of Radiation

1.3.2.1 Introduction

As discussed earlier, the deposition of energy resulting from the interaction of radiation with matter leads to the excitation and ionization of the atoms within biological tissues. The measurement of the absorbed energy is known as dosimetry. In SI units, the unit of absorbed dose is the Gray (Gy):

$$1 \text{ Gray (Gy)} = 1 \text{ Joule absorbed/kg} \quad (1.26)$$

The effects of dissipating radiation energy in the biological material include the following:

- Excitation and ionization (applications to the detection and measurement of radiation are discussed in Section 1.2.4)
- Breaking of chemical bonds
- Biological effects

1.3.2.2 Biological Effects

DNA is the primary target for the induction of biological effects from radiation in all living organisms. There are broad similarities in radiation responses in different organisms, and yet wide differences in radiation sensitivity. The range in lethality from acute exposure to radiation varies by three to four orders of magnitude among organisms, and mammals are among the most sensitive and viruses are among the most radioresistant (Figure 1.9; Whicker and Schultz, 1982).

As previously described, damage from radiation is initiated by ionization, which occurs if the radiation has sufficient energy to eject one or more orbital electrons from the atom in which it interacts. Ionizing radiation is characterized by a large release of energy (approximately 34 eV per event), an amount that is more than enough to break strong chemical bonds (e.g., only 4.9 eV is required to break a C=C bond) (Hall and Giaccia, 2006).

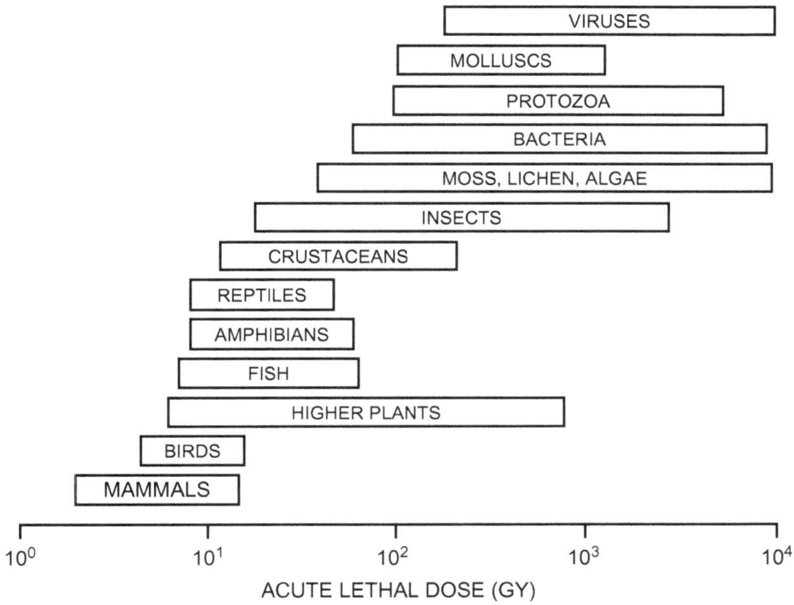

FIGURE 1.9 Lethal dose range. Approximate acute lethal dose ranges for various taxonomic groups. *From Whicker and Schultz (1982); UNSCEAR (2008), with permission.*

The ionization process and resulting charged particles can subsequently produce significant damage to biological cells. Such damage is often referred to as direct effects. Much of the biological damage from radiation, however, is due to indirect effects from free radicals (Figure 1.10). Free radicals are the fragments of atoms that remain after ionization. Free radicals have an unpaired or odd number of orbital electrons, resulting in a high degree of chemical instability. Such free radicals can easily break chemical bonds and are a main cause of damage from radiation exposure. Free radicals react with cellular molecules within fractions of a second after their formation, thus they have a short life. The OH• free radical, formed by the ionization of cellular water, is among the most common because of the abundance of water in all biological tissues (about 80% of the mass of a living cell is water). To appreciate the quantity of free radicals produced, consider their concentration (expressed in terms of a G value, defined as the number of radicals produced per 100 eV of energy absorbed in the medium). The G value for the OH• radical is 2.6 (IAEA, 2010). Thus, if a 5 MeV alpha particle were to dissipate all of its energy within cellular water, approximately 130,000 free radicals of OH• could theoretically be produced.

Free radicals are not unique to radiation, but are produced in response to many stressors—smoking, air pollution, exposure to solar UV radiation, tissue inflammation, even normal metabolic processes—all of which produce

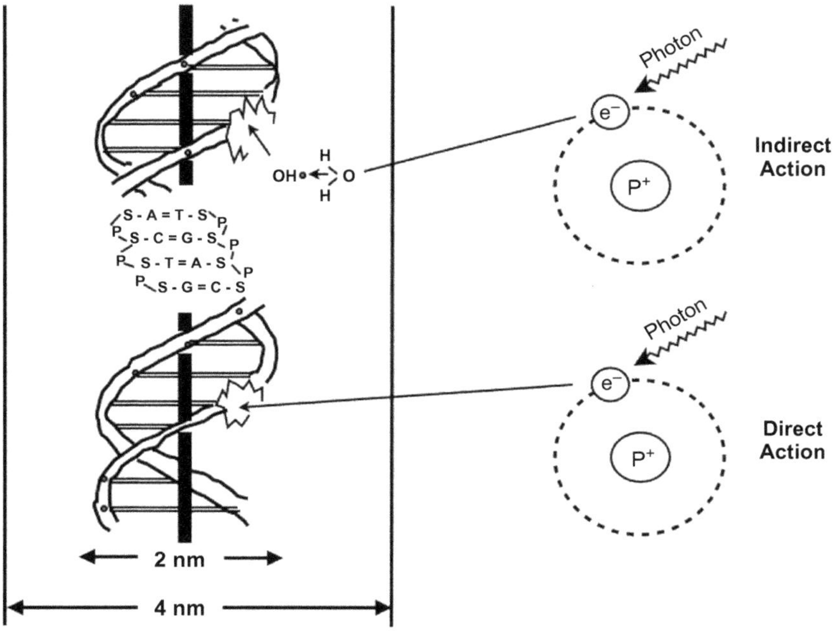

FIGURE 1.10 Biologic effects of radiation. Direct versus indirect effects of radiation on DNA. *From Hall and Giaccia (2006,) with permission.*

damaging free radicals. Such abundant free radical production results in humans experiencing approximately 10^4 to 10^5 endogenous oxidative damages per cell per day among the 3×10^9 bases in the genome (IAEA, 2010). Damage caused by the free radicals is so abundant that very efficient repair mechanisms have evolved within all biological species to counter their effects.

Radiation and the free radicals produced can damage DNA by causing several different types of lesions (e.g., single-strand breaks (SSBs), double-strand breaks (DSBs), base changes, and interstrand cross links). The number of DNA lesions caused by a dose of 1–2 Gy is ~1000 base damages, 1000 SSBs, and ~40 DSBs (IAEA, 2010). DSBs are central to radiation-induced damage and their numbers correlate with radiosensitivity and the probability of cell survival. There are efficient DNA repair processes specific to each type of lesion. For DSBs the two primary repair pathways are nonhomologous end joining (NHEJ) and homologous recombination (UNSCEAR, 2000). The mechanisms of the two repair pathways are such that NHEJ is much more prone to errors during the repair process.

Errors in repair do occur and can result in apoptosis, chromosome aberrations, or mutations. Apoptosis (cell death through a programmed self-destruction) is by far the most likely outcome following repair errors. Occasionally, mutations form and their impacts within a population are dependent on the cell type in which they occur. The two general types of cells are germ and somatic. Germ cells refer to the primordial cells from which eggs or sperm are derived. All other tissues (bone, muscle, blood, etc.) are derived from somatic cells. A mutation within a somatic cell can lead to cell death, or if the DNA damaged cell has undergone mis-repair, such that the cell is still viable, then the mutation in the somatic cell can lead to cancer. Mutations in reproductive germ cells can decrease the number of gametes, increase embryo lethality, or be inherited, resulting in the alteration of the offspring. For humans, the risk of hereditary effects in offspring of exposed individuals is about 10% of the cancer risk to the exposed parents (UNSCEAR, 2000). The risk of nonfatal cancer for humans has been estimated at 1×10^{-5} per mSv (ICRP, 1991). Thus, the risk of hereditary damage in humans is about one chance in a million per millisievert absorbed by the parents. For nonhuman biota, the risk of hereditary effects is currently unknown.

Most mutations are deleterious, offer no advantage to the individual that possesses them, and are subsequently removed from the population. Some mutations are neutral, have no apparent effect on the individuals that possess them, and can persist over many generations within a population. Rarely, a mutation might offer a selective advantage (e.g., increase the efficiency of water absorption in the roots of a plant that contains the mutation). Such selective advantages would spread in a population.

The deleterious effects of ionizing radiation to biological systems have been known from the earliest days to be primarily dose dependent. Effort has therefore been expended over the years in defining the effective dose to a biological system. The issue is complex because the effective dose depends not only on the gross energy deposited, but also on the quality of the radiation and the radiation sensitivity of the affected tissue. Some comments on the formal approach adopted by the radiation protection community are presented in the following paragraphs.

1.3.2.2.1 The Sievert

In SI units, the effective dose to humans is the Sievert (Sv), which is the absorbed dose (Gy) adjusted by two dimensionless weighting factors: (1) the radiation weighting factor, (w_R Table 1.5A), to account for the biological effectiveness of the absorbed radiation, and (2) the tissue weighting factor, (w_T Table 1.5B), to account for differences in the radiation sensitivities of different organs of the body. For example, mammalian nerve and muscle cells (which are highly specialised) have lower radiosensitivities than reproductive germ cells or gastrointestinal stem cells; the latter are mitotically active and comparatively undifferentiated. That young, rapidly growing tissues are more radiosensitive was first postulated in 1906 (Bergonie and Tribondeau, 1906). The w_R is a balanced consequence of the deposition of energy within tissue resulting from different types of radiation, as described in Section 1.3.1.

The equivalent dose and the effective dose are both expressed as Sieverts (Sv) and are related to the absorbed dose (Gy) by the following equations:

$$\text{equivalent dose (Sv)} = \text{absorbed dose (Gy)} \times w_R \qquad (1.27)$$

$$\begin{aligned}\text{effective dose (Sv)} &= \text{absorbed dose (Gy)} \times w_R \times w_T \\ &= \text{equivalent dose (Sv)} \times w_T \qquad (1.28)\end{aligned}$$

The weighting factors are listed in Table 1.5.

In modern literature, the biological effects of radiation are classified as deterministic or stochastic effects. Readers are referred to a detailed discussion in ICRP 103 (2007).

1.3.2.2.2 Deterministic (Prompt) Effects

Deterministic effects of irradiation arise when large numbers of cells are destroyed by radiation and harmful tissue reactions ensue. They may be called somatic effects, as they only appear in the exposed person and do not affect genetic material passed on to future generations, or prompt effects, as they can occur within days, or hours, or even less. The severity of the effect depends on the dose above a threshold of about 100 mGy. Examples of deterministic effects in humans include cataracts and skin erythema. Infertility is also considered a somatic effect.

1.3.2.2.3 Stochastic (Delayed) Effects

Stochastic effects have no dose threshold and the probability of having the effect is proportional to the absorbed dose. Radiation-induced cancer and genetic effects are stochastic. In the case of cancers, there is evidence of radiation risk at doses

TABLE 1.5A Radiation Weighting Factor (w_R)

Radiation		w_R
β, X, γ rays		1
α		20
Neutrons		
	Thermal	5
	0.5 MeV	7.5
	0.5 MeV	20
	0.1–2 MeV	20
	2–20 MeV	5

From ICRP (1991), with permission.

TABLE 1.5B Tissue Weighting Factors (w_T, $\gamma w_T = 1$)

Tissue	Fraction of Risk	Tissue	Fraction of Risk
Gonads	0.20	Liver	0.05
Colon	0.12	Oesophagus	0.05
Lungs	0.12	Thyroid	0.05
Stomach	0.12	Skin	0.01
Bladder	0.05	Bone surface	0.01
Breast	0.05	Others	0.17

below 100 mGy and they also appear among members of the public exposed only to background radiation. The main stochastic effects are initiated by damaging chromosomes in a somatic cell and may occur months or years after exposure. On the basis of the available evidence, the population-weighted average lifetime risk of death from cancer following an acute dose equivalent to all body organs of 0.1 Sv is estimated to be 0.8%, although the lifetime risk varies considerably with age at the time of exposure. For low LET radiation (see Section 1.3.1.1), accumulation of the same dose over weeks or months, however, is expected to reduce the lifetime risk appreciably, possibly by a factor of 2 or more.

Although genetic defects caused by damage to chromosomes in reproductive cells can also occur, "heritable effects of radiation have yet to be clearly demonstrated in man, but the absence of a statistically significant increase in genetically related disease in the [30,000] children [studied] of atomic bomb survivors . . . is not inconsistent with the animal data" (BEIR V Report, 1990).

Considerable and continuous debate occurs over the effects of low doses of radiation. The underlying problem is that a relatively high natural background cancer rate in humans (~20%) makes it difficult to discern small increases in cancer incidence following exposure to low doses. The derivation of cancer risk is based on populations of humans exposed to high doses of medical radiation, or survivors from the atomic bombings of Japan during World War II (WWII). Because of the lack of sufficient data, some form of extrapolation from the high-dose region is required to estimate risks from exposures to low doses (Figure 1.11). A linear-no-threshold model (LNT) has been used by the United States National Research Council, the National Council on Radiation Protection and Measurements, and the United Nations Scientific Committee on the Effects of Atomic Radiation (UNSCEAR, 1996, 2000). The LNT is a conservative approach and can be interpreted to mean that all radiation, regardless of how small the dose, has the potential to cause cancer. Because insufficient human data exist to counter the LNT model, it continues to be used by regulatory agencies for human radiation exposure (BEIR V Report, 1990).

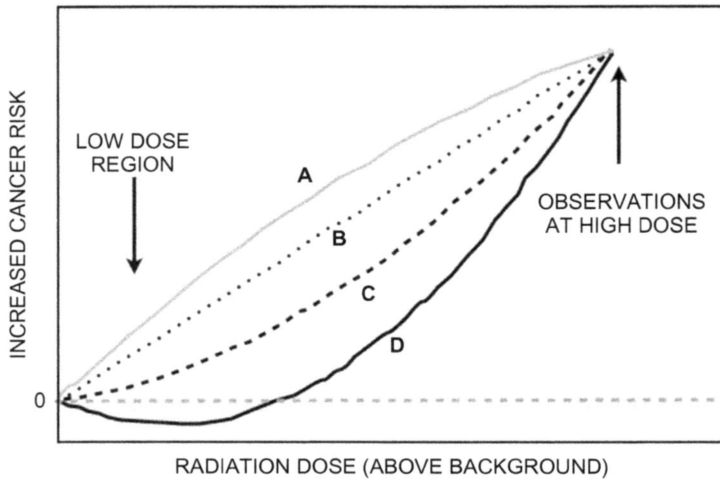

FIGURE 1.11 Cancer risk versus radiation dose. Alternative assumptions for the extrapolation of the cancer risk versus radiation dose to low dose levels given a known risk at a high dose: (A) supralinearity, (B) linear, (C) linear-quadratic, and (D) hormesis. *From http://en.wikipedia.org/wiki/Radiation_hormes, with permission.* Accessed May 25, 2011.

However, alternative hypotheses exist. The French Academy of Sciences – National Academy of Medicine stated in their 2005 report that many laboratory studies have observed radiation hormesis when animals are exposed to low-level radiation (Duport, 2003; Calabrese, 2004). Radiation hormesis is the hypothesis that low doses of ionizing radiation, slightly above natural background levels, are actually beneficial because they stimulate the activation of repair mechanisms (Luckey, 1991). These repair mechanisms are not activated in the absence of ionizing radiation, and importantly, they then protect against other types of stresses or disease. The upregulated repair mechanisms following exposure to low doses of radiation are thought by some to be sufficiently effective that when stimulated they not only cancel the detrimental effects of ionizing radiation, but also inhibit disease unrelated to radiation exposure (Kaiser, 2003; Feinendegen, 2005; Tubiana et al., 2009).

1.3.2.3 Human Radiological Protection

Radiological protection is predicated on the premise that the deleterious effects of radiation are determined principally by the effective dose resulting from exposure to external sources or from the ingestion or inhalation of radionuclides.

1.3.2.3.1 External Gamma Sources

Estimates of the absorbed dose rate from external gamma sources can be made for point sources, as well as for flat or volumetric sources of infinite extent, using tabulated parameters.

In the case of the point source, the relevant parameter is the specific gamma ray constant, which is the dose rate (mSv/h) from a source of 1 MBq at a distance of 1 m. Allowance is made for the separation distance from the source using the inverse square law and for the effect of shielding. Selected gamma ray constants, together with the thickness of lead shielding to reduce the dose rate by 95%, are listed in Table 1.4. Unger and Trubey (1982) published a comprehensive table.

1.3.2.3.2 Ingested and Inhaled Radionuclides

Dose to radiation workers (and the general public) is also accumulated by the ingestion or inhalation of radionuclides. The biological impact of the absorbed dose depends on their metabolic distribution and biokinetics within the body (different elements will be distributed and excreted according to their similarity to any nutrient analogues that may exist; see Chapter 6 for more detailed discussion), on the quality of the radiation components (α, β, and the energy distribution of the X and γ radiation), and on their relative sensitivities to different organs of the body (Eqs. 1.27 and 1.28). For practical human radiological protection, these parameters have been subsumed within a single parameter, the dose conversion factor for each radionuclide, and tabulated in the IAEA Safety Series 115 (IAEA, 1996). Important examples are listed in Table 1.6.

The International Commission on Radiological Protection (ICRP) maintains three fundamental principles for radiation protection that also underpin radioecological research: (1) justification, (2) optimisation, and (3) the applications of dose limits.

TABLE 1.6 Committed Effective Dose per Unit Intake (Sv/Bq) (Ingestion, Infants and Adults)

Radionuclide	Half-life	Infants (1–2 y) Sv/Bq	Adults (>17 y) Sv/Bq
Tritiated water	12.3 y	4.8 E-11	1.8 E-11
^{14}C	5730 y	1.6 E-09	5.8 E-10
^{60}Co	5.27 y	2.7 E-08	3.4 E-09
^{90}Sr	29.1 y	7.3 E-08	2.8 E-08
^{99m}Tc	6.02 h	1.3 E-10	2.2 E-11
^{131}I	8.04 d	1.8 E-07	2.2 E-08
^{137}Cs	30.0 y	1.2 E-08	1.3 E-08
^{198}Au	2.69 d	7.2 E-09	1.0 E-09
^{238}U	4.47 + 09 y	1.2 E-07	4.5 E-08

1.3.2.3.3 Justification

Generally speaking, no planned exposure should be introduced unless it produces sufficient net benefit to the exposed individuals or to society to offset the radiation detriment it causes. The justification may need to be re-examined as new information or technology becomes available.

1.3.2.3.4 Optimisation

The process of optimisation is intended for those situations that have been deemed to be justified and is central to the process of protection. A source-related process is optimised if the number of people exposed and the magnitude of individual doses are as low as reasonably achievable, taking economic and societal factors into account. This is what the As Low As Reasonably Achievable (ALARA) principle is all about. (ICRP Publications 60, 1991; 103, 2007).

1.3.2.3.5 Application of Dose Limits

International Radiation Safety Standards are based on an approximated overall fatal risk coefficient of 5% per Sv. Based on these types of considerations, the dose limits listed in Table 1.7 are recommended (ICRP 103, 2007).

1.3.2.4 Environmental Radiological Protection

There are fundamental differences in determining the risk to humans following exposure to radiation and the risks to a radioactively contaminated environment (Figure 1.12). Human risk analysis largely focuses on cancer risks to individuals, and dose-response relationships are sufficiently established so that risk factors (i.e., the probability of lethality from cancer per unit of dose) are established. In contrast, ecological risk to nonhuman biota is seldom concerned with individuals; instead it is concerned with populations of plants

TABLE 1.7 Recommended Dose Limits in Planned Exposure Situations

Type of Limit	Occupational	Public
Effective dose	20 mSv per year, averaged over defined periods of 5 years	1 mSv in a year
Annual equivalent dose in:		
Lens of the eye	150 mSv	15 mSv
Skin	500 mSv	50 mSv
Hands and feet	500 mSv	–

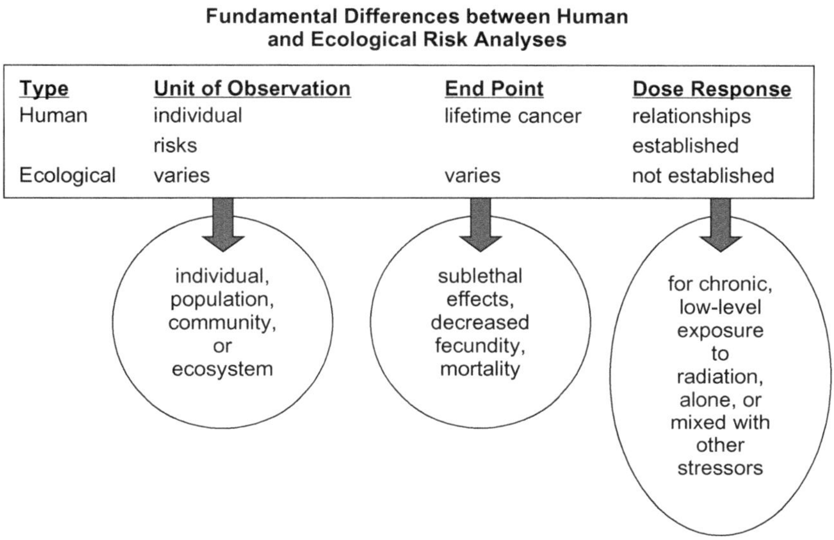

FIGURE 1.12 Fundamental differences between human and ecological risk analyses. (from Hinton, et al. 2004)

and animals. Management of the environment centres on a viable population of organisms, not on single individuals within the population. End points for ecological risks are not cancer oriented; instead, they include a wide assortment of effects ranging from chromosomal damage to reduced reproductive success. The dose-response relationships for these end points are not yet established (but studies are underway; see the following material); therefore, there are no risk factors that equate dose to the probability of an outcome for nonhuman biota.

The criteria for determining if an ecosystem is at risk from radioactive contamination are currently changing. Historically, the paradigm for protecting the environment was that if humans were protected, then so was the rest of the environment (ICRP, 1977; IAEA, 1992). That is, the protection criterion for humans (1 mSv per year) was considered to be sufficiently restrictive that populations of plants and animals living in the same environment would be sufficiently protected. Recently, that perspective has begun to change (e.g. Hinton et al. 2004). The ICRP recognised the need to provide advice on environmental protection, and that a clear framework was required to assess the relationships between exposure and dose, dose and effects, and any consequences of effects (ICRP, 2007). The ICRP has stated that the framework they are developing for environmental protection should complement the approach used for the protection of humans (ICRP, 2009). Consequently, the ICRP has suggested a similar reference model approach as used for humans (i.e., "Reference Man") for nonhuman biota. They have, therefore, proposed a small set of reference animals and plants (RAPs) for which reference dosimetric models have been developed and knowledge on radionuclide uptake and radiation effects collated. The background to, and development of, a framework for environmental protection is available within relevant documents such as ICRP Publication 91 (ICRP, 2003) and ICRP Publication 103 (ICRP, 2007).

The end points considered to be most relevant in determining risks to non-human biota are increased mortality, increased morbidity, and decreased reproductive output. Of the three, changes in reproduction are thought to be the most sensitive to radiological exposures. Much more data are needed, however, before we can confidently predict population level impacts to non-human biota as a function of radiological exposures. Data are particularly scarce for chronic, low-level exposures; exposures over several generations; and when radiological exposures are combined with other types of contaminants or stressors.

The ability to predict population-level effects under such scenarios are complicated by the large natural variation in sensitivities to radiation among the individuals within a population, indirect effects, compensating mechanisms, and adaptation to the stress. An example of an indirect effect is the greater abundance of resources (i.e., food, water, light, etc.) available to radioresistant individuals when radiosensitive individuals decline within a population. The same analogy holds relative to a greater abundance of resources available to radioresistant populations within a community when radiosensitive populations decline (i.e., one species of insect declines leaving more resources to a radioresistant insect species occurring within the same community). Such interactions are extremely difficult to predict. Likewise, compensating mechanisms have been documented in populations of exposed animals that complicate the prediction of effects. An example of a compensating mechanism is provided by Blaylock (1969). He documented an increased mortality of fish embryos exposed to a dose rate of 4 mGy per day of radiation in a contaminated lake. This effect, however, was compensated for when the fish produced larger brood sizes, with the net result that no effect to the population was observed.

Several organisations and research groups are actively trying to improve our knowledge of radiation impacts on the environment and to derive benchmarks of acceptable dose rates that will be considered protective of the structure and function of ecosystems. Consolidation of data within a common database is augmenting their efforts. A radiation-effects database, called FREDERICA, has been developed and is freely available online at www.frederica-online.org (Copplestone et al., 2008).

Bearing in mind that this volume is focussed on tropical radioecology, it should be noted that research facilities within tropical regions that may specifically be used to evaluate biotic responses to chronic or acute radiation exposure include the large-scale gamma-irradiation facility at Kajang in Malaysia (www.nuclearmalaysia.gov.my). Also, some of the earliest studies on the effects of radiation on the tropical environment were carried out in the rainforests of Costa Rica using a large ^{137}Cs source that was raised remotely to expose the surrounding forest (Odum and Pigeon, 1970).

1.4. ENVIRONMENTAL RADIOACTIVITY

1.4.1 Classes of Radioactivity

Broadly speaking, radionuclides may be classified into those of anthropogenic origin or those that are natural in the environment. Both are extensively used in environmental science. Anthropogenic (induced by human activity) radionuclides are fission products or are produced by the activation of target materials with neutrons or high-energy particles from reactors or accelerators or eluted from radioisotope generators. Natural environmental radionuclides are isotopes of primordial elements or are continuously produced by natural processes.

1.4.2 Naturally Occurring Radioactive Materials

Typically, environmental radionuclides are grouped under the generic headings cosmogenic and primordial, which pertain to their mode of production. They cover all forms of radioactive disintegration and a wide range of half-lives. Naturally occurring radioactive materials (NORMS) have always been present within the Earth's crust and are present in all biota to varying degrees.

1.4.2.1 Cosmogenic Radioactivity

Cosmogenic radionuclides are continuously produced from the bombardment of stable nuclides by cosmic radiation. The sun is the closest source of such radiation, but the cosmic ray flux is primarily galactic, modulated by the sun and by solar wind (Gosse and Phillips, 2001). The collision process, whereby the cosmic particle shatters or interacts with the stable element to form a radionuclide, is called spallation. This process mostly occurs in the upper atmosphere and lower troposphere, but also on the Earth's surface. It should be noted that the production rate is lower in the tropics than at the poles (in the atmosphere or at the crust) primarily due to the effect of the Earth's magnetic field (Gosse and Phillips, 2001).

Nuclides produced in the upper troposphere and lower stratosphere have a wide range of half-lives ranging from minutes or less (e.g., ^{39}Cl, 56 minutes) to millions of years (e.g., ^{10}Be, 1.6×10^6 years), but most range from tens to thousands of years. Common cosmogenic radionuclides include ^{26}Al, ^{14}C, ^{36}Cl, ^{3}H, ^{32}Si, ^{10}Be, and ^{7}Be. They subsequently diffuse to the surface of the land and the oceans, with rainfall or associated with particulates, from whence they disperse into ecosystems. The depositional processes show seasonal variability while favouring surface deposition in temperate zones over tropical and subtropical regions. Many of these nuclides, including ^{3}H, ^{14}C, and ^{36}Cl, were also formed during atmospheric nuclear weapons testing. The fallout component still dominates in many areas.

Nuclides produced on the Earth's surface can be used, for example, to determine how long any particular rock or soil has been exposed at the surface. This information can be very useful in interpreting glaciation histories (e.g., in the Bolivian Andes; Smith et al., 2005) or for other geomorphology studies (e.g., in Sri Lanka; Hewawasam et al., 2003) and in helping to clarify different environmental processes such as climate change (Leduc et al., 2006).

1.4.2.2 Primordial Radioactivity

The primordial radionuclides refer to those generated during the formation of the universe and whose half-lives are so long that they are still present today. They include potassium-40 (^{40}K), ^{238}U, ^{235}U, and ^{232}Th. One kilogram of soil typically contains 400 Bq ^{40}K (typical range 140–850 Bq), 35 Bq ^{238}U (typical range 16–110 Bq), and 30 Bq ^{232}Th (typical range 11–64 Bq) (UNSCEAR, 2000, Table 5, Annex B). These values are global averages and some soils vary substantially from these norms. Some of the areas with high levels of natural radioactivity (Cullen and Penna Franca, 1977) are in tropical zones. They include Kerala in southern India (Nair et al., 1999, Christa et al., 2011) and Guarapari in Brazil (Pfeiffer et al., 1981), both areas are influenced by high concentrations of Th associated with rare earth sand deposits, including the minerals monazite and xenotime. Tropical areas rich in uranium mineralisation include the Alligator Rivers region of northern Australia and Oklo in Gabon, Africa (see Chapter 3).

As we have already seen, ^{238}U, ^{235}U, and ^{232}Th also generate several radioactive progeny via their decay chains through a series of shorter lived nuclides (Section 1.2.3.2 and Figure 1.4). These decay chain products are considered secondary primordial radionuclides, despite their shorter half-lives, as they have been continuously created from their parents since the time of formation.

NORMS are also very useful in helping to better understand many natural processes, particularly in the case of the U- and Th-series radionuclides. The differing half-lives and chemistries of the radioactive elements within each series can lead to varying degrees of radioactive disequilibrium between sequential members in the series. The degree of disequilibrium established will depend on the length of time the material containing the parent radionuclides has been formed (which effects the potential ingrowth of the progeny nuclides), the stability or porosity of the material (which controls the relative mobility of the radioactive constituents), the type of radiation by which the nuclide decays (e.g., as distinct from beta and gamma emissions, alpha particles can cause physical defects in solids giving rise to potential escape paths or even ejecting the daughter radionuclide from the surface of the material), and the environmental conditions to which the material is exposed (e.g., inundation–episodic or perennial, subsurface burial, or exposure to wind or heat from any source).

Studies involving disequilibria have been widely discussed in more depth elsewhere, including within the Environmental Radioactivity series (e.g., Ivanovich and Harmon, 1982; Carroll and Lerche, 2003; McLaughlin et al., 2005; Shaw, 2007; Krishnaswami and Cochran, 2008; Froehlich, 2010) and will only be mentioned in brief in this chapter (Section 1.4.4).

1.4.3 Anthropogenic Radionuclides

Radioactivity was discovered in 1896 by Henri Becquerel (1852–1908) when studying compounds containing uranium, one year after the identification of X-rays by Wilhelm Roentgen (1845–1923). Soon after, Pierre Curie (1859–1906) and his with Marie Curie (1867–1934), working in the Sorbonne in Paris, found that thorium compounds also contained radioactivity with three types of emissions: (1) positive, (2) negative, and (3) neutral (later named alpha, beta, and gamma rays by Ernest Rutherford, 1871–1937). The first man-made element, technetium, was produced by deuterium bombardment of molybdenum in a cyclotron in Italy in 1937 (www.webelements.com). All isotopes of technetium are radioactive.

Many hundreds of nuclides have subsequently been produced artificially. Some are naturally occurring radionuclides released into the environment by industrial or mining processes (the so-called technologically enhanced nuclides, or TENORMS; see Section 1.4.3.2 for more discussion). Many are also produced as by-products of fission within nuclear reactors or testing of nuclear weapons (e.g., ^{137}Cs and ^{90}Sr), or as activation products from the neutron reactions that occur within nuclear reactors (e.g., ^{3}H and ^{60}Co). Such radionuclides are also used for medical purposes and as tracers of natural phenomena (see Section 1.4.4). All these nuclides have half-lives and decay properties that make them potentially useful in tracer studies, some of which are listed in Table 1.8 and discussed further in Section 1.4.4.

TABLE 1.8A Gamma Ray Emitters Extensively Employed for Applications

Nuclide	Decay Mode	$T_{1/2}$	E_γ keV (γ%)[a]	Application
^{7}Be	EC	53.28 d	478(10.6)	Environmental (see Table 1.9)
^{22}Na	β^+	2.602 y	511(180), from β^+, 1275 (100)	Biological tracer (e.g., ion transport)
^{24}Na	β^-	14.96 h	1369(100), 2754(100)	Industrial tracer
^{40}K	β^-/EC	1.28 10^8 y	1461 (11)	K/Ar dating geological samples

continued

TABLE 1.8A Gamma Ray Emitters Extensively Employed for Applications—Cont'd

Nuclide	Decay Mode	$T_{1/2}$	E_γ keV (γ%)[a]	Application
^{46}Sc	β^-	83.3 d	889(100), 1121(100)	Industrial tracer
^{51}Cr	EC	27.70 d	320(9.85)	^{51}Cr edetate, a radiopharmaceutical
^{54}Mn	EC	312.2 d	835(100)	tracer
^{59}Fe	β^-	44.51 d	1099(56), 1292(44).	
^{60}Co	β^-	5.27 y	1173(100), 1332(100)	Radiotherapy; radiation sterilisation
^{64}Cu	β^-/β^+	12.70 h	511(34.8)	
^{65}Zn	EC	243.9 d	1116(51)	
^{82}Br	β^-	35.34 h	554(71), 619(43), 698 (29), 777(84), 828(24), 1044(27)	Environmental tracer
^{95}Zr	EC	64.00 d	724(44), 757(55)	
^{99}Mo	β^-	65.92 h	141(91), 740(14)	$^{99}Mo/^{99m}Tc$ generators
^{99m}Tc	IT	6.007 h	141(89)	Radiopharmaceuticals
^{109}Cd	EC	462.6 d	88(3.7) in equilibration with Ag-109 (40s)	Nondestructive analysis sources
^{131}I	β^-	8.021 d	364(81)	Radiotherapy
^{137}Cs	β^-	30.07 y	662 (85)	Environmental (see Table 1.9) Gamma sources
^{192}Ir	β^-/EC	73.83 d	296(29), 308(30), 316 (83), 468(48)	Nondestructive testing
^{198}Au	β^-	2.696 d	412(96)	Cancer therapy; environmental and industrial tracing
^{226}Ra	α	1600 y	A wide range of X-rays and gamma rays from ^{226}Ra and progeny	Cancer therapy (Brachytherapy)
^{241}Am	α	433 y	59.5(36)	Smoke detectors

[a]*The percentage yield of the X-rays or gamma rays. Only higher energy radiations with yields exceeding 1% are listed. For a complete tabulation, readers are referred to http://ie.lbl.gov/toi/perchart.htm and thence to the individual isotopes.*

TABLE 1.8B Beta Emitters Extensively Employed for Applications

Nuclide	Decay mode E_{max} β^-	$T_{1/2}$	Applications
$^{3}H(T)$	β^- (100%) 18.6 keV	12.33 y	Environmental, (see Table 1.9)
^{10}Be	β^- (100%) 555.8 keV	1.51 106 y	Environmental, (see Table 1.9)
^{14}C	β^- (100%) 156.5 keV	5730 y	Environmental, (see Table 1.9)
^{32}Si	β^- (100%) 224 keV	150 y	Potential for dating the recent past (Fifield and Morgenstern, 2009)
^{32}P	β^- (100%) 1711 keV	14.3 d	Nuclear medicine, especially aspects of cancer diagnosis and treatment
^{36}Cl	β^- (98.1%) 708.6 keV EC + β^+ (1.9%)	3.01 10^5	Environmental, (see Table 1.9)
^{63}Ni	β^- (100%) 66.9 keV	100.1 y	Sources for electron capture technology (e.g., detection of explosives at airports)
^{85}Kr	β^- (100%) 687 keV γ (514 .keV) 0 .43 %	10.8 y	Industrial β source
^{90}Sr	β^- (100%) 546 keV	28.8 y	Extensive uses in medicine (superficial radiotherapy) and industry (thickness gauges)
^{210}Pb	β^- (100%) 63.5 keV	22.3 y	Environmental, (see Table 1.9)

1.4.3.1 Atmospheric Nuclear Testing

Many anthropogenic radionuclides were produced during the atmospheric testing of nuclear weapons from the 1950s to the 1970s (Figure 1.13). The Trinity test carried out at Alamogordo, New Mexico, on July 16, 1945, was the first such test on the planet. That test was quickly followed by the use of atomic bombs on Hiroshima and Nagasaki on August 6 and 9, respectively, of the same year and subsequent development and testing from that time over a period of more than two decades. Underground nuclear testing continued through the 1990s and in one case, in 2009. Most, but not all, of the tests were made in the Northern Hemisphere; hence, the fallout predominates north of the equator.

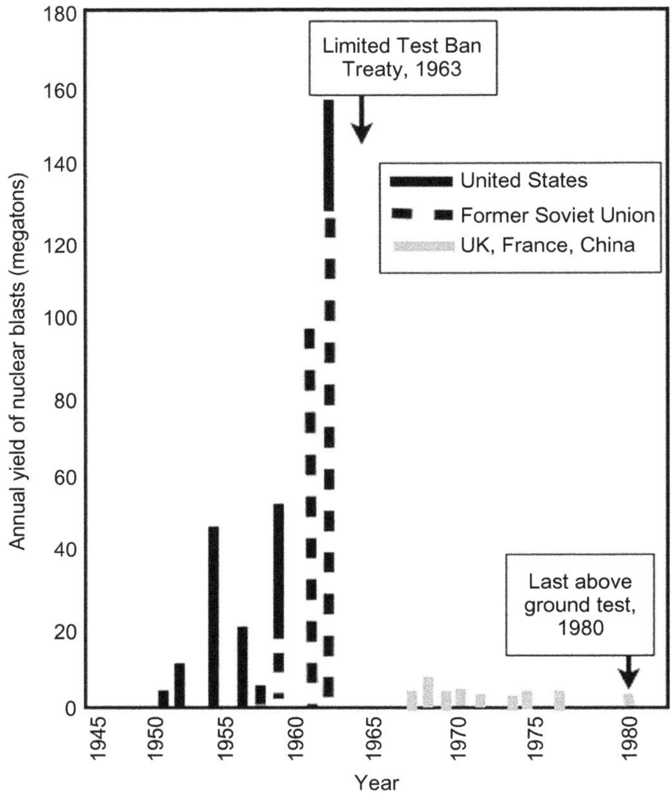

FIGURE 1.13 Atmospheric nuclear testing. The annual yield (megatons) of above ground nuclear tests (1945–1980).

Nuclear explosions produce many radioactive elements as a consequence of uranium and plutonium fission and neutron capture reactions. Most of these have short half-lives, but two radionuclides of particular concern have been ^{137}Cs and ^{90}Sr because of their relatively long half-lives of 30 and 28 years, respectively. ^{60}Co was of concern for some time, having a half-life of about 5 years, but this has, by now, mostly decayed. ^{14}C and ^{3}H, naturally occurring radionuclides with half-lives of about 5730 and 12 years, respectively, were also considerably increased in concentration in the environment as a consequence of nuclear testing. The so-called "bomb pulses" of these two nuclides have proved very useful in a wide range of biogeographic investigations including helping to determine the age of groundwaters (Aggarwal et al., 2007) and for forensic analyses of growth in humans and other biota (Stenström et al., 2010). Global fallout from nuclear weapons testing accounts for approximately 1000 times the radioactive emissions to the environment from nonmilitary sources (Shaw, 2007; www.epa.gov/rpdweb00/rert/

nuclearblast.html). By way of illustration, reference will be made to the global distribution of tritium (^3H). Tritium has three principal sources: (1) cosmogenic, (2) weapons fallout, and (3) industrial. The fallout component dominated from the mid-1960s until quite recently. The impact in the Southern Hemisphere was much less than in the Northern Hemisphere, as shown in Figure 1.14A, which depicts the annual average tritium levels in precipitation in Vienna and Invercargill (New Zealand). Data on which Figure 1.14 is based are downloaded from the IAEA WISER database at http://nds121.iaea.org/wiser/index.php. Other environmental nuclides, such as ^{14}C (Figure 1.15; http://web.science.uu.nl/AMS/Radiocarbon.htm) and ^{36}Cl, also exhibit a

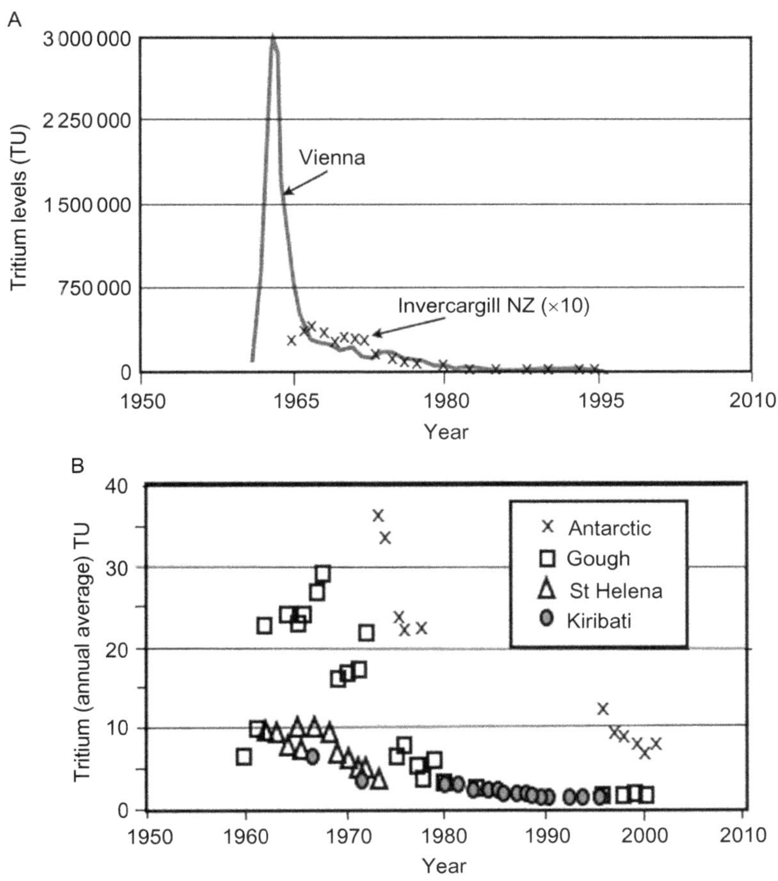

FIGURE 1.14 Tritium in precipitation. (A) Mean annual tritium levels in precipitation in Vienna and Invercargill, New Zealand. (Note that data from Invercargill has been scaled by a factor of 10.) (B) This figure illustrates that the level of environmental isotopes (tritium) tends to decrease with latitude as well as with time (after the bomb peak). *From the IAEA WISER database, http://nds121.iaea.org/wiser/gui/gnip_all_index.php, with permission.*

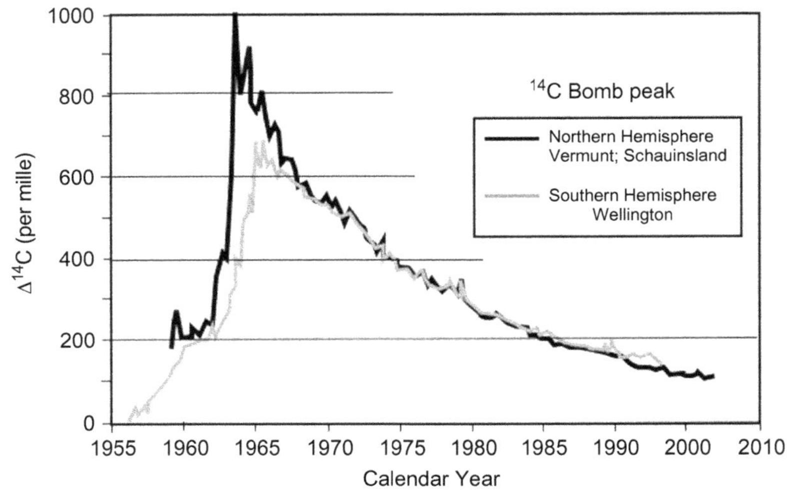

FIGURE 1.15 Atmospheric ^{14}C post-1956. Variation of the levels of ^{14}C in the Northern and Southern Hemispheres after 1956. *From the University of Utrecht AMS Lab, http://www.phys.uu.nl/ams/Radiocarbon.htm, with permission.*

well-defined bomb pulse, as well as a steady cosmogenic input. Applications vary depending on whether the study focuses on the recent (postnuclear) bomb pulse or the long-term (steady-state) cosmogenic input. A number of examples are listed in Table 1.9.

Since this book is dealing with tropical radioecology, it is worthwhile noting that the levels of fallout products from atmospheric weapons testing is lowest in the tropics and increases with increasing latitude toward the poles (see also Chapter 2). This has been illustrated in Figure 1.14B for atmospheric tritium by data from four IAEA stations located from near the equator to the Antarctic. The stations include: Kiribati (Lat. 1.87), St. Helena (Lat. 24.14), Gough Island (Lat. 40.32), and the Antarctic (Lat. 75.25). Despite the considerable variability in the data, the trends can be readily seen.

1.4.3.2 Technologically Enhanced Radioactive Materials

Humans initiated low-level radioactive contamination of the environment at least 2000 years ago, as a consequence of Roman silver and lead mining (Shaw, 2007; citing Rosman et al., 1997). In the 1920s, the first nuclear industry was based on the extraction of radium (^{226}Ra, a daughter of ^{238}U) for a range of medical, industrial, and scientific purposes (Williams and Kirchmann, 1990). However, it was the development of nuclear weapons toward the end of WWII that saw the start of the Atomic Age.

From a tropical and subtropical perspective, most anthropogenic sources have resulted from atmospheric weapons testing, as noted in the previous

TABLE 1.9 Environmental Radioisotopes and their Applications

Isotope	Source	Half-life	Application	Comments
^{3}H (tritium)	Atmospheric testing; cosmogenic	12.25 y	Groundwater recharge Oceanographic mixing	Environmental tritium is dominantly sourced by atmospheric testing; hence, tritium in water implies that a component of the water is from 'post nuclear' (i.e., post-1950) precipitation. Environmental tritium is used to identify groundwater recharge areas and to study oceanographic surface mixing processes.
^{7}Be	Cosmogenic	53 d	Sediment accumulation and redistribution over the previous half year	The presence of ^{7}Be in a sediment core indicates that the material has been at the surface over the past six months. Be-7 is correlated with ^{137}Cs in sediment cores, which provide information on the rate of accumulation of sediment in post-nuclear times (i.e., over the past 40 years).
^{14}C	Atmospheric testing	5730 y		In post-nuclear times, atmospheric testing is the major source of ^{14}C. It therefore exhibits a typical 'bomb' pulse, which has been used to study the uptake of CO_2 by the oceans; to investigate mixing processes in the upper layers of the oceans in post-nuclear times; and to provide evidence of the source of carbon (modern vegetation or mineral) in commercial products.
	Cosmogenic		Prehistory; evolution of coastal and other ecological systems in recent geological time; global climate change studies	^{14}C is used for dating carbon-containing materials up to 50,000 y. This technique has been widely used for dating artefacts, bones, and charcoal and thereby has made a major contribution to an understanding of prehistory and the evolution of ecological systems; studying groundwater flow patterns; investigating coastal processes through the dating of marine corals, of shell grit in dunes, and of sediments in estuaries and lakes; better understanding global oceanographic circulation patterns; and dating of tree rings, ice cores, and corals as a contribution to global climate change studies.
^{36}Cl	Atmospheric testing	310,000 y	Salinity, groundwater quality	The ^{36}Cl bomb pulse is used to study chloride migration (i.e., salinity processes in the unsaturated zone and in the modern groundwater).
	Cosmogenic		Dating of old groundwater	^{36}Cl is used to date groundwater up to 1 M years. More generally, it is used to study aspects of the chloride cycle (i.e., the evolution of groundwater quality).

continued

TABLE 1.9 Environmental Radioisotopes and their Applications—Cont'd

Isotope	Source	Half-life	Application	Comments
^{137}Cs	Atmospheric testing	30.1 y	Sedimentation, soil erosion	^{137}Cs is used to measure the rate of sedimentation and erosion in post-nuclear times.
^{232}Th ^{238}U ^{235}U	Primordial Primordial Primordial	$1.4\ 10^{10}$ y $4.47\ 10^{9}$ y $7.0\ 10^{8}$ y	Dating geological samples	^{232}Th, ^{238}U, and ^{235}U are the progenitors of ^{208}Pb, ^{206}Pb, and ^{207}Pb, respectively, and are used to date geological material older than 109 years. Ion microprobe methods are available to date individual inclusions within natural materials and have been widely applied to the dating of zircons.
^{234}U	U-238 series	245,500 y	Dating sediments and corals; dating sedimentary uranium deposits	$^{234}U/^{238}U$ ratios are measured in geochemical samples and natural waters. Together with other data they are used to study the hydrogeological stability of uranium deposits.
^{230}Th		75,380 y		$^{230}Th/^{234}U$ ratios are used for assessing the accumulation of sediments in estuaries and for the dating of corals.
^{226}Ra		1601 y		^{226}Ra is the parent of ^{222}Rn.
^{220}Rn		3.83 d		^{222}Rn in houses is extensively monitored as a source of radiation dose to the general public. It is also used in studies of large-scale atmospheric migration. Air masses, which have traversed land, are significantly higher in radon than those that have traversed only the sea.
^{210}Pb		22.3 y	Sediment dating over the past 100y	The ^{210}Pb method has been extensively applied to the dating of sediments in lakes and estuaries over the past 100 years. The sediment is cored and individual sections assayed for ^{210}Pb. There are two components of ^{210}Pb: the unsupported and the supported. The unsupported Pb, which is used in measuring the accumulation rates, is adsorbed on the surface from the decay of the ^{222}Rn dissolved in the associated water. The supported ^{210}Pb is derived from the decay of the ^{238}U within the sediment minerals. Experimentally, the supported lead is calculated from the measured uranium levels and is subtracted from the total to obtain the unsupported component.

section. Apart from such testing, the nuclear industry also makes controlled and/or limited releases of radioactivity into the broader environment by a number of means including:

- Tracer experiments to evaluate environmental processes
- Gaseous and liquid discharges from nuclear installations
- Wastes from hospitals and research facilities
- Materials seeping from radioactive waste repositories

Generally speaking, the impact of controlled releases of radionuclides from peaceful industrial and medical applications and waste disposal are small and restricted to the vicinity of the release points. The situation can be different for uncontrolled releases or accidents. For example, radioactivity from the accident in the nuclear power plant at Chernobyl in 1986 has been detected in the Antarctic and Arctic and all latitudes in between (Shaw, 2007). The effects of Fukushima seem to be more localised, but some trace levels of radioactivity have been detected across the Northern Hemisphere, as well as at some sites south of the equator including Australia, Fiji, and Papua New Guinea (http://www.ctbto.org/press-centre/highlights/2011/fukushima-related-measurements-by-the-ctbto/fukushima-related-measurements-by-the-ctbto-page-1/). The releases into the marine environment from the Fukushima Daiichi plant, consequent to the earthquake and tsunami in March 2011, have been detected 600 km off the Japanese shore (Buesseler et al., 2012) and in bluefin tuna caught on the other side of the Pacific, in California waters (Madigan et al. 2012). The activity concentrations in the wider aquatic environment are to be assessed in an IAEA coordinated research program currently under development (IAEA, 2011). That being said, away from obvious known sources, most concentrations of technogenic radionuclides arising from the nuclear industry are extremely low.

Other significant sources of TENORMS include coal and oil-fuelled power stations (releasing U- and Th-series radionuclides via gases and fly ash); the mining industry (e.g., uranium, tin, and phosphate); the oil industry (through the pumping of brines and groundwaters); general industry (e.g., in the manufacture of ^{241}Am smoke detectors), including housing and construction (e.g., through the accumulation of radon in enclosed spaces); and the recycling industries.

1.4.4 Applications of Environmental Radionuclides

Detailed studies of the dispersion of environmental radionuclides reflect the dynamics of natural processes that have occurred in the interval between the formation of the nuclide and sampling. Applications have been very widespread and include the following:

- Establishing an absolute time frame for many geological processes
- Dating of artefacts
- Studying the dynamics of groundwater and oceanic processes

- Estimating the rates of erosion and accumulation of fertile soils and sediments
- Understanding the dynamics of the transport of radionuclides in the geosphere and biosphere

Pre-WWII applications of radioisotope tracing were on a relatively small scale and restricted to the fields of chemistry, biology, and agriculture. In the immediate postwar period, there was considerable interest in extending the techniques to industry and the environment. Progress was rapid and was reported in the proceedings of two famous international conferences on the Peaceful Uses of Atomic Energy sponsored by the United Nations (UN, 1956; UN, 1958). Most classes of applications are covered in these seminal reports. Subsequent progress has resulted from enormous advances in the associated technologies of radiation detector science and microprocessors, as well as in data handling, storage, and visualisation. The following few paragraphs provide, in brief, some examples of radionuclide applications in tropical and subtropical studies. The list is certainly not exclusive in terms of the extent of approaches used; however, it exemplifies the range of nuclear techniques that are available to radioecologists to better understand their particular fields of endeavour.

The simplest environmental example is the use of tritiated water, HTO, as a tracer for the transport of water through subtropical ecosystems (e.g., Twining et al., 2011) or to determine the age of shallow groundwaters (Aggarwal et al., 2007). Tritium has also been used in conjunction with other tracers to map the dispersion and dilution of sewage effluent from tropical systems (IAEA, in press).

More complex geomorphological studies have been undertaken in the tropics and elsewhere. In one example from Carroll and Lerche (2003), multiple sites within the sedimentation zone of the Amazon River off the coast of Brazil were assessed using the ^{210}Pb excess technique to provide a detailed picture of the time-variable deposition in the region of interest. The fallout nuclide ^{137}Cs, often together with ^{210}Pb (a Rn daughter falling with rain) and/or ^{7}Be (a particle-reactive cosmogenic nuclide), has been applied in innumerable studies of soil erosion and deposition across the equatorial regions (Pfitzner et al., 2004), as well as in more temperate areas.

Atmospheric studies have sometimes used the same three radionuclides to evaluate vertical mixing and transport models (Rehfeld and Heimann, 1995; El-Hussein et al., 2001; Alonso Hernández et al., 2004). Studies with ^{222}Rn (Zahorowski et al., 2005) and ^{137}Cs (Kido et al., 2011) have been used to assess the movement of air masses across the tropics.

In some sophisticated biospheric examples, radiotracers may be used to study the uptake and dispersion of selected heavy metals. For example, some years ago, ^{65}Zn and ^{54}Mn were injected into the Magela Creek in the Alligator Rivers region of the Northern Territory during the annual wet (flood) season to assess their dispersion and bioavailability (Figure 1.16; IAEA, in press). Combined radionuclides have also been used in multitracer experiments to study biokinetics, biomonitoring, and bioaccumulation in, for example, the tropical lagoons of New Caledonia (Metian and Warnau, 2008; Metian et al., 2008;). Also, in conjunction with inductively-coupled plasma mass spectrometry (ICP-MS) analyses, they have been used to assess the history of exposure to contaminants by crocodilians (Twining et al., 1999) and to identify the home ranges of those animals (Markich et al., 2002).

In a terrestrial example, ^{65}Zn, ^{85}Sr, and ^{134}Cs were applied to two types of agricultural soils in the Northern Territory of Australia to assess bioaccumulation into human food crops (Twining et al., 2004). Similar studies have been carried out in a range of tropical countries (see Chapters 3 and 5 for more details on these experiments).

Despite the demonstrated usefulness of radiotracers for environmental studies, it is becoming increasingly difficult to justify to regulators large-scale studies involving their release into the environment. Readers will recall that the ICRP criterion of justification involves balancing the benefit of the new knowledge against any radiation detriment to the workers, the community, and the environment. Assessments by regulators include whether the knowledge can be obtained by other means, including advanced numerical modelling or new experimental techniques. The challenge for the radiotracer practitioner is to couple ever-decreasing levels of radionuclide release with other techniques to obtain insights into the behaviour of the ecosystems that would otherwise be unavailable. Of critical interest is the use of tracers to evaluate numerical models, which are the bases for prediction and therefore for management decisions designed to ensure the sustainability of ecosystems for future generations.

1.5. CONCLUDING COMMENTS

Bearing in mind the likelihood that nuclear energy will be a substantial component of social and economic development across the tropics in future decades, this chapter has sought to introduce readers to the scientific principles that underpin a profound understanding of tropical radioecology. Attempts were also made to share some insights into the skills and knowledge required of practitioners in the field. To this end, the chapter included an introduction to the science of radioactivity and radiation and covered the principles of measurement and the assessment of experimental uncertainties.

Reference was made to the fundamentals of radiation protection. However, there is a modern recognition that the assessment of radiation effects on critical human groups is an inadequate basis for generalisation to the environment at large. A brief introduction to environmental radiological protection was therefore included. However, readers are also directed to more detailed discussions of dose assessment in Chapter 7.

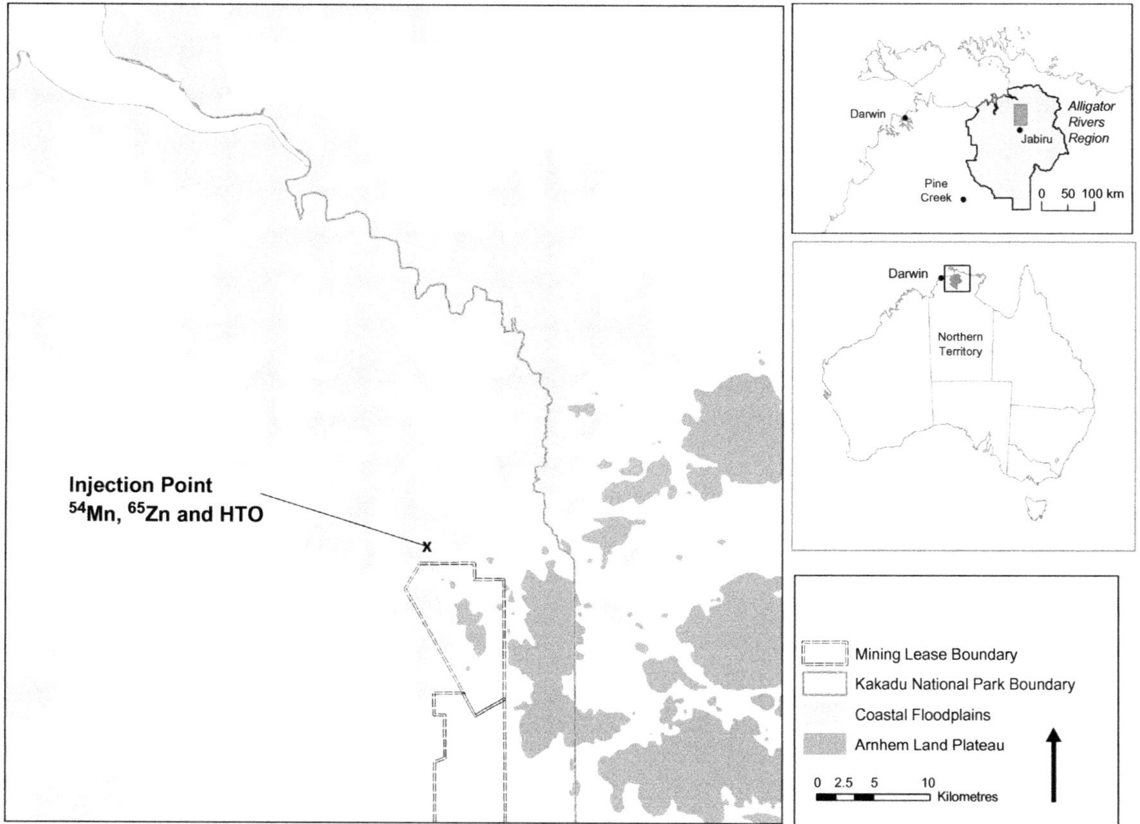

FIGURE 1.16 Radioactive tracer study of the transport of heavy metals (zinc and manganese) across the Magela Plains within the Alligator Rivers Region of the Northern Territory of Australia: Location map

Tropical radioecology is a modern discipline firmly rooted in a broad spectrum of fundamental sciences that produce extensive and accessible databases, many of which are maintained by national laboratories and a range of international agencies. Extensive use of a number of these databases is made in this chapter. Their contribution to public domain science and therefore to general understanding of our fragile planet is difficult to overstate.

In essence, this book is concerned with the sources of radionuclides in the tropics and the transport, behaviour, fate, and impact of those radionuclides throughout regional ecosystems. Whilst much of the information is generic to any environment, and most of the studies to date have been undertaken in temperate regions, the significance of tropical and subtropical biodiversity and endemic ecologies is a given, as are the very high human populations across the tropics. These factors, coupled with massive and burgeoning aquaculture, agriculture, forestry, and mining industries, which in turn lead to the remobilisation and redistribution of radionuclides, underline the importance of the study of tropical radioecology, which cannot be overestimated. The following chapters in this volume will lead the reader progressively through the physical, chemical, and biotic factors affecting radioecology, and the modelling tools available to integrate those factors in a range of tropical systems. They provide an excellent basis from which future scientific investigations and appropriate environmental management can be undertaken to balance the costs and benefits of nuclear developments in tropical and subtropical areas of our planet.

REFERENCES

Aggarwal, P.K., Gat, J.R., Froehlich, K.F., 2007. Isotopes in the Water Cycle: Past, Present, and Future of a Developing Science. Springer, Berlin, p. 381.

Airey, P.L., Howard, C.J., 2003. Comment on a General Property of Radionuclides. Journal of Radioanalytical and Nuclear Chemistry 257 (1–3), p. 45–46(2).

Alonso Hernández, C.M., Cartas Águila, H., Díaz Asencio, M., Muñoz Caravaca, A., 2004. Reconstruction of ^{137}Cs signal in Cuba using ^{7}Be as tracer of vertical transport processes in the atmosphere. J. Environ. Radioact. 75 (2), 133–142.

Audi, G., Wapstra, A.H., 1993. The 1993 Atomic Mass Evaluation (I) Atomic Mass Table. Nuclear Physics A565, 1–65.

Bergonie, J., Tribondeau, L., 1906. De quelques résultats de la radiotherapie et essai de fi xation d'une technique rationnelle. Comptes-Rendus des Séances de l'Académie des Sciences 143, 983–985.

BEIR V Report, 1990. Committee on the Biological Effects of Ionizing Radiation (BEIR V), *National Research Council 1990*, Executive Summary, p 4.

Blaylock, B.G., 1969. Fecundity of a *Gambusia affinis* population exposed to chronic environmental radiation. Radiat. Res. 37, 1088–1094.

Bryson, B., 2003. A Short History of Nearly Everything. Broadway Books, New York.

Buesseler, K., Jayne, S., Fishe, N., Rypina, I., Baumann, H., Baumann, Z., Breier, C., Douglass, E., George, J., Macdonald, A., Miyamoto, H., Nishikawa, J., Pike, S.,

Yoshida, S., 2012. Fukushima-dervied radionuclides in the ocean and biota off Japan. Proceedings of the National Academy of Sciences. www.pnas.org/cgi/doi/10.1073/pnas.1120794109.

Calabrese, E.J., 2004. Hormesis: From marginalization to mainstream: A case for hormesis as the default dose-response model in risk assessment. Toxicol. Appl. Pharmacol. 197, 125–136.

Carroll, J., Lerche, I. (Eds.), 2003. Sedimentary processes: Quantification using radionuclides. In: Radioactivity in the Environment Book Series, vol. 5. Elsevier, Amsterdam, p. 282. ISBN 978 0080443003.

Copplestone, D., Hingston, J., Real, A., 2008. The development and purpose of the FREDERICA radiation effects database. J. Environ. Radioact. 99, 1456–1463.

Christa, E.P., Jojo, P.J., Vaidyan, V.K., Anilkumar, S., Eappen, K.P., 2011. Radiation dose in the high background radiation area in Kerala, India. Radiat. Prot. Dosimetry. 148 (4): p. 482–486. doi:10.1093/rpd/ncr198.

Cullen, T.L., Penna Franca, E. (Eds.), 1977. Proceedings of the International Symposium on areas of high natural radioactivity: Pocos de Caldas. Brazil, June 16–20, 1975. Academia Brasileira de Ciencias, Rio de Janeiro.

Duport, P., 2003. A database of cancer induction by low-dose radiation in mammals: Overview and initial observations. International Journal of Low Radiation 1, 120–131.

El-Hussein, A., Mohamemed, A., Abd El-Hady, M., Ahmed, A.A., Ali, A.E., Barakat, A., 2001. Diurnal and seasonal variation of short-lived radon progeny concentration and atmospheric temporal variations of ^{210}Pb and ^{7}Be in Egypt. Atmos. Environ. 35 (25), 4305–4313.

Feinendegen, L.E., 2005. Evidence for beneficial low-level radiation effects and radiation hormesis. Br. J. Radiol. 78 (925), 3–7.

Fifield, K., Morgenstern, U., 2009. Silicon-32 as a tool for dating the recent past. Review Article. Quaternary Geochronology 4 (5), 400–405.

Firestone, R.B., Shirley, V.S. (Eds.), 1998. Table of Isotopes. Eighth ed. John Wiley and Sons Inc, New York. ISBN 0471-14918-7.

Freedman, W.L., Kolb, E.W., 2006. Chapter 1. Cosmology. In: Fraser, G. (Ed.), The New Physics for the Twenty First Century. Cambridge University Press, Cambridge, pp. 13–40. ISBN-13: 978 0 521 81600 9.

Froehlich, K. (Ed.), 2010. Environmental radionuclides: Tracers and timers of terrestrial processes. In: Radioactivity in the Environment Book Series, vol. 16. Elsevier, Amsterdam, p. 438. ISBN 978 0 08 043873 3.

Geyh, M.H., Sleicher, H., 1990. Absolute Age Determinations: Physical and Chemical Methods and their Applications. Springer Verlag, Berlin, Heidelberg.

Gosse, J.C., Phillips, F.M., 2001. Terrestrial *in situ* cosmogenic nuclides: Theory and application. Quaternary Science Reviews 20, 1475–1560.

Hall, E.J., Giaccia, A.J., 2006. Radiobiology for the Radiologist. Lippincott, Williams and Wilkins ISBN 0-7817-4151-3.

Hewawasam, T., von Blanckenburg, F., et al., 2003. Increase of human over natural erosion rates in tropical highlands constrained by cosmogenic nuclides. Geology 31 (7), 597–600.

Hinton, T.G., 1998. Risks from exposure to radiation. In: Newman, M.C. (Ed.), Fundamentals of Ecotoxicology. Ann Arbor Press, Chelsea, MI, USA ISBN- 1575040131 (Chapter 14).

Hinton, T.G., Beford, J.S., Congdon, J.C., Whicker, F.W., 2004. Effects of radiation on the environment: A need to question old pardigms and enhance collaboration among radiation biologists and radiation ecologists. Radiat. Res. 162, 332–338.

Hubbell, J.H., Seltzer, S.M., 1996. Tables of X-Ray Mass Attenuation Coefficients and Mass Energy-Absorption Coefficients from 1 keV to 20 MeV for Elements Z=1 to 92 and 48

Additional Substances of Dosimetric Interest NISTIR 5632. http://www.nist.gov/pml/data/xraycoef/index.cfm.

IAEA, 1992. Effects of Ionizing Radiation on Plants and Animals at Levels Implied by Current Radiation Protection Standards. Technical Reports Series No. 332. International Atomic Energy Agency, Vienna.

IAEA, 1996. International Basic Safety Standards for Protection against Ionizing Radiation and for the Safety of Radiation Sources. IAEA Safety Series 115 (STI/PUB/996).

IAEA, 2004. Quantifying Uncertainty in Nuclear Analytical Measurements. IAEA-TECDOC-1401.

IAEA, 2010. Radiation biology: A Handbook for Teachers and Students. IAEA-TCS-42. Vienna, Austria. ISSN 1018-5518.

IAEA, 2011. Regional Benchmarking for the Pacific and Adjacent Marine Regions of the Extent and Impact of the Releases of Radioactivity from the Fukushima Daiichi Nuclear Power Plant into the Marine Environment. IAEA Technical Cooperation Program 2012-2013 (Under development).

IAEA, 2012. Use of Intentionally Discharged Radioactive Tracers to Study Surface Water Processes. IAEA TECDOC, in press.

ICRP, 1977. Recommendations of the International Commission on Radiological Protection, Publication 26. Pergamon Press, Oxford and New York.

ICRP, 1991. The 1990 Recommendations of the International Commission on Radiation Protection. Publication 60, 21, 1–3. Elsevier Science Ltd, Kidlington, Oxford, UK.

ICRP, 2003. A Framework for Assessing the Impact of Ionising Radioation on Non-Human Species. Publication 91. Elsevier Science Ltd, Kidlington, Oxford, UK. ISBN 13: 978-0-08-044310-2.

ICRP, 2007. The 2007 Recommendations of the International Commission on Radiological Protection. Publication 103. Ann. ICRP 37 (2–4) Elsevier.

ICRP, 2009. Environmental Protection: The Concept and Use of Reference Animals and Plants. Publication 108 Elsevier ISBN-13 978-0-444-52934-3.

ICRU, 2001. International Commission on Radiation Units and Measurements, Report 65, Quantities, Units, and Terms in Radioecology. Journal of the ICRU 1, 48.

Ivanovich, M., Harmon, R.S. (Eds.), 1982. Uranium Series Disequilibrium: Applications to Environmental Problems. Clarendon Press, Oxford.

Kaiser, J., 2003. HORMESIS: Sipping from a poisoned chalice. Science 302, 376–379.

Kido, H., et al., 2011. The simulation of long-range transport of Cs-137 by yellow sand phenomena in East Asia. Progress in Nuclear Science and Technology 1, 420–423.

Krishnaswami, S., Cochran, J.K., 2008. U-Th series nuclides in aquatic systems. In: Radioactivity in the Environment Book Series, vol. 13. Elsevier, Amsterdam, p. 458. ISBN 978 0 08 045012 4.

Leduc, G., Thouveny, N., et al., 2006. Authigenic Be-10/Be-9 signature of the Laschamp excursion: A tool for global synchronisation of paleoclimatic archives. Earth and Planetary Science Letters 245 (1–2), 19–28.

Lowenthal, G.L., Airey, P.L., 2001. Practical applications of radioactivity and nuclear radiations. In: Cambridge University Press 181–231. ISBN-13 978-0-521-01980-4 (Chapter 7).

Luckey, T.D., 1991. Radiation Hormesis. CRC Press, Boca Raton, FL. ISBN 0-8493-6159-1.

Madigan, D., Baumann, Z., Fisher, N., 2012. Pacific bluefin tuna transport Fukushima-dervided radionuclides from Japan to California. Proceedings of National Academy of Sciences. www.pnas.org/cgi/doi/10.1073/pnas.1204859109.

Markich, S.J., Jeffree, R.A., Harch, B.D., 2012. Catchment-specific element signatures in estuarine crocodiles (*Crododylus porosus*) from the Alligator Rivers Region, northern Australia. Sci. Total Environ. 287, 83–95.

McLaughlin, J.P., Simopoulos, S.E., Steinhausler, F., 2005. The natural radiation environment VII. In: Radioactivity in the Environment Book Series, vol. 7. Elsevier, Amsterdam, p. 1191. ISBN 0 08 044137 8.

Metian, M., Warnau, M., 2008. The tropical brown alga *Lobophora variegata* (Lamouroux) Womersley: A prospective bioindicator for Ag contamination in tropical coastal waters. Bull. Environ. Contam. Toxicol. 81 (5), 455–458.

Metian, M., Bustamante, P., Hédouin, L., Warnau, M., 2008. Accumulation of nine metals and one metalloid in the tropical scallop *Comptopallium radula* from coral reefs in New Caledonia. Environ. Pollut. 152 (3), 543–552.

Mendeleev, D., 1869. Zeitscrift für Chemie 12, 405–406. A translation of On the Relationship of the Properties of the Elements to their Atomic Weights by Carmen Giunta is available at http://web.lemoyne.edu/~GIUNTA/EA/MENDELEEVann.HTML.

Nair, M.K., Nambi, K.S., Amma, N.S., Gangadharan, P., Jayalekshmi, P., Jayadevan, S., et al., 1999. Population study in the high natural background radiation area in Kerala, India. Radiat. Res. 152 (Suppl. 6), S145–S148.

Odum, H.T., Pigeon, R.F. (Eds.), 1970. A tropical rainforest. A study of irradiation and ecology at El Verde, Puerto Rico, Nat. Tech. Info. Service 1678 Springfield VA.

Pfeiffer, W.C., Penna-Franca, E., Ribeiro, C.C., Nogueira, A.R., Londres, H., Oliveira, A.E., 1981. Measurements of environmental radiation exposure dose rates at selected sites in Brazil. An. Acad. Bras. Cienc. 53 (4), 683–691.

Pfitzner, J., Brunskill, G., et al., 2004. ^{137}Cs and excess ^{210}Pb deposition patterns in estuarine and marine sediment in the central region of the Great Barrier Reef Lagoon, north-eastern Australia. J. Environ. Radioact. 76 (1–2), 81–102.

Pirard, B., D'uston, C., Maurice, S., Gasnault, O., 2005. Performance limits of new generation scintillators for planetary gamma ray spectroscopy. Lunar and Planetary Science XXXVI, http://www.lpi.usra.edu/meetings/lpsc2005/pdf/2187.pdf.

Rehfeld, S., Heimann, M., 1995. Three dimensional atmospheric transport simulation of the radioactive tracers ^{210}Pb, ^{7}Be, ^{10}Be, and ^{90}Sr. Journal of Geophysical Research 100 (D12), 26 141–26, 161.

Rosman, K.J.R., Chisholm, W., Hong, S., Candelone, J.P., Boutron, C.F., 1997. Lead from Carthaginian and Roman Spanish mines isotopically identified in Greenland ice dated from 600 B.C. to 300 A.D. Environ. Sci. Technol. 31, 3413–3416.

Shaw, G., 2007. Radioactivity in the terrestrial environment. Radioactivity in the Environment book series, vol. 10. Elsevier (M. Baxter, series editor).

Smith, J.A., Seltzer, G.O., Farber, D.L., Rodbell, D.T., Finkel, R.C., 2005. Early local last glacial maximum in the tropical Andes. Science 308, 678–681.

Stenström, K., Skog, G., Nilsson, C.M., Hellborg, R., Svegborn, S.L., Georgiadou, E., et al., 2010. Local variations in ^{14}C – How is bomb-pulse dating of human tissues and cells affected. Nuclear Instruments and Methods in Physics B: Beam Interactions with Materials and Atoms 268 (7–8), 1299–1302.

Tubiana, M., Feinendegen, L.E., Yang, C., Kaminski, J.M., 2009. The linear no-threshold relationship is inconsistent with radiation biologic and experimental data. Radiology 251, 13–22.

Twining, J.R., Payne, T.E., Itakura, T., 2004. Soil-water distribution coefficients and plant transfer factors for ^{134}Cs, ^{85}Sr and ^{65}Zn under field conditions in tropical Australia. J. Environ. Radioact. 71, 71–87.

Twining, J.R., Hughes, C.E., Harrison, J.J., Hankin, S., Crawford, J., Johansen, M., et al., 2011. Biotic, temporal, and spatial variability of tritium concentrations in transpirate samples collected in the vicinity of a near-surface low-level nuclear waste disposal site and nearby research reactor. J. Environ. Radioact. 102, 551–558.

Twining, J.R., Markich, S.J., Prince, K.E., Jeffree, R.A., 1999. Osteoderms of estuarine crocodiles record their enhanced Pb exposure in Kakadu National Park. Environ. Sci. Technol. 33, 4396–4400.

UN, 1956. Proceedings of the International Conference on the Peaceful Uses of Atomic Energy. Geneva 1955 (United Nations New York).

UN, 1958. Proceedings of the International Conference on the Peaceful Uses of Atomic Energy. Geneva 1958 (United Nations New York).

Unger, L.M., Trubey, D.K., 1982. Specific Gamma Ray Dose Constants for Nuclides Important for Dosimetry and Radiological Assessment. Report ORNL/RSIC-45/R1. http://www.orau.org/documents/ivhp/health-physics/ornl-rsic-45.pdf.

UNSCEAR, 1996. Sources and Effects of Ionizing Radiation. United Nations Scientific Committee on the Effects of Atomic Radiation, NY Scientific Annex within 1996 Report to the General Assembly, United Nations Publications, New York.

UNSCEAR, 2000. Sources and Effects of Ionizing Radiation, UNSCEAR 2000. Report to the General Assembly with Scientific Annexes. United Nations Publications, New York.

UNSCEAR, 2008. Effects of Ionizing Radiation on Non-human Biota, United Nations Scientific Committee on the Effects of Atomic Radiation. United Nations Publications, New York.

Whicker, F.W., Schultz, V., 1982. Radioecology: Nuclear Energy and the Environment. CRC Press, Boca Raton, FL.

Williams, A.R., Kirchmann, R.J., 1990. Radium: A historical introduction. Ch1-1 in The Environmental Behaviour of Radium. IAEA Vienna. vol. 1. IAEA Technical Report Series No. 310.

Zahorowski, W., Chambers, S., Wang, T., Kang, C.H., Uno, I., Poon, S., et al., 2005. Radon-222 in boundary layer and free tropospheric continental outflow events at three ACE-Asia sites. Tellus 57B, 124–140.

Chapter 2

Radionuclide Behaviour and Transport in the Tropical Atmospheric Environment

Paul Martin[1] and John L. McBride[2]
[1]*Australian Radiation Protection and Nuclear Safety Agency, Yallambie, Victoria, Australia*
[2]*School of Earth Sciences, University of Melbourne, Parkville, Australia*

2.1. INTRODUCTION

The behaviour of radionuclides in the atmosphere is of importance for several reasons. They may be used for investigations of environmental processes, for example for testing of models for atmospheric mixing or for estimation of residence times of aerosols. They may be used as a tracer or marker for anthropogenic activity, such as tracking of industrial pollution or detection of nuclear weapons testing. Their study may also be aimed at estimation of the radiological impact of the presence of the radionuclides themselves, i.e. for estimation of dose to humans and/or to non-human biota.

The distribution of a radionuclide in the Earth's atmosphere is determined by five main factors. First, there is the origin of the radionuclide (i.e., the mechanism that introduces it into the atmosphere and the location/distribution of that introduction). Second, there is the degree to which the radionuclide is involved in chemical reactions and/or becomes associated with airborne particles. Third, horizontal and vertical mixing mechanisms (primarily turbulent) cause dilution and dispersion to parts of the atmosphere distant from its place of origin. Fourth, wet and dry deposition to the Earth's surface may result in its physical removal. Finally, radioactive decay, quantified by the radionuclide's half-life, acts as another removal mechanism.

The main source types are:

Cosmic radiation. Interaction of cosmic rays with the atmosphere yields several radionuclides, such as ^{3}H, ^{7}Be, ^{10}Be, ^{14}C, ^{36}Cl, and ^{32}Si. These are primarily produced in the stratosphere and, to a lesser extent, in the troposphere.

Radioactivity in the Environment, Vol. 18. http://dx.doi.org/10.1016/B978-0-08-045016-2.00002-3

Resuspension of dust. Airborne dust generated by physical means, such as by the action of wind on the Earth's surface or by industrial processes, contains radionuclides such as ^{137}Cs and members of the uranium and thorium decay series.

Exhalation of gases from the Earth's surface. This is the case for the three naturally occurring isotopes of radon: ^{222}Rn, ^{220}Rn, and ^{219}Rn.

Decay of a radioactive parent in the atmosphere. An example is decay of ^{222}Rn, resulting in formation of several radioactive progeny in a decay series.

Releases from operations of the nuclear fuel cycle, for example, uranium mining, fuel fabrication, reactor operation, and fuel reprocessing. This refers to both planned and accidental releases.

Nuclear weapons production and testing. The set of radionuclides released to the atmosphere depends on whether the test is above or below ground.

Radionuclides originating from the Earth's surface may be referred to as "terrigenic," and those from cosmic radiation as "cosmogenic." Radionuclides produced as a result of human activities are referred to as "anthropogenic" (Hu et al., 2010). Some radionuclides have both cosmogenic and anthropogenic source terms, such as ^{3}H and ^{36}Cl.

In some cases, the radionuclide will undergo chemical reactions with other components of the atmosphere. For example, ^{14}C enters the carbon cycle, while ^{3}H enters the hydrogen cycle. Their behaviour has been described in other reviews (Happell et al., 2004; Cook et al., 2010) and will not be discussed further here.

Isotopes of the noble gases radon, xenon, and krypton do not take part in chemical reactions, and they do not become associated with particles. Rather, they are dispersed in the atmosphere until they are removed by radioactive decay. In the case that a steady-state condition applies, the total activity present in the atmosphere will essentially depend upon the rate of introduction and the half-life of the radionuclide.

The behaviour of other radionuclides is dependent upon their association with particles or aerosols. In this case, their behaviour may also be influenced by the size distribution of those aerosols. Examples include ^{7}Be, ^{10}Be, ^{210}Pb, ^{210}Bi, and ^{210}Po.

This chapter reviews and discusses observations and current understanding of the distribution of radionuclides in the tropical atmosphere. Of course, the tropical atmospheric environment cannot be considered in isolation; however, there are significant differences in meteorological systems between the tropics and the higher latitudes and, consequently, in radionuclide transport mechanisms. Therefore, the chapter focuses primarily on such differences based on information from studies published in the open literature. The major transport processes are described in Section 2.2, cosmogenic and anthropogenic radionuclides are discussed in Section 2.3, and terrigenic radionuclides are discussed in Section 2.4.

2.2. THE LARGE-SCALE STRUCTURE AND TRANSPORT PROCESSES IN THE TROPICS

2.2.1 Long-term Mean, Zonally Averaged Structure

The transport of radionuclides and other substances in the Earth's atmosphere is governed by the large-scale flow fields referred to by meteorologists as the large-scale circulation, or general circulation. The major forcing mechanism for this flow is the radiative imbalance between the tropics and poles illustrated in Figure 2.1. This imbalance requires a transport of energy by atmospheric and oceanic currents to maintain energy balance. Approximately 75% of the meridional heat transport is carried out by atmospheric circulations and the remaining 25% by ocean currents.

In response to this transport requirement, the atmosphere has a direct circulation, known as the Hadley cell, extending from the tropics to subtropical latitudes. This is illustrated in Figure 2.2. The figure is schematic in character but is a useful foundation for interpretation of the actual flow in subsequent figures. The heavy solid lines represent the direct thermal cells known as the Hadley cells or circulations. These emanate at the surface in the subtropics and consist of equatorward flow in the easterly trade winds. The source or beginning of the trade flow is the axis of the subtropical ridge of high pressure circling the globe at approximately 30° latitude in either hemisphere. An important aspect of the trade wind flow for transport is the steadiness of the flow, usually defined as the monthly vector mean wind speed divided by

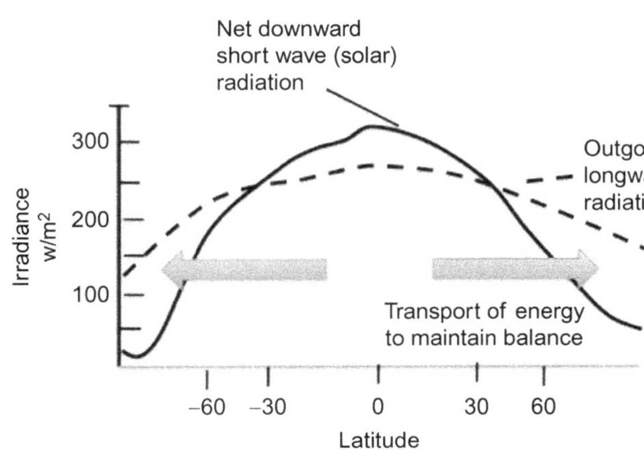

FIGURE 2.1 Annual mean radiative imbalance at the top of the atmosphere. The solid line is the absorbed short wave radiation from the sun, its latitudinal variation reflecting the latitudinal variation of solar insolation. The dashed line is the outgoing longwave, or the earth, radiation. Its latitudinal variation is proportional to σT^4, where σ is the Stefan-Boltzmann constant and T is the effective radiative temperature at that latitude. The horizontal scale is such that the spacing between latitudes is proportional to the area between them.

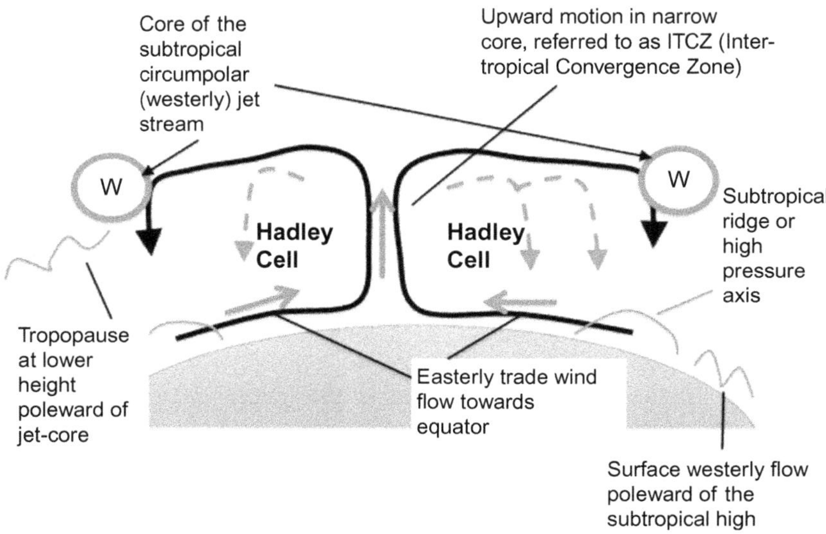

FIGURE 2.2 Schematic diagram of the "Hadley cell," which represents the zonally averaged atmospheric circulation in the tropics.

the scalar mean wind. In the main regions of the Atlantic and Pacific trades, the steadiness usually exceeds 90% (Garstang, 1967; Brummer et al., 1974; Enfield, 1981). Consequently, this means that radionuclides and other chemicals can cross the width of the oceans.

The trade flow from the Northern and Southern Hemispheres meets in a narrow zone of precipitation or upward vertical motion known as the Intertropical Convergence Zone (ITCZ). At upper levels, the return flow is away from the equator. Conservation of angular momentum with respect to the Earth's axis dictates that the poleward flow becomes westerly with increasing latitude. The poleward axis of the direct, or Hadley, cell is marked by the subtropical jet stream, whose core lies approximately over the location of the surface subtropical ridge. For practical purposes, the tropical circulation could be defined as occupying the region between the two subtropical highs at the surface, or equivalently, the two (one in each hemisphere) subtropical jet cores at upper altitudes. As will be shown in the following sections, these zones travel north and south following the seasonal cycle.

These schematic features are fundamental to understanding long-range transport and dispersion in the tropics. Also important is the vertical motion. The upward motion in the tropics is effectively confined to the upward motions within the narrow convective cores of deep cumulonimbus clouds. There are energy conservation reasons for the large imbalance between the areas of upward and downward motion. The upward vertical cells are called "vertical hot towers," a phrase coined by Riehl and Malkus (1958). A consequence from this confinement of the upward motion to the "hot towers" for transport of radionuclides is that their transport vertically from the surface also generally only occurs within the convective regions, which is also the location where wet deposition occurs.

Ninety percent or more of the tropics is a region of slow downward vertical motion or subsidence. The downward motion is from low to high atmospheric pressure, which causes compressional warming. From energy balance considerations, this counters the net radiative flux divergence or radiational cooling of the free troposphere. The net radiational cooling is of the order of 1 degree per day, which requires an ~30 hPa per day downward motion. This defines the effective time scale of the Hadley circulation, as the downward branch takes approximately 30 days to descend from the tropopause (~100 hPa) to the surface (~1000 hPa). This has implications for the distributions of radionuclides with an upper tropospheric source, such as ^{7}Be. The vertical distributions over most of the tropics in the nonconvective or trade wind regimes generally reflect this slow downward transport.

As previously described, the upward transport in the tropics occurs primarily in convective towers that penetrate from near the surface to the upper troposphere. This has implications for the modelling of transport processes in global circulation models. The horizontal resolution of such models is generally significantly larger than the size of a convective updraft. Hence the rainfall processes, as well as vertical transports of momentum, water vapour, and aerosols, are not explicitly resolved by the grid-scale dynamics of such models. Rather, these are carried out by the convective parameterisations of the models. Thus there is extensive literature on the sensitivity of global chemical transports to the convective parameterisations used. Examples in the context of the global simulations of transports of ^{222}Rn (radon) and ^{210}Pb are given in the literature (Jacob and Prather, 1990; Mahowald et al., 1997; Zhang et al., 2008; Feng et al., 2010).

Some of the circulation characteristics just described can be seen in Figure 2.3, which is a depiction of the long-term average flow next to the surface during the month of January (Figure 2.3a) and July (Figure 2.3b). The dashed line stretching across the globe in both panels is the surface location of the axis of the ridge of high pressure, or the subtropical high belt. As can be seen, the "subtropical ridge" is continuous in both hemispheres, circling the globe on this long-term average depiction. On individual days, it is represented by individual high-pressure weather systems. On the equatorward side of the subtropical ridge, the easterly trade winds flow across the tropical oceans toward the equator. In the upper panel of Figure 2.3a is a zoomed insert illustrating the trade wind flow between the two high-pressure belts.

Referring back to the schematic diagram of the flow in Figure 2.2, the upper level, poleward branch of the Hadley circulation occurs in the troposphere immediately below the tropopause. Given the primarily stratospheric origin of cosmogenic radionuclides and the importance of radionuclides with

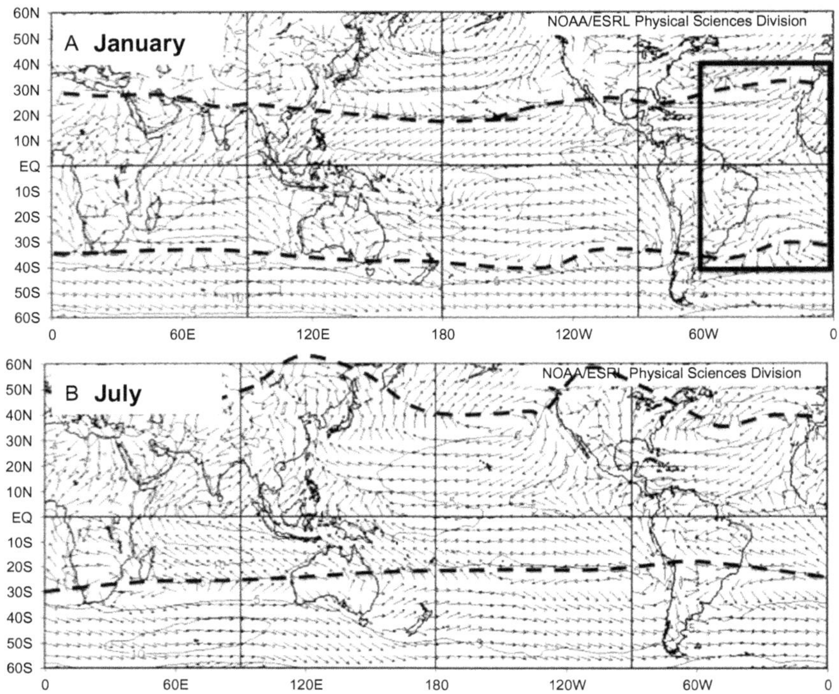

FIGURE 2.3 Surface flow in (a) January and (b) July with the location of the subtropical surface-pressure ridge marked by a dashed line.

a primary terrigenic source (e.g., radon and its progeny), an important component of the circulation is the distribution and frequency of exchange of mass between the troposphere and stratosphere, known as stratosphere–troposphere exchange (STE). Global scale aspects of STE are reviewed by Holton et al. (1995) and Stohl et al. (2003). The zonally averaged picture is a global scale meridional cell with upward motion from the troposphere into the stratosphere in the tropics, poleward transport within the stratosphere, and downward motion through the tropopause into the troposphere in mid and high latitudes. The timescale of this "Brewer-Dobson" circulation is several years and the forcing mechanism is believed to be through eddies in the extratropical upper stratosphere and lower mesosphere associated with breaking Rossby waves.

Despite the slow timescale of the Brewer-Dobson circulation, the actual exchange events occur on a timescale of several days. In the deep tropics, the exchange is upward and is believed to be associated with an overshoot of convective towers into the stratosphere in the ITCZ. Despite the short (several hours) timescale of a convective tower, the persistence of deep convection in the ITCZ results in a slow mean ascent (Plumb, 1996; James et al.,

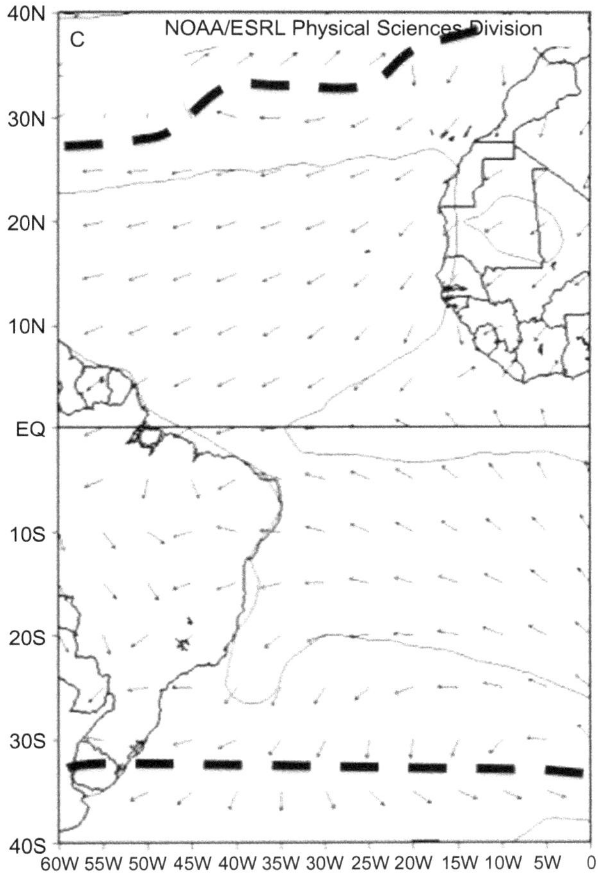

FIGURE 2.3—cont'd (c) Zoomed insert of the outlined box in (a). It is included to better show the easterly trade wind flow toward the equator. Base Images provided by the NOAA/ESRL Physical Sciences Division, Boulder Colorado, http://www.esrl.noaa.gov/psd/. The longterm average flow is derived from the NCEP reanalysis data as described by Kalnay et al. (1996).

2003). Conversely, the exchange events in the mid-latitudes are associated with tropopause folds or stratospheric intrusions, which are part of the structure of developing subtropical and mid-latitude cyclones. These events mix air in both directions from stratosphere to troposphere and vice versa, with the net exchange being downward. The timescale is that of a developing cyclone, which is several days, but the intrusion of stratospheric air can be quite deep, bringing it to within 3–5 km of the surface (Beekmann et al., 1997; Elbern et al., 1998). The subsequent mixing with surrounding tropospheric air occurs on a longer timescale.

The geographical distribution and seasonal and interannual variability of STE is the topic of active research. For our purposes, it suffices to note that the downward exchange through tropopause folds occurs in the region of the subtropical and polar jet streams and in the mid-latitude regions of baroclinic storm activity or "storm tracks." Referring back to the global plan view of Figure 2.3, the intrusions occur everywhere poleward of the subtropical ridge marked by the dashed line.

2.2.2 Deviations from the Zonally Averaged Structure: Monsoon, Gyres, El Niño Southern Oscillation

The primary deviation from the zonal mean structure described earlier is associated with the presence of monsoon flows. In simple terms, the monsoons are associated with the ITCZ convection being located over the tropical portions of the continents in summer. The solar heating of the landmasses interacts with the radiative force illustrated in Figure 2.1 such that the ITCZ is drawn poleward toward the heated landmasses. The consequences of this for global circulation are that in the vicinity of the continents there is a very large seasonality of the ITCZ rainfall with a summer–wet, winter–dry seasonal cycle. This brings a strong seasonality in wet deposition and removal of aerosols from the troposphere. The preferential location of the ITCZ convection over the tropical landmasses in summer also has implications for the transport of terrigenic radionuclides, such as radon from the land surface to the upper troposphere in the convective updrafts.

The word monsoon generally refers to the annually reversing wind system, which is an intrinsic component of the monsoon circulation. For example, the monsoon currents bring easterly winds over the northern Indian Ocean during the northern winter and westerly winds over the northern summer. Historically, these currents had a major influence on trade, which followed a yearly cycle with trade in one direction during the westerly regime and return voyages flowing during the easterly regime. Similar considerations apply to the transport of aerosols and radionuclides. Some examples of the seasonal reversal of the currents are shown in Figure 2.4.

An important aspect of the monsoon is the presence of transients within the monsoon current known as Active and Break monsoon events. The major event is the monsoon onset, marking a sudden beginning of the annual wet season. The onset brings about a sudden drop in the atmospheric aerosol content. One consequence is the short-term change in soil moisture, which has a large impact on radon exhalation, as discussed in Section 2.4.1. A comprehensive review of observations and mechanisms of the global monsoon is given by Webster et al. (1998). A review of the transients, including the mechanisms of onset, particularly for the Southern Hemisphere component of the monsoon, is given by Wheeler and McBride (2011).

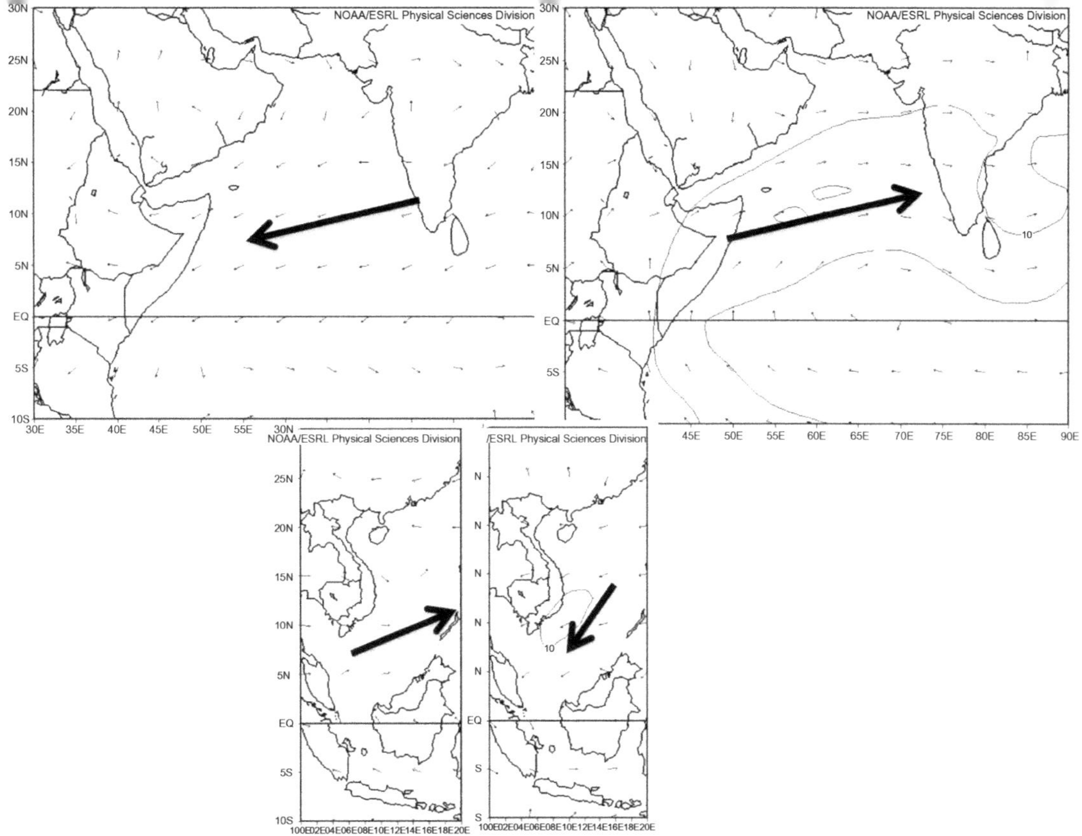

FIGURE 2.4 Examples of the seasonal reversal of winds associated with the monsoon. The upper panels show the flow at 850 hPa (approximately 1.5 km above the surface) for the Indian Ocean. The left panel is for January and the right panel for July. The plotted data are from the NCEP reanalysis and the bold arrows schematically show the direction of flow. The lower panels are for the South China Sea, with the September flow at left and December flow at right. *From base maps provided by the NOAA/ESRL Physical Sciences Division, Boulder, Colorado, http://www.esrl.noaa.gov/psd/.*

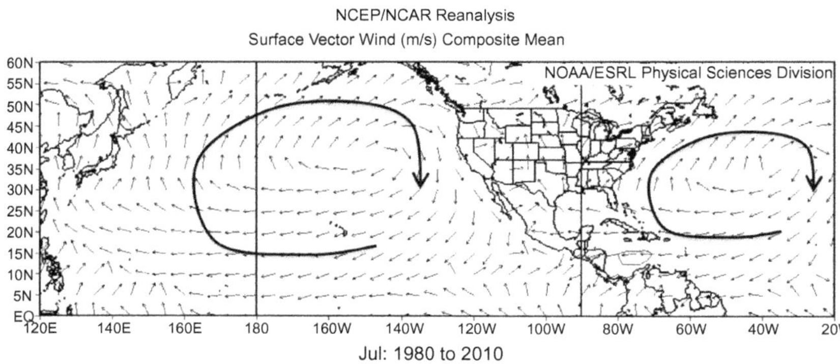

FIGURE 2.5 The mean flow at the surface during July, illustrating the anticyclonic gyres important for tropical transport. *From base map provided by the NOAA/ESRL Physical Sciences Division, Boulder, Colorado, http://www.esrl.noaa.gov/psd/.*

A second deviation from zonal symmetry, important for long-range atmospheric transport, is the presence of the atmospheric gyres associated with the anticylones making up the subtropical ridge. Over the subtropical oceans, semi permanent anticylones, or high-pressure cells, are located. These are present all year, but are particularly well marked in the northern Atlantic and northern Pacific during the northern summer months. These Northern Hemisphere gyres are shown in Figure 2.5. The equatorward side of the gyres is where the trade wind flow occurs; the gyres are also characterised by "steadiness" and thus are capable of ocean-wide transport.

A recent example of ocean-wide transport in the atmospheric gyres occurred during the 2011 release of radionuclides from the Fukushima Daiichi nuclear power plant in Japan following an earthquake and tsunami event. Trace activities of ^{133}Xe ($t_{1/2}$ = 5.25 days) were subsequently observed at Darwin in the Southern Hemisphere (Stohl et al., 2012). John McBride asked the Australian National Meteorological and Oceanographic Centre to run daily forward trajectories from a starting location of Fukushima using an atmospheric transport model. These confirmed that on certain days during March 2011, trajectories entered the Pacific gyre, circled it, and crossed into the Southern Hemisphere, taking ~10–15 days to complete the circuit.

The third deviation from zonal symmetry discussed in this chapter is the El Niño Southern Oscillation phenomenon (ENSO). ENSO is a global-scale phenomenon involving "coupling" between processes in the atmosphere and ocean centred on the tropical Pacific Ocean. The atmospheric structure of ENSO is monitored by the difference in mean sea level pressure between the western Pacific (designated by the pressure at Darwin) and the eastern Pacific (designated by the pressure at Tahiti). Pressure at these sites varies out of phase on a timescale of months to years. The accompanying signal in the ocean is monitored by the sea surface temperature averaged over a number of regions along the equatorial eastern Pacific. The most common index is that for the Niño 3.4 region 5°N–5°S, 120°–170°W. ENSO is the dominant mode of interannual variability in the tropical atmosphere and has global impacts extending well out of the tropics. It is a slow phenomenon with a maximum in the power spectra of approximately four years. The two extremes of ENSO are known as El Niño and La Niña. The individual events have lifetimes of about a year in length. There is extensive literature on ENSO. Major reviews include Philander (1990) and Diaz et al. (2001). The major impact of ENSO on radionuclide behaviour will be its modulation of the location and strength of the regions of tropical convection.

2.3. COSMOGENIC AND ANTHROPOGENIC RADIONUCLIDES

The interactions of cosmic ray particles with the Earth's atmosphere produce a number of radionuclides (Lal and Peters, 1967; Masarik and Beer, 1999). Their production occurs primarily in the stratosphere, but a significant fraction occurs in the troposphere (mainly the upper troposphere). Some radionuclides are also produced in the crust (e.g., ^{36}Cl is formed in the oceans from neutron capture by ^{35}Cl).

The cosmogenic radionuclides are relatively unimportant in terms of radiological dose to humans, but they are widely used as investigative tools in atmospheric and hydrological studies and for dating purposes.

Production varies over the ~11 year solar cycle; high solar activity deflects cosmic rays and results in a decrease in cosmogenic radionuclide production. The solar cycle is asymmetric, and the length and amplitude vary from cycle to cycle (Hathaway et al., 1994). The last four cycle minima occurred in 1976, 1986, 1996, and 2008-2009.

The production rate of cosmogenic radionuclides is lower in the tropics than at higher latitudes due to deflection of incoming particles toward the poles by the Earth's geomagnetic field. This deflection is energy dependent, so the average energy of incoming particles over the tropics is higher, and therefore more penetrating, than over the higher latitudes (Li et al., 2007). This, combined with a higher altitude tropopause, means that although the production rate is lower in the tropics, the proportion produced in the troposphere is somewhat higher.

These latitudinal variations in production rates should be reflected in the concentrations of the very short-lived cosmogenic radionuclides as these originate from the local tropospheric production. For example, concentrations of ^{39}Cl ($t_{1/2}$ = 56.2 minutes) in rainwater in northern Greece are significantly higher than in Israel (Papastefanou, 2007).

In the case of long-lived radionuclides, such as ^{10}Be ($t_{1/2} = 1.51 \times 10^6$ years) and ^{36}Cl ($t_{1/2} = 3.01 \times 10^5$ years), relatively rapid removal from the troposphere, primarily by wet deposition, means that the main source term for observed concentrations in the air will be the stratosphere. Hence, STE and subsequent behaviour in the troposphere will be the main drivers. For radionuclides with intermediate half-lives, such as ^{7}Be ($t_{1/2} = 53.2$ days), the situation is more complex.

Several of the cosmogenic radionuclides, such as ^{36}Cl, have also been produced as a result of anthropogenic activities. In addition, for long-lived fission products, such as ^{137}Cs ($t_{1/2} = 30$ years) and ^{90}Sr ($t_{1/2} = 28.8$ years), the main stock of existing activity in the atmosphere is in the stratosphere, so their behaviour parallels that of the long-lived cosmogenic radionuclides.

2.3.1 ^{7}Be, ^{10}Be, ^{137}Cs, and ^{90}Sr

^{7}Be is one of the most widely reported of the cosmogenic radionuclides largely due to its relative ease of measurement in air filter, rainwater, and surface soil samples by high-resolution gamma ray spectrometry and its application to atmospheric mixing and surface soil erosion studies. In contrast, ^{10}Be is difficult to measure (requiring accelerator mass spectrometry) and relatively few data have been reported.

Globally, approximately two-thirds of ^{7}Be and ^{10}Be production occurs in the stratosphere. Stratospheric ^{7}Be concentrations are significantly lower in the tropics than at higher latitudes because of the lower production rate, and the effect of the relatively short ^{7}Be half-life on transport from the higher latitudes. Consequently, the ^{10}Be/^{7}Be ratio is relatively high in the tropical stratosphere (Koch and Rind, 1998).

Once formed, ^{7}Be and ^{10}Be rapidly attach to aerosols and are subject to the removal and transport processes that affect them. Average residence times for aerosols are $\sim$1 year in the stratosphere and several days in the troposphere. ^{7}Be produced in the stratosphere is removed primarily by radioactive decay and reaches the Earth's surface only when advection of stratospheric air into the troposphere occurs (Turekian et al., 1983). The distribution pattern of its production, combined with its relatively rapid removal in the troposphere, results in a gradient in ^{7}Be across the tropopause and in the troposphere. Jordan et al. (2003) reported an average concentration for the lower stratosphere of $\sim$185 mBq m^{-3}. For ground-level stations, average concentrations are generally in the range of 2–5 mBq m^{-3}, although average concentrations are $<$2 mBq m^{-3} for many tropical locations.

Long-term variability in surface air ^{7}Be concentrations is observed on timescales corresponding to the solar cycle ($\sim$11 years) and ENSO ($\sim$3–5 years). For a given production rate, the following are the major factors influencing surface air concentrations (Rehfeld and Heimann, 1995):

- Mass exchange between stratosphere and troposphere
- Vertical transport in the troposphere
- Wet deposition
- Horizontal transport from subtropics and mid-latitudes into the tropics and polar regions

Since these factors vary on a seasonal basis, it is not surprising that ^{7}Be surface concentrations show an annual cyclical behaviour at most locations. Unfortunately, it is difficult to separate the influence of the above four factors at one location based on ^{7}Be data alone, and there is conflicting evidence for the importance of each in the literature. This especially applies to the influence of STE in spring/summer, with some authors attributing a significant influence and others finding contrary evidence (Baskaran, 1995; Doering and Akber, 2008).

This question could be investigated using ^{10}Be measurements, as this isotope is an excellent indicator of the stratospheric contribution. Dibb et al. (1994) used this approach to show that STE is a significant source of Be isotopes in the Arctic troposphere. Unfortunately, few studies use this approach because of the high cost of ^{10}Be determinations, and none seem to have been reported for tropical locations.

The long-lived fission products ^{137}Cs and ^{90}Sr are also tracers of stratospheric air contribution. This is because the primary contribution of these radionuclides in the troposphere is from the existing stock in the stratosphere, in the absence of significant activity from new anthropogenic inputs or resuspended dust (Hernandez et al., 2007; Masson et al., 2010).

Figure 2.6 shows mean latitudinal deposition density to the Earth's surface of ^{90}Sr from nuclear weapons testing, based on measurement data. The deposition

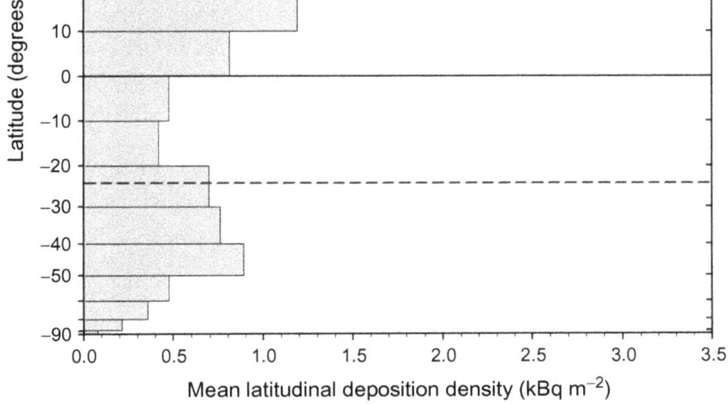

FIGURE 2.6 Mean latitudinal deposition density of ^{90}Sr from nuclear weapons testing, based on measurements. *Data derived from Table 8 of UNSCEAR (2000).*

density was substantially greater in the mid-latitudes than in the tropics and polar regions because of the relative influence of STE. This conclusion applies to long-lived weapons testing products, such as ^{137}Cs and ^{90}Sr, since most of the deposited activity originated from debris injected into the stratosphere. However, in the case of shorter lived radionuclides, the stratospheric contribution was much lower because of decay, and the latitudinal variation in deposition was more dependent on the location of injection (UNSCEAR, 2000).

Rangarajan and Gopalakrishnan (1970) reported the results of a comparative study of ^{7}Be and ^{137}Cs in surface air from Bombay and several other stations in India from 1965 to 1966. Increased concentrations were observed for both radionuclides in spring with the peak in March/April. Since ^{137}Cs could only be sourced in significant quantities from the stratosphere, its peak indicated the occurrence of STE in the spring period. Although ^{137}Cs and ^{7}Be were positively correlated, the $^{7}Be/^{137}Cs$ ratio reached minimum values during the March/April period because of the contribution from troposphere-origin ^{7}Be. On the basis of these data, Rangarajan and Gopalakrishnan (1970) concluded that the observed ^{7}Be was about 60% of stratospheric origin in the spring, decreasing to 25% in the autumn.

High positive correlations for ^{137}Cs and ^{7}Be for Cienfuegos, Cuba (22°N), were also reported by Alonso Hernández et al. (2004); see Figure 2.7a. Wet deposition results in removal of these radionuclides and so will tend to reduce airborne concentrations (Alonso Hernández et al., 2006). Figure 2.7b shows an example of the negative correlation between ^{7}Be and rainfall for Cienfuegos.

Koch and Mann (1996) examined long-term data from six sites covering a wide range of latitudes and concluded that the ^{7}Be variability due to ENSO was primarily caused by changes in precipitation patterns. High concentrations can be expected in the subtropics, particularly dry continental areas such as central Australia and the Sahara, due to low precipitation and sinking air motion (Rehfeld and Heimann, 1995). In contrast, lower concentrations can generally be expected nearer the equator due to the lower production rate, less descending air masses, and removal of ^{7}Be around the ITCZ because of convective rainfall (Uematsu et al., 1994). Horizontal transport will bring some ^{7}Be into the tropics, but losses during transport from wet deposition and radioactive decay will reduce this input term.

These trends are confirmed by comparisons of data from stations over the North American and South American continents, which show ^{7}Be concentration minima at the tropical and high latitude stations, and maxima in the mid-latitudes (Viezee and Singh, 1980). Doering and Akber (2008) examined ^{7}Be data from 13 locations in the southwestern Pacific region and confirmed that lower concentrations are prevalent at locations between the equator and the Tropic of Capricorn. They found that peak concentrations occur in early spring in the tropics, early summer at subtropical latitudes, and mid-to-late summer at middle latitudes. From analysis of the data, they concluded that STE and vertical transport of air in the troposphere were the primary drivers of the seasonality.

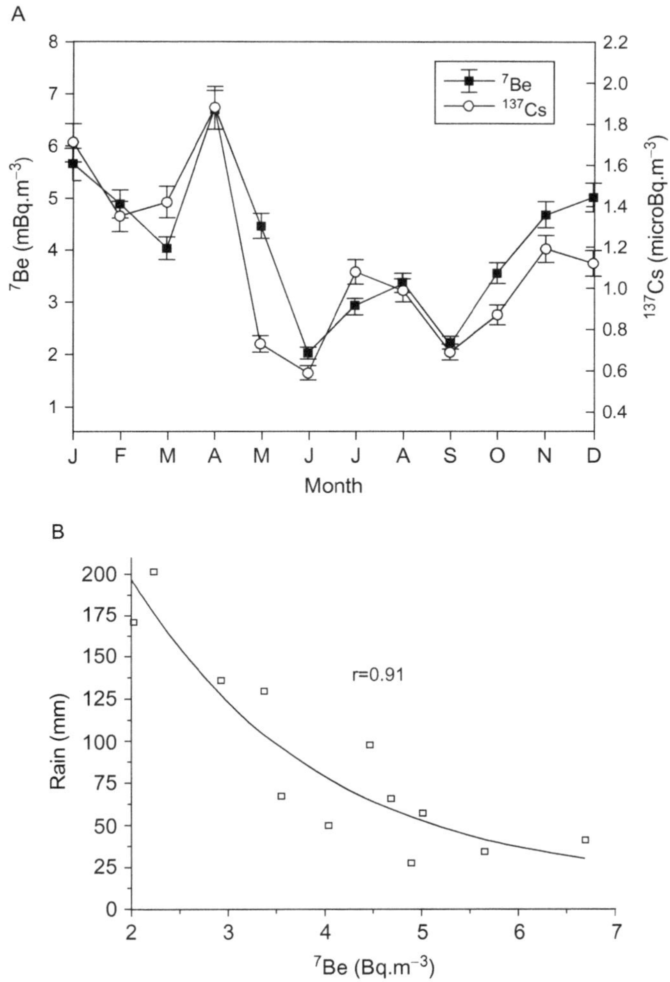

FIGURE 2.7 (a) Mean monthly ^{7}Be and ^{137}Cs concentrations in air and (b) rainfall versus mean monthly ^{7}Be concentrations in air at Cienfuegos, Cuba (22°N), July 1994 to December 1997. *From Alonso Hernández et al. (2004), with permission.*

Figure 2.8 shows mean monthly ^{7}Be surface air concentrations at six locations derived from data reported by EML (1999). The three stations in Figure 2.8a are on the western coast of South America and have low rainfall due to the influence of the Humboldt current. This is especially the case for Antofagasta, which has an annual rainfall of only 4 mm with high year-to-year rainfall variability. The average concentration at Guayaquil is much lower than that at the other two stations, which is expected because of its proximity to the equator. All three stations show relatively low seasonal variability for ^{7}Be.

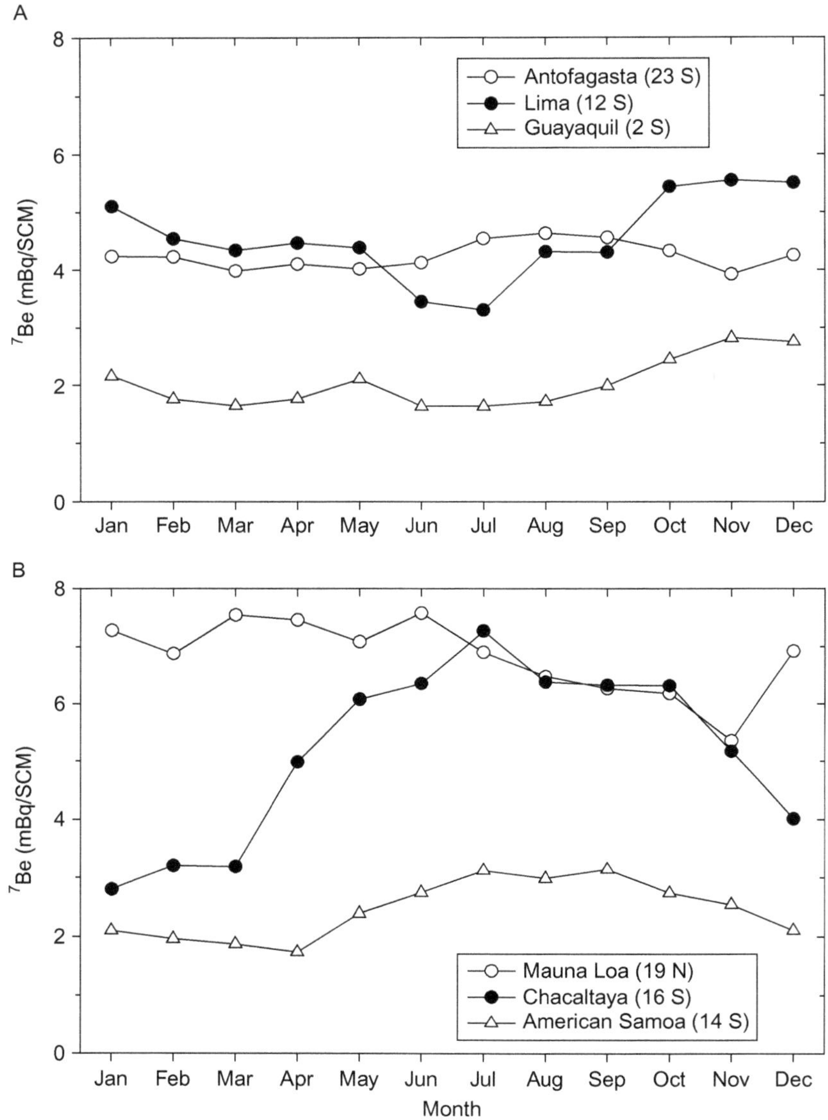

FIGURE 2.8 Mean monthly ^{7}Be concentrations in air (mBq/SCM) at (a) Antofagasta, Chile (23°S, 31 m); Lima, Peru (12°S, 13 m); and Guayaquil, Ecuador (2°S, 7 m); and (b) Mauna Loa, Hawaii (19°N, 3401 m); Chacaltaya, Bolivia (16°S, 5220 m); and Tutuila, American Samoa (14°S, 77 m). Sampling period: 1970–1999. *Data from EML (1999).*

The stations at Mauna Loa and Chacaltaya are elevated (3401 and 5220 m, respectively), and higher quantities of ^{7}Be are observed as they are sampling mid-troposphere air (Figure 2.8b). Chacaltaya has lower concentrations in the summer; Feely et al. (1989) attributed this to washout of ^{7}Be by rain showers as air is transported up the eastern slopes of the Andes prior to arrival at

Chacaltaya. In American Samoa, high mid-year concentrations more or less coincide with the period of low rainfall (Feely et al., 1989).

However, the degree of seasonal variability in ^{7}Be concentrations at a location cannot be attributed only to the influence of rainfall. For example, Uematsu et al. (1994) reported that in 1985 the ^{7}Be concentrations at the island of Nauru (1°S) averaged 1.4 mBq m^{-3} with a standard deviation of only 0.4 mBq m^{-3}; the annual rainfall was 2060 mm.

In summary, the behaviour of the radionuclides discussed here is affected by a number of factors, and the relative importance of these factors varies considerably with sampling location.

2.3.2 ^{36}Cl

^{36}Cl is produced in the atmosphere by spallation of ^{40}Ar, and in the oceans by neutron capture by ^{35}Cl. Nuclear weapons testing over the oceans in the 1950s to early 1970s resulted in a 'bomb pulse' of anthropogenic ^{36}Cl, but over the following decades concentrations returned to essentially pretest levels (Suter et al., 1987). Measurement of ^{36}Cl fallout requires accelerator mass spectrometry, and relatively few studies have been published to date.

Due to its long half-life and relatively rapid removal from the troposphere, ^{36}Cl fallout can be expected to be driven primarily by STE. Consequently, concentrations should be higher in the Northern Hemisphere than in the Southern Hemisphere (due to stronger STE), and to peak in the mid-latitudes (Heikkilä et al., 2009). Mid-latitude Northern Hemisphere studies have given fallout values of ~30–40 atoms m^{-2} s^{-1}. Keywood et al. (1998) published data for fallout at several sites in Australia, including three sites in the wet–dry tropics. They concluded that the mean fallout for the Southern Hemisphere is about 15 atoms m^{-2} s^{-1}. Of interest were the fallout data for the three tropical locations, which were ~3–4 times higher than expected based on the temperate latitude data. This enhancement was most pronounced during the summer monsoon period. Keywood et al. (1998) discussed hypotheses for these observations based on enhanced STE in the tropics and sweeping of ^{36}Cl from arid areas to the tropics, but concluded that these were unable to provide a convincing explanation.

2.4. TERRIGENIC RADIONUCLIDES

2.4.1 Radon and Short-lived Radon Progeny

The element radon has three naturally occurring isotopes: ^{222}Rn of the uranium series, ^{220}Rn of the thorium series, and ^{219}Rn of the actinium series. These are often referred to as radon, thoron, and actinon, respectively, and in the remainder of this section the term "radon" will be used for ^{222}Rn only.

Radon is formed upon decay of its parent ^{226}Ra in soils, house-building materials, and so forth. As it is a noble gas, radon may escape from these matrices and make its way into the atmosphere. The process of escape from the ground or building material surface is known as radon "exhalation." Once

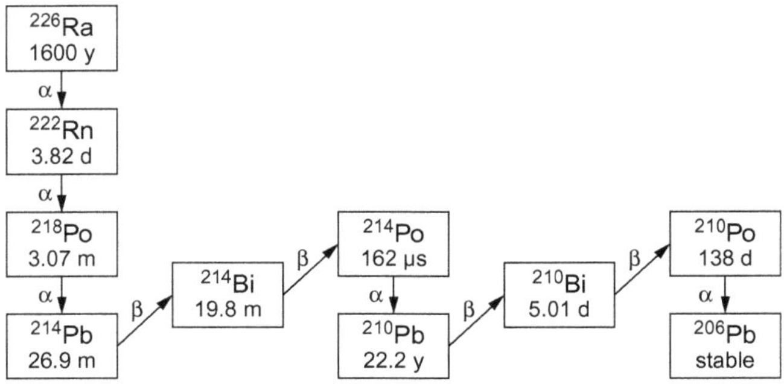

FIGURE 2.9 Portion of the uranium decay chain headed by ^{226}Ra.

in the atmosphere, radioactive decay to its progeny (Figure 2.9) is the only significant removal mechanism for radon. Upon their formation in the atmosphere, its progeny rapidly bind to submicron aerosols.

A large number of studies have been undertaken worldwide on the topic of concentrations of radon and radon progeny (and to a lesser extent of thoron and thoron progeny) indoors, especially in temperate and high-latitude regions. This is due to their important contribution to radiological dose to humans, which is estimated to have a worldwide average of about 50% of the natural dose (UNSCEAR, 2000). However, radon is less important for radiological dose in the tropics and subtropics. The reasons for this include the generally greater ventilation in houses in hotter climates, less use of underground rooms (i.e., cellars) where high radon concentrations can build up, and in many (especially rural) areas, a greater proportion of the day spent outdoors (Langroo et al., 1991; Iimoto et al., 2001; Arogunjo et al., 2004; Rochedo and Lauria, 2008). Nevertheless, the correlation of radon concentrations indoors with latitude shows considerable scatter (UNSCEAR, 2000), and some regions with very high natural background radiation may have a specific problem requiring investigation. In addition, many mining operations of uranium-bearing ores take place in tropical locations, and in this case radon surveys and/or monitoring may be required as part of a radiological protection regime (Jha et al., 2000b; Martin et al., 2004; Fernandes et al., 2006; Tripathi et al., 2008).

Radon and its progeny have been used extensively in the study of atmospheric processes. For radon itself, this includes study of the diurnal variability of the planetary boundary layer (Yoshioka and Iida, 2005), transport processes including convection (Feng et al., 2010), and distinguishing between continental and maritime air sources (Zahorowski et al., 2004). It has been used widely to evaluate tracer transport in global models (Jacob et al., 1997; Stevenson et al., 1998; Stockwell and Chipperfield, 1999; Taguchi et al., 2002; Josse et al., 2004; Zhang et al., 2008). The progeny (in particular ^{210}Pb) have been used to study wet scavenging of aerosols, dry deposition to the Earth's surface, and large-scale transport and convection.

Figure 2.10 shows the estimated mean annual latitudinal radon exhalation flux density plotted against latitude. The latitude axis is scaled so that the area enclosed by the curve between any two latitudes is proportional to the total radon flux (Bq s^{-1}) from that latitudinal band (Martin, 2011). The main factor influencing the mean flux density at each latitude is the proportion of the surface with ice-free land. The Southern Hemisphere contributes only about 26% of the Earth's total radon flux, due to its large areas of ocean and ice-covered land.

The region between the Tropic of Cancer and Tropic of Capricorn contributes about 38% to the total flux. In the Northern Hemisphere, the tropical component of the flux is 28% of the total for this hemisphere. However, in the Southern Hemisphere the tropical component is much greater (66% of the hemisphere flux). Given these contributions, an understanding of the factors affecting radon exhalation fluxes in the tropics is important if we are to improve modelling of radon behaviour, especially for the Southern Hemisphere.

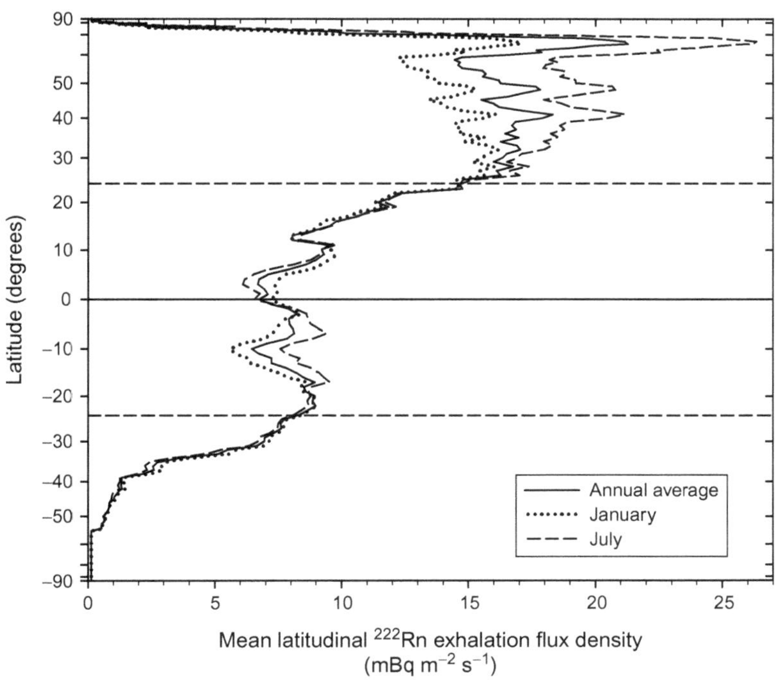

FIGURE 2.10 Estimated mean annual latitudinal radon exhalation flux density. Data have a latitudinal resolution of one degree. *Data from http://www.nmt.edu/~schery/mapdata.html (Schery and Wasiolik, 1998).*

The radon exhalation flux density for ice-free soil is affected by a number of factors including the ^{226}Ra activity concentration; the ^{226}Ra distribution in the soil grains (i.e., absorbed on the surface or contained within the soil matrix); soil structural characteristics such as porosity, grain size, and grain shape; and the soil moisture and temperature, atmospheric pressure, and wind speed.

Of the temporally variable factors, soil moisture has the greatest influence (Jha et al., 2000a). The relationship is complex, but in general, rainfall results in a reduction in exhalation. Most studies of seasonal variability show reductions in the wetter months: usually winter in temperate zones (Whittlestone et al., 1998) and wet monsoonal in the tropics (Lawrence et al., 2009).

This dependence on soil moisture, combined with increased ice and snow cover in winter in the high latitudes, means that the tropical contribution to a hemisphere's total flux should be higher in winter than in summer. This effect is reflected in the model predictions by Schery and Wasiolek (1998) for January and July (shown in Figure 2.10).

For locations with long dry seasons of very low rainfall and/or wet seasons with highly variable daily rainfall, it is likely that the wet season flux will not only be lower than the dry season flux, but also much more variable. An example from northern Australia illustrating both lower and more variable wet season fluxes is shown in Figure 2.11. Over eight sites, the average dry season flux densities were higher by a factor of 2.8 ± 0.4 (Lawrence et al., 2009).

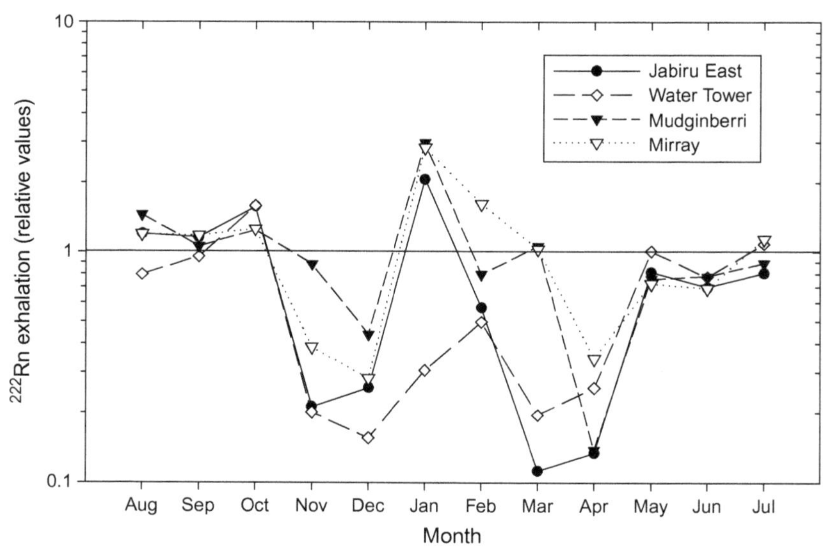

FIGURE 2.11 ^{222}Rn flux density measurements (normalised to the dry season averages) at four sites (13°S) in northern Australia between August 2002 and July 2003. Wet season is approximately November to April. *From Lawrence et al. (2009), with permission.*

The effect of rain on radon exhalation flux may be highly site specific. In a study at a site in Nepal, Girault et al. (2009) found that the effect of rainfall was transient, with similar average fluxes during the monsoon and dry winter seasons. However, in this case, advective discharge of gas occurred rather than radon within the soil column being transported predominantly by diffusion.

Low atmospheric pressure and high wind speed are likely to increase the exhalation flux density, which would imply an increase under cyclonic conditions. However, the rainfall accompanying such events seems more likely to lead to a strong reduction in radon flux densities. Cyclones and vertical hot towers should be efficient mechanisms for transporting radon to the upper troposphere, but the radon is most likely to originate from the period prior to high rainfall or from the region outside the rainfall-affected area of the tower.

Once it has been introduced into the atmosphere, radon's behaviour is determined only by atmospheric mixing processes and removal by radioactive decay. The majority of available atmospheric concentration data have been obtained close to ground level (usually ~ 1 m height), which is not only close to the source, but within the Planetary Boundary Layer (PBL). It is generally accepted that a "typical" average outdoor concentration is 10 Bq m^{-3}, but this figure must be treated with caution as actual concentrations vary considerably both temporally and geographically. In particular, island and coastal locations tend to have lower outdoor concentrations due to the influence of maritime air (Mishra et al., 1980; Tanji et al., 1992; Akber et al., 1996; Bhugwant et al., 2001; Zahorowski et al., 2004).

The existence of diurnal cycles in ground level radon and short-lived progeny concentrations is well established and easily explainable in terms of greater mixing during the daytime due to solar insolation. The difference between daytime and nighttime concentrations varies considerably from day to day. On nights when temperature inversions occur, leading to a low PBL and still conditions, radon concentrations typically begin rising after sunset and decrease at sunrise. Average peak-to-trough concentration ratios are commonly a factor of between 2 and 5, with peak-to-trough ratios of more than an order of magnitude over a single 24 hour period being common (Jha et al., 2000b; Martin et al., 2004). Concentrations of short-lived radon progeny (^{218}Po to ^{214}Po) follow a similar diurnal pattern, but with a delay dependent on their half-lives (Akber and Pfitzner, 1994; Dueñas et al., 1994; Magalhães et al., 2003). Dry and/or wet deposition to the surface is also a factor resulting in lower progeny concentrations close to the ground.

The exhalation flux density in the near vicinity of the measurement point appears to influence the measured radon and short-lived progeny concentrations in outdoor air, especially under very still conditions. For example, Bollhöfer and Ryan (2007) measured radon concentrations at 37 sites over the former uranium mine site, Nabarlek (12°S), with a total area of about 100 ha. They obtained variations of up to an order of magnitude between sites, with the higher concentrations from sites with higher local radon exhalation flux densities. Mahesh et al. (2005) reported positive correlations between local radon exhalation flux densities and radon and radon progeny concentrations at seven sites in southwest India (13–15°N). Another example is the importance of the local influence on concentrations at the Mauna Loa Observatory (20°N) in Hawaii. Radon global simulations with chemical transport models (CTMs) underestimate the concentrations at this station, in part due to the local signal (Whittlestone et al., 1996), although other factors, such as limited vertical resolution in the models, are also important (Taguchi et al., 2002).

Both experimental data and CTM predictions for monthly average radon concentrations at tropical monsoonal locations show a strong seasonal dependence. Figure 2.12 shows an example from Jaduguda, India, where the winter season concentrations obtained were approximately a factor of five greater than those for the rainy season. Comparable results have been reported for locations in southern China (Zahorowski et al., 2005), northern Australia (Akber and Pfitzner, 1994; Martin et al., 2004), and Rio de Janeiro, Brazil (Magalhães et al., 2003). The influence of rainfall on radon exhalation fluxes discussed earlier is certainly one cause of this seasonal variation. However, there are several other influencing factors, including variation in solar insolation, predominant wind direction, and atmospheric stability conditions between dry and rainy seasons. Mishra et al. (1980) reported measurements of radon over the Arabian Sea during the summer monsoon, found decreasing

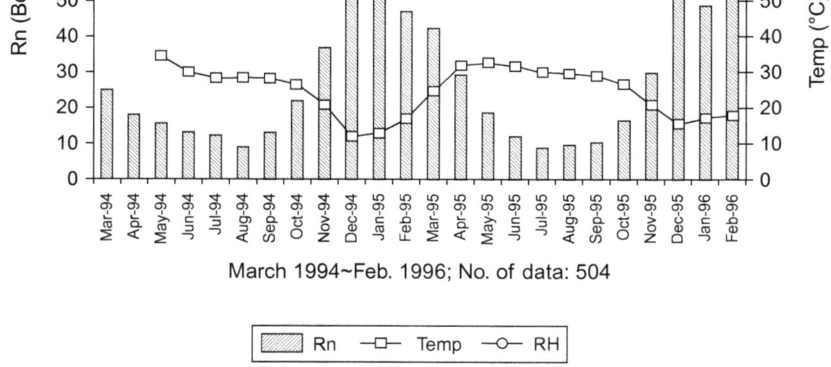

FIGURE 2.12 Mean ^{222}Rn concentrations, and temperature and relative humidity (at 8 a.m.), at Jaduguda (23°N), India. Summer is March–June, rainy season is July–October, and winter is November–February. *From Jha et al. (2000b), with permission.*

levels from 20°N toward the equator, and concluded that there was a progressive reduction in the continental component of the monsoon air.

Because radon is not removed by wet or dry deposition, it is a powerful tracer for vertical mixing processes. Vertical profiles of radon concentration have been obtained primarily from airborne sampling with some time series data from towers (Liu et al., 1984; Kritz et al., 1993; Ramonet et al., 1996; Zahorowski et al., 2006, 2011).

2.4.2 ^{210}Pb–^{210}Bi–^{210}Po

The main source of ^{210}Pb in the atmosphere is decay of ^{222}Rn following its exhalation from the Earth's surface. A secondary source is suspension of dust due to wind; however, this source is minor except in some particular locations (e.g., close to uranium mining operations) or in particular conditions (e.g., dust storms). ^{210}Pb associated with suspended dust affects the PBL because the dust settles toward the Earth's surface following its suspension. Another source is volcanic activity, which may have a local influence on concentrations, but it is a minor contributor globally (Lambert et al., 1979; Baskaran, 2011).

The degree of disequilibrium between ^{210}Pb and its radioactive progeny, ^{210}Bi and ^{210}Po, are sometimes used to calculate atmospheric residence times for ^{210}Pb. However, the available data indicate that the $^{210}Po/^{210}Pb$ ratio is unsuitable for this application because there are sources of ^{210}Po other than decay of atmospheric ^{210}Pb (Turekian et al., 1977; Rangarajan and Eapen, 1990). Suggested alternate sources include resuspended dust, volcanic activity (Lambert et al., 1979; Nho et al., 1997), organic matter from zooplankton in wind-driven sea spray, volatile compounds of polonium emitted from the ocean (Kim et al., 2000), and biomass burning (Tateda and Iwao, 2008). The situation is further complicated by the fact that some sources, such as volcanic activity (Sato, 2003) and industrial lead emissions (Settle et al., 1982), also emit ^{210}Pb.

^{210}Po has been shown to be enriched in fire plumes (Lambert et al., 1991) due to the volatility of elemental Po and Po compounds at temperatures in the low hundreds of degrees Celsius. The flux of ^{210}Po from savannah burning has been estimated to be $\sim$10% of global production (Nho et al., 1996). In the case of large burning events it can certainly have a local or regional effect. From a study at Lamto, Ivory Coast (6°N), Nho et al. (1996) concluded that biomass burning was an important contributor to ^{210}Po concentrations (although the situation was complex since a winter peak in ^{210}Po and ^{210}Pb was due primarily to the transport of dust from the Sahara). Paatero et al. (2009) reported airborne $^{210}Po/^{210}Pb$ ratios in Finland increasing from the usual 0.03–0.05 to a maximum of 0.35 during the spring 2006 biomass fires in Eastern Europe.

Dibb et al. (1999) observed elevated levels of ^{210}Pb throughout the free troposphere over the South Pacific as part of the NASA PEM-Tropics airborne sampling campaign, and attributed the most likely cause to fires in Africa. However, their conclusions did not relate to ^{210}Pb released directly by the fires; instead it related to ^{222}Rn. Under convective conditions, radon is effectively transported to higher levels in the troposphere because it is not removed from the ascending air column by precipitation. Radon in the upper troposphere decays to ^{210}Pb, which has a longer residence time than ^{210}Pb in lower tropospheric levels and can be transported considerable distances. For these reasons, the vertical distribution of ^{210}Pb in the troposphere is complex, and concentrations in the upper troposphere are commonly as high, or higher, than those in the mid-level troposphere (Sykora and Froehlich, 2010).

Figure 2.13 shows the available data on ^{210}Pb surface air concentrations from the database of Preiss et al. (1996). These data have limitations because of the small number of locations represented, in particular the lack of data from Africa and eastern South America and the preponderance of coastal and near-coastal sites. Nevertheless, the strong correlation of the ^{210}Pb latitudinal averages with the radon source term (Figure 2.10) is readily apparent. In Figure 2.13, island-based (including Australian) data are indicated by open symbols and continental data by closed symbols. In the figure, the lower

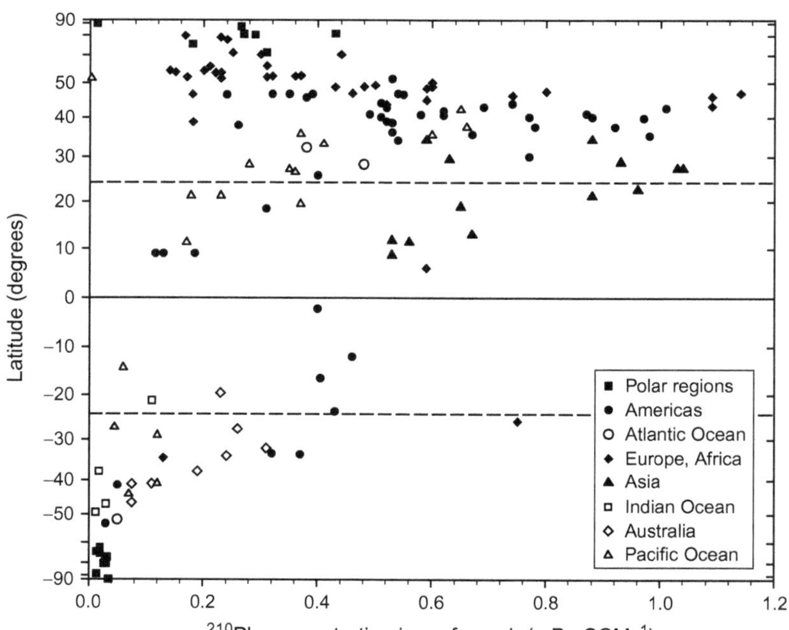

FIGURE 2.13 Mean annual ^{210}Pb concentration in surface air as a function of latitude. Each point represents a separate measurement location. *Data from Preiss et al. (1996), http://www-lgge.Obs.ujf-grenoble.fr/axes/radioactivite/Pb210_database/airconc.shtml.*

average concentrations on islands is apparent, with the exception of several Japanese sites, which experience a strong influence from the Asian land mass during the winter monsoon (Yamamoto et al., 2006). Preiss et al. (1996) proposed that asymmetry is to be expected for ^{210}Pb concentrations between eastern and western parts of the continents due to the different dominant air masses, and that this asymmetry is reversed between temperate and tropical latitudes (i.e., concentrations should be greater on the western parts of the continents in the tropics).

The primary removal mechanism for ^{210}Pb from the air column is wet deposition in both large-scale systems and convective updrafts. Figure 2.14 shows data reported by Mélières et al. (2003) for ^{210}Pb deposition and air concentration as a function of rainfall at Petit Saut (French Guiana). Deposition was positively correlated with rainfall at low rainfall rates, as has been observed in other studies, although fluctuations increased at higher rates. Air concentrations showed only weak negative correlation with rainfall, indicating that the replenishment of ^{210}Pb-bearing aerosol in the air column is fast in comparison with removal by rainfall in this region.

Martin (2003) reported uranium series radionuclide concentrations in successive samples collected during convective storms in northern Australia. Uranium, thorium, and radium isotope concentrations in the rainwater decreased to one-half or less from their initial value after the first 1–2 millimetres of rainfall. However, a (relatively small) reduction in ^{210}Pb was only observed for three of the six storms, implying that most of the ^{210}Pb activity was sourced from in-cloud rainout rather than below-cloud washout.

As with radon, ^{210}Pb is an important trace constituent for the evaluation of climate models. The radiative forcing by anthropogenic aerosols is a major

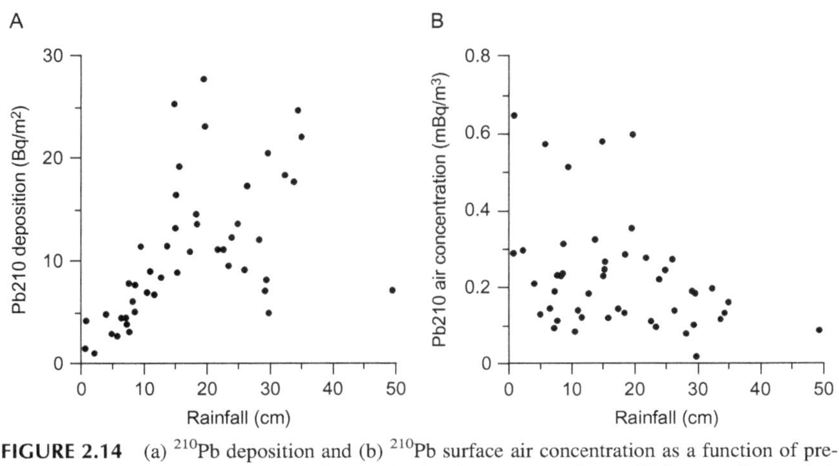

FIGURE 2.14 (a) ^{210}Pb deposition and (b) ^{210}Pb surface air concentration as a function of precipitation at Petit Saut (5°N), French Guiana. Each value corresponds to a 15 day sampling period. *From Mélières et al. (2003), with permission.*

uncertainty in current assessments (Liu et al., 2001). Thus, it is important to evaluate the climate model ability to simulate the coupling of aerosols to the hydrological cycle arising from the short lifetime of aerosols against wet deposition (Koch et al., 1999; Kiehl et al., 2000; Liu et al., 2001). The quick attachment of ^{210}Pb and ^{7}Be to aerosols makes them important tracers to evaluate the ability of climate models to simulate the wet deposition process.

2.4.3 Radionuclides on Resuspended Dust

The suspension of surface soil particles into the atmosphere occurs primarily as a result of the action of wind, and airborne radionuclide concentrations may increase substantially during heavy storms as a result (Hernandez et al., 2007). Another process leading to suspension is rainsplash, especially when combined with wind. Raindrop size distributions in tropical storms are often larger than those in temperate regions, which could, in theory, lead to a greater influence of rainsplash on dust suspension and translocation of surface-bound radioactivity to vegetation. Other factors, such as vehicular traffic and agricultural and mining operations, may be important for dust suspension on a local or regional scale.

Almost all of the mass of airborne dust is present in particles of diameter greater than ~1 μm. As a result, in addition to wet deposition, sedimentation and dry deposition can be expected to play an important role in their subsequent removal from the atmosphere.

There are two basic scenarios for suspension of radioactivity on dust: radionuclides that are approximately well mixed in the soil column, and radionuclides that have been deposited on the soil surface (often having an approximately exponentially decreasing activity concentration with depth in the soil). Typically, the uranium and thorium series radionuclides and ^{40}K are examples of the former, while many cosmogenic and anthropogenic radionuclides, such as ^{7}Be, ^{137}Cs, and ^{90}Sr, may be present mainly on the soil surface.

Production of airborne dust with contaminants that had been previously deposited on the soil surface is referred to as 'resuspension' (Sehmel, 1980; Nicholson, 1988). Resuspension of activity is commonly estimated using an empirical resuspension factor, K, which is the airborne concentration (Bq m^{-3}) divided by the soil contamination density (Bq m^{-2}). The resuspension factor is often given as a time-dependent quantity (Jourdain, 2009), and reported factors range over several orders of magnitude. Site-specific data are preferable given the number of factors that can influence resuspension, including age of the contamination on site, vegetation cover, rainfall amount and intensity, and airborne activity due to biomass burning (Moroney, 1992). Little data have been reported for the tropical environment; in one data study from the Ranger Uranium Mine site in northern Australia, Akber (1992) obtained a value for K of 10^{-8} m^{-1}, which is within the range of values reported elsewhere.

The suspension of dust may, in some cases, be a significant contributor to concentrations of ^{210}Pb and its progeny in the atmosphere (Nho et al., 1996).

As discussed in Section 2.4.2, this causes difficulties with their use for calculation of atmospheric residence times; one approach to this problem is to determine a member of the uranium decay chain above ^{210}Pb to apply a correction (Martin, 2003; Carvalho, 1995).

2.5. CONCLUSIONS

This chapter has reviewed the literature on radionuclide behaviour in the tropical atmosphere with an emphasis on differences from the higher latitude regions. Some key points include the following:

- Ground-level concentrations of stratospheric-origin radionuclides are lower in the tropics than in the mid-latitudes, primarily due to lower STE.
- Good time-series datasets are available for ^{7}Be, but interpretation can be difficult due to the number of factors involved. One approach to address this problem would be concurrent measurements of ^{10}Be.
- One study in northern Australia showed ^{36}Cl concentrations in rainwater significantly higher than expected. Further tests to confirm and explain this result would be of value.
- The tropics are a significant source of ^{222}Rn and its progeny to the atmosphere, particularly in the Southern Hemisphere. More information on radon exhalation flux densities, and especially the influence of soil moisture, would be needed to improve the usefulness of radon in testing of global circulation and CTMs.
- There is evidence that vertical transport of ^{222}Rn in tropical convective systems is an important source of ^{210}Pb to the upper troposphere.
- Biomass burning may be an important source of ^{210}Po in the tropical atmosphere, at least on a local and regional scale.
- Wet–dry contrasts over the monsoon can be useful in the study of radionuclide behaviour in the atmosphere. Examples include changes in radon exhalation and in ^{210}Pb residence times between wet and dry seasons.
- Unfortunately, there is a general lack of data available from tropical Africa.

This chapter included a basic description of the major atmospheric circulations in the tropics, and it was used to aid interpretation of published studies. The review was structured into two major sections covering cosmogenic and anthropogenic radionuclides and terrigenic radionuclides. It is hoped this synthesis of the meteorology with the published radionuclide literature will provide a useful summary for both climate scientists and radioecologists.

ACKNOWLEDGMENTS

The authors thank John Twining for his advice and encouragement. They also thank Alan Wain and Jim Fraser of the Australian Bureau of Meteorology National Meteorological and Oceanographic Centre for helpful discussions and for carrying out trajectory calculations related to the Fukushima Daiichi accident.

REFERENCES

Akber, R.A., 1992. Suspension of dust in the vicinity of Ranger Uranium Mine: An estimate for the land application area. In: Proceedings of the Workshop on Land Application of Effluent Water from Uranium Mines in the Alligator Rivers Region. AGPS, Canberra, pp. 267–286.

Akber, R.A., Pfitzner, J.L., 1994. Atmospheric concentrations of radon and radon daughters in Jabiru East. Technical Memorandum 45. Supervising Scientist, Darwin NT.

Akber, R., Martin, P., Martin, T., Waggitt, P., 1996. Radon and gamma dose rate measurements at selected sites on Christmas Island, Indian Ocean. Internal Report 224. Supervising Scientist, Darwin NT.

Alonso Hernández, C.M., Cartas Águila, H., Díaz Asencio, M., Muñoz Caravaca, A., 2004. Reconstruction of ^{137}Cs signal in Cuba using ^{7}Be as tracer of vertical transport processes in the atmosphere. J. Environ. Radioact. 75, 133–142.

Alonso Hernández, C.M., Cartas-Águila, H., Díaz-Asencio, M., Muñoz Caravaca, A., Martín-Pérez, J., Sibello-Hernández, R., 2006. Atmospheric deposition of ^{137}Cs between 1994 and 2002 at Cienfuegos, Cuba. Cuba. J. Environ. Radioact. 88, 199–204.

Arogunjo, A.M., Ohenhen, H.O., Olowookere, S.P., 2004. A re-evaluation of the occupancy factors for effective dose estimate in tropical environment. Radiat. Prot. Dosimetry 112, 259–265.

Baskaran, M., 1995. A search for the seasonal variability on the depositional fluxes of ^{7}Be and ^{210}Pb. J. Geophys. Res. 100, 2833–2840.

Baskaran, M., 2011. Po-210 and Pb-210 as atmospheric tracers and global atmospheric Pb-210 fallout: A review. J. Environ. Radioact. 102, 500–513.

Beekmann, M., Ancellet, G., Blonsky, S., DeMuer, D., Ebel, A., Elbern, H., et al., 1997. Regional and global tropopause fold occurrence and related ozone flux across the tropopause. Journal of Atmospheric Chemistry 28, 29–44.

Bhugwant, C., Rivière, E., Keckhut, P., Leveau, J., 2001. Variability of carbonaceous aerosols, ozone, and radon at Piton Textor, a mountain site on Réunion island (south-western Indian Ocean). Tellus 53B, 546–563.

Bollhöfer, A., Ryan, B., 2007. Radiological impact assessment of the rehabilitated Nabarlek site. Supervising Scientist Report 193. In: eriss Research Summary 2005–2006. Supervising Scientist, Darwin NT, pp. 143–146. http://www.environment.gov.au/ssd/publications/ssr/193.html.

Brummer, B., Augstein, A., Riehl, H., 1974. On the low-level wind structure in the Atlantic trade. Quart. J. Roy. Meteor. Soc. 100, 109–121.

Carvalho, F.P., 1995. Origins and concentrations of ^{222}Rn, ^{210}Pb, ^{210}Bi and ^{210}Po in the surface air at Lisbon, Portugal, at the Atlantic edge of the European continental landmass. Atmos. Environ. 29, 1809–1819.

Cook, G.T., Scott, E.M., Harkness, D.D., 2010. Radiocarbon as a tracer in the global carbon cycle. In: Environmental Radionuclides: Tracers and Timers of Terrestrial Processes. Radioactivity in the Environment, vol. 16. Elsevier, pp. 89–138.

Diaz, H.F., Hoerling, M.P., Eischeid, J.K., 2001. ENSO variability, teleconnections, and climate change. International Journal of Climatology 21, 1845–1862.

Dibb, J.E., Meeker, L.D., Finkel, R.C., Southon, J.R., Caffee, M.W., Barrie, L.A., 1994. Estimation of stratospheric input to the Arctic troposphere: ^{7}Be and ^{10}Be in aerosols at Alert, Canada. J. Geophys. Res. 99, 12855–12864.

Dibb, J.E., Talbot, R.W., Meeker, L.D., Scheuer, E.M., Blake, N.J., Blake, D.R., et al., 1999. Constraints on the age and dilution of Pacific Exploratory Mission-Tropics biomass burning plumes from the natural radionuclide tracer ^{210}Pb. J. Geophys. Res. 104, 16233–16241.

Doering, C., Akber, R., 2008. Describing the annual cyclic behaviour of ^{7}Be concentrations in surface air in Oceania. J. Environ. Radioact. 99, 1703–1707.

Dueñas, C., Pérez, M., Fernández, M.C., Carretero, J., 1994. Disequilibrium of radon and its short-lived daughters near the ground with atmospheric stability. J. Geophys. Res. 99, 12865–12872.

Elbern, H., Hendricks, J., Ebel, A., 1998. A climatology of tropopause folds by global analyses. Theor. Appl. Climatol. 59, 181–200.

EML, 1999. Surface air sampling program database. http://www.eml.st.dhs.gov/databases/sasp.

Enfield, D.B., 1981. Annual and nonseasonal variability of monthly low-level wind fields over the Southeastern tropical Pacific. Mon. Wea. Rev. 109, 2177–2190.

Feely, H.W., Larsen, R.J., Sanderson, C.G., 1989. Factors that cause seasonal variations in Beryllium-7 concentrations in surface air. J. Environ. Radioact. 9, 223–249.

Feng, W., Chipperfield, M.P., Dhomse, S., Monge-Sanz, B.M., Yang, X., Zhang, K., et al., 2010. Evaluation of cloud convection and tracer transport in a three-dimensional chemical transport model. Atmos. Chem. Phys. Discuss. 10, 22953–22991.

Fernandes, H.M., Simoes Filho, F.F.L., Perez, V., Franklin, M.R., Gomiero, L.A., 2006. Radioecological characterization of a uranium mining site located in a semi-arid region in Brazil. J. Environ. Radioact. 88, 140–157.

Garstang, M., 1967. Sensible and latent heat exchange in low latitude synoptic scale systems. Tellus 3, 492–508.

Girault, F., Koirala, B.P., Perrier, F., Richon, P., Rajaure, S., 2009. Persistence of radon-222 flux during monsoon at a geothermal zone in Nepal. J. Environ. Radioact. 100, 955–964.

Happell, J.D., Östlund, G., Mason, A.S., 2004. A history of atmospheric tritium gas (HT) 1950–2002. Tellus 56B, 183–193.

Hathaway, D.H., Wilson, R.M., Reichmann, E.J., 1994. The shape of the sunspot cycle. Solar Physics 151, 177–190. http://solarscience.msfc.nasa.gov/predict.shtml.

Heikkilä, U., Beer, J., Feichter, J., Alfimov, V., Synal, H.A., Schotterer, U., et al., 2009. ^{36}Cl bomb peak: comparison of modeled and measured data. Atmos. Chem. Phys. 9, 4145–4156.

Hernandez, F., Karlsson, L., Hernandez-Armas, J., 2007. Impact of the tropical storm Delta on the gross alpha, gross beta, ^{90}Sr, ^{210}Pb, ^{7}Be, ^{40}K, and ^{137}Cs activities measured in atmospheric aerosol and water samples collected in Tenerife (Canary Islands). Atmos. Environ. 41, 4940–4948.

Holton, J.R., Haynes, P.H., McIntyre, M.E., Douglass, A.R., Rood, R.B., Pfister, L., 1995. Stratosphere-troposphere exchange. Reviews of Geophysics 33, 403–439.

Hu, Q.H., Weng, J.Q., Wang, J.S., 2010. Sources of anthropogenic radionuclides in the environment: A review. J. Environ. Radioact. 101, 426–437.

Iimoto, T., Kosako, T., Sugiura, N., 2001. Measurements of summer radon and its progeny concentrations along with environmental gamma dose rates in Taiwan. J. Environ. Radioact. 57, 57–66.

Jacob, D.J., Prather, M.J., 1990. Radon-222 as a test of convective transport in a general circulation model. Tellus 42B, 118–134.

Jacob, D.J., Prather, M.J., Rasch, P.J., et al., 1997. Evaluation and intercomparison of global atmospheric transport models using ^{222}Rn and other short-lived tracers. J. Geophys. Res. 102, 5953–5970.

James, P., Stohl, A., Forster, C., Eckhardt, S., Selbert, P., Frank, A., 2003. A 15-year climatology of stratosphere-troposphere exchange with a Lagrangian particle dispersion model: Methodology and validation. J. Geophys. Res 108 (D12), 8519. doi:10.1029/2002JD002637.

Jha, S., Khan, A.H., Mishra, U.C., 2000a. A study of the ^{222}Rn flux from soil in the U mineralised belt at Jaduguda. J. Environ. Radioact. 49, 157–169.

Jha, S., Khan, A.H., Mishra, U.C., 2000b. Environmental Rn levels around an Indian U complex. J. Environ. Radioact. 48, 223–234.

Jordan, C.E., Dibb, J.E., Finkel, R.C., 2003. $^{10}Be/^{7}Be$ tracer of atmospheric transport and stratosphere–troposphere exchange. J. Geophys. Res 108 (D8), 3-1–3-14.

Josse, B., Simon, P., Peuch, V.H., 2004. Comprehensive nuclear-test-ban treaty. Tellus 56B, 339–356.

Jourdain, F., 2009. Resuspension. In: Quantification of Radionuclide Transfer in Terrestrial and Freshwater Environments for Radiological Assessments. IAEA, Vienna, pp. 63–68. IAEA TECDOC-1616.

Kalnay, E., Kanamitsu, M., Kistler, R., Collins, W., Deaven, D., Gandin, L., et al., 1996. The NCEP/NCAR reanalysis 40-year project. Bull. Amer. Meteor. Soc. 77, 437–471.

Keywood, M.D., Fifield, L.K., Chivas, A.R., Cresswell, R.G., 1998. Fallout of chlorine 36 to the Earth's surface in the southern hemisphere. J. Geophys. Res. 103, 8281–8286.

Kiehl, J.T., Schneider, T.L., Rasch, P.J., Barth, M.C., Wong, J., 2000. Radiative forcing due to sulfate aerosols from simulations with the National Center for Atmospheric Research Community Climate Model. Version 3. J. Geophys. Res. 105, 1441–1457.

Kim, G., Hussain, N., Church, T.M., 2000. Excess ^{210}Po in the coastal atmosphere. Tellus 52B, 74–80.

Koch, D., Rind, D., 1998. Beryllium 10/beryllium 7 as a tracer of stratospheric transport. J. Geophys. Res. 103, 3907–3917.

Koch, D.M., Mann, M.E., 1996. Spatial and temporal variability of ^{7}Be surface concentrations. Tellus 48B, 387–396.

Koch, D.M., Jacob, D.J., Tegen, I., Rind, D., Chin, M., 1999. Tropospheric sulfur simulation and sulfate direct radiative forcing in the Goddard Institute for Space Studies general circulation model. J. Geophys. Res. 104, 23,799–23,822.

Kritz, M.A., Rosner, S.W., Kelly, K.K., Loewenstein, M., Chan, K.R., 1993. Radon measurements in the lower tropical stratosphere: Evidence for rapid vertical transport and dehydration of tropospheric air. J. Geophys. Res. 98, 8725–8736.

Lal, D., Peters, B., 1967. Cosmic ray produced radioactivity on the Earth. In: Sitte, K. (Ed.), Encyclopedia of Physics. Springer-Verlag, New York, pp. 551–612.

Lambert, G., Buisson, A., Sanak, J., Ardouin, B., 1979. Modification of the atmospheric polonium 210 to lead 210 ratio by volcanic emissions. J. Geophys. Res. 84 (C11), 6980–6986.

Lambert, G., Le Cloarec, M.F., Ardouin, B., Bonsang, B., 1991. Long-lived radon daughters signature of Savanna Fires. In: Levine, J.S. (Ed.), Global Biomass Burning. MIT Press, London, pp. 181–184.

Langroo, M.K., Wise, K.N., Duggleby, J.C., Kotler, L.J., 1991. A nationwide survey of ^{222}Rn and γ radiation levels in Australian homes. Health Phys. 61, 753–761.

Lawrence, C.E., Akber, R.A., Bollhöfer, A., Martin, P., 2009. Radon-222 exhalation from open ground on and around a uranium mine in the wet–dry tropics. J. Environ. Radioact. 100, 1–8.

Li, S.W., Li, Y.S., Tsui, K.C., 2007. Radioactivity in the atmosphere over Hong Kong. J. Environ. Radioact. 94, 98–106.

Liu, S.C., McAfee, J.R., Cicerone, R.J., 1984. Radon 222 and tropospheric vertical transport. J. Geophys. Res. 89 (D5), 17291–17297.

Liu, H., Jacob, D.J., Bey, I., Yantosca, R.M., 2001. Constraints from ^{210}Pb and ^{7}Be on wet deposition and transport in a global three-dimensional chemical tracer model driven by assimilated meteorological fields. J. Geophys. Res. 106 (D11), 12109–12128.

Magalhães, M.H., Amaral, E.C.S., Sachett, I., Rochedo, E.R.R., 2003. Radon-222 in Brazil: An outline of indoor and outdoor measurements. J. Environ. Radioact. 67, 131–143.

Mahesh, H.M., Avadhani, D.N., Siddappa, K., 2005. Concentration, distribution, and transportation of ^{222}Rn and its decay products in the environment of coastal Karnataka and Kaiga in southwest India. Radioactivity in the Environment, vol. 7, Elsevier, pp. 542–553.

Mahowald, N.M., Rasch, P.J., Eaton, B.E., Whittlestone, S., Prinn, R.G., 1997. Transport of ^{222}radon to the remote troposphere using the Model of Atmospheric Transport and Chemistry and assimilated winds from ECMWF and the National Center for Environmental Prediction/NCAR. J. Geophys. Res. 102 (D23), 28,139–28,151.

Martin, P., 2003. Uranium and thorium series radionuclides in rainwater over several tropical storms. J. Environ. Radioact. 65, 1–18.

Martin, P., 2011. A note on the contribution of tropical regions to the Earth's radon flux. In: Sources and Measurements of Radon and Radon Progeny Applied to Climate and Air Quality Studies. IAEA, Vienna, pp. 75–76. IAEA Proceedings Series.

Martin, P., Tims, S., Ryan, B., Bollhöfer, A., 2004. A radon and meteorological measurement network for the Alligator Rivers Region, Australia. J. Environ. Radioact. 76, 35–49.

Masarik, J., Beer, J., 1999. Simulation of particle fluxes and cosmogenic nuclide production in the Earth's atmosphere. J. Geophys. Res. 104, 12099–12111.

Masson, O., Piga, D., Gurriaran, R., D'Amico, D., 2010. Impact of an exceptional Saharan dust outbreak in France: PM_{10} and artificial radionuclides concentrations in air and in dust deposit. Atmos. Environ. 44, 2478–2486.

Mélières, M.A., Pourchet, M., Richard, S., 2003. Surface air concentration and deposition of lead-210 in French Guiana: Two years of continuous monitoring. J. Environ. Radioact. 66, 261–269.

Mishra, U.C., Rangarajan, C., Eapen, C.D., 1980. Natural radioactivity of the atmosphere over the Indian land mass, inside deep mines, and over adjoining oceans. In: Natural Radiation Environment III. US Department of Energy, Washington DC, pp. 327–346. US DOE, Special Symposium Series 51, CONF 780422.

Moroney, J.R., 1992. Pathway analysis concepts for radiological impact assessment. In: Proceedings of the Workshop on Land Application of Effluent Water from Uranium Mines in the Alligator Rivers Region. AGPS, Canberra, pp. 192–225.

Nho, E.Y., Ardouin, B., Le Cloarec, M.F., Ramonet, M., 1996. Origins of ^{210}Po in the atmosphere at Lamto, Ivory Coast: Biomass burning and Saharan dusts. Atmos. Environ. 30, 7065–7075.

Nho, E.Y., Le Cloarec, M.F., Ardouin, B., Ramonet, M., 1997. ^{210}Po, an atmospheric tracer of long-range transport of volcanic plumes. Tellus 49B, 429–438.

Nicholson, K.W., 1988. A review of particle resuspension. Atmos. Environ. 22, 2639–2651.

Paatero, J., Vesterbacka, K., Makkonen, U., Kyllönen, K., Hellen, H., Hatakka, J., et al., 2009. Resuspension of radionuclides into the atmosphere due to forest fires. Journal of Radioanalytical and Nuclear Chemistry 282, 473–476.

Papastefanou, C., 2007. Chlorine-39 in rainfall at a temperate latitude (40 °N). J. Environ. Radioact. 92, 175–182.

Philander, S.G., 1990. El Niño, La Niña, and the Southern Oscillation. Academic Press, location, p. 289.

Plumb, R.A., 1996. A "tropical pipe" model of stratospheric transport. J. Geophys. Res. 101, 3957–3972.

Preiss, N., Mélières, M.A., Pourchet, M., 1996. A compilation of data on lead-210 concentration in surface air and fluxes at the air-surface and water-sediment interfaces. J. Geophys. Res. 101, 28847–28862.

Ramonet, M., Le Roulley, J.C., Bousquet, P., Monfray, P., 1996. Radon-222 measurements during the Tropoz II campaign and comparison with a global atmospheric transport model. Journal of Atmospheric Chemistry 23, 107–136.

Rangarajan, C., Eapen, C.D., 1990. The use of natural radioactive tracers in a study of atmospheric residence times. Tellus 42B, 142–147.

Rangarajan, C., Gopalakrishnan, Smt.S, 1970. Seasonal variation of beryllium-7 relative to caesium-137 in surface air at tropical and sub-tropical latitudes. Tellus 12, 116–121.

Rehfeld, S., Heimann, M., 1995. Three dimensional atmospheric transport simulation of the radioactive tracers ^{210}Pb, ^{7}Be, ^{10}Be, and ^{90}Sr. J. Geophys. Res. 100, 26141–26161.

Riehl, H., Malkus, J.S., 1958. On the heat balance in the equatorial trough zone. Geophysica 6, 503–538.

Rochedo, E.R.R., Lauria, D., 2008. International versus national regulations: Concerns and trends. Appl. Radiat. Isot. 66, 1550–1553.

Sato, J., 2003. Natural radionuclides in volcanic activity. Appl. Radiat. Isot. 58, 393–399.

Schery, S.D., Wasiolek, M.A., 1998. Modelling ^{222}Rn flux from the Earth's surface. In: Katase, A., Shimo, M. (Eds.), ^{222}Rn and Thoron in the Human Environment. World Scientific, pp. 207–217.

Sehmel, G.A., 1980. Particle resuspension: A review. Environ. Int. 4, 107–127.

Settle, E.M., Patterson, C.C., Turekian, K.K., Cochran, J.K., 1982. Lead precipitation fluxes at tropical oceanic sites determined from ^{210}Pb measurements. J. Geophys. Res. 87, 1239–1245.

Stevenson, D.S., Collins, W.J., Johnson, C.E., Derwent, R.G., 1998. Intercomparison and evaluation of atmospheric transport in a Lagrangian model (STOCHEM), and an Eulerian model (UM), using ^{222}Rn as a short-lived tracer. Q. J. Roy. Meteor. Soc. 124, 2477–2491.

Stockwell, D.Z., Chipperfield, M.P., 1999. A tropospheric chemical transport model: Development and validatation of the model transport schemes. Q. J. Roy. Meteor. Soc. 125, 1743–1783.

Stohl, A., Bonasoni, P., Cristofanelli, P., Collins, W., Feichter, J., Frank, A., et al., 2003. Stratosphere-troposphere exchange: A review, and what we have learned from STACCATO. J. Geophys. Res. 108 (D12), 8516. doi:10.1029/2002JD002490.

Stohl, A., Seibert, P., Wotawa, G., Arnold, D., Burkhart, J.F., Eckhardt, S., et al., 2012. Xenon-133 and caesium-137 releases into the atmosphere from the Fukushima Daiichi nuclear power plant: Determination of the source term, atmospheric dispersion, and deposition. Atmos. Chem. Phys. 12, 2313–2343.

Suter, M., Beer, J., Bonani, G., Hoffman, H.J., Michel, D., Oeschger, H., et al., 1987. ^{36}Cl studies at the ETH/SIN AMS facility. Nuc. Instrum. Meth. Phys. Rev B29, 211–215.

Sykora, I., Froehlich, K., 2010. Radionuclides as tracers of atmospheric processes. Radioactivity in the Environment 16, 51–88 Chapter 3. Elsevier.

Taguchi, S., Iida, T., Moriizumi, J., 2002. Evaluation of the atmospheric transport model NIRE-CTM-96 by using measured radon-222 concentrations. Tellus 54B, 250–268.

Tanji, T., Okino, M., Sugioka, I., Mochizuki, S., 1992. Radon and its daughters in the atmosphere over the equatorial Pacific Ocean. Radiat. Prot. Dosimetry 45, 399–401.

Tateda, Y., Iwao, K., 2008. High ^{210}Po atmospheric deposition flux in the subtropical coastal area of Japan. J. Environ. Radioact. 99, 98–108.

Tripathi, R.M., Sahoo, S.K., Jha, V.N., Khan, A.H., Puranik, V.D., 2008. Assessment of environmental radioactivity at uranium mining, processing and tailings management facility at Jaduguda, India. Appl. Radiat. Isot. 66, 1666–1670.

Turekian, K.K., Nozaki, Y., Benninger, L.K., 1977. Geochemistry of atmospheric radon and radon products. Ann. Rev. Earth Planet. Sci. 5, 227–255.

Turekian, K.K., Benninger, L.K., Dion, E.P., 1983. ^{7}Be and ^{210}Pb total deposition fluxes at New Haven, Connecticut, and at Bermuda. J. Geophys. Res. 88, 5411–5415.

Uematsu, M., Duce, R.A., Prospero, J.M., 1994. Atmospheric beryllium-7 concentrations over the Pacific Ocean. Geophysical Research Letters 21, 561–564.

UNSCEAR, 2000. Sources and effects of ionizing radiation. Volume I: Sources. UNSCEAR 2000 Report to the General Assembly. UNSCEAR, Vienna.

Viezee, W., Singh, H.B., 1980. The distribution of beryllium-7 in the troposphere: Implications on stratospheric/tropospheric air exchange. Geophysical Research Letters 7, 805–808.

Webster, P.J., Magana, V.O., Palmer, T.N., Shukla, J., Tomas, R.A., Yanai, M., et al., 1998. Monsoons: Processes, predictability, and the prospects for prediction. J. Geophys. Res. 103, 14451–14510.

Wheeler, M.C., McBride, J.L., 2011. Australasian monsoon. In: Lau, W.K.M., Waliser, D.E. (Eds.), Intraseasonal Variability in the Atmosphere-Ocean Climate System. 2nd ed. Springer, pp. 147–198.

Whittlestone, S., Schery, S.D., Li, Y., 1996. Pb-212 as a tracer for local influence on air samples at Mauna Loa Observatory. Hawaii. J. Geophys. Res. 101, 14777–14785.

Whittlestone, S., Zahorowski, W., Schery, S.D., 1998. Radon flux variability with season and location in Tasmania, Australia. Journal of Radioanalytical and Nuclear Chemistry 236, 213–217.

Yamamoto, M., Sakaguchi, A., Sasaki, K., Hirose, K., Igarashi, Y., Kim, C.K., 2006. Seasonal and spatial variation of atmospheric ^{210}Pb and ^{7}Be deposition: Features of the Japan Sea side of Japan. J. Environ. Radioact. 86, 110–131.

Yoshioka, K., Iida, T., 2005. The diurnal change in the vertical distribution of atmospheric ^{222}Rn due to the growth and rise of the stable stratification height in the atmospheric boundary layer. Radioactivity in the Environment 7, 489–496.

Zahorowski, W., Chambers, S.D., Henderson-Sellers, A., 2004. Ground based radon-222 observations and their application to atmospheric studies. J. Environ. Radioact. 76, 3–33.

Zahorowski, W., Chambers, S., Wang, T., Kang, C.H., Uno, I., Poon, S., et al., 2005. Radon-222 in boundary layer and free tropospheric continental outflow events at three ACE-Asia sites. Tellus 57B, 124–140.

Zahorowski, W., Chambers, S., Williams, A.G., 2006. Radon-222 as a tracer of atmospheric transport phenomena on different spatial and temporal scales. In: Proceedings of the 15th Pacific Basin Nuclear Conference, Sydney. http://www.pacificnuclear.org/pnc/homeframe_pnc.html.

Zahorowski, W., Chambers, S., Crawford, J., Williams, A.G., Cohen, D.D., Vermeulen, A.T., et al., 2011. ^{222}Rn observations for climate and air quality studies. In: Sources and Measurements of Radon and Radon Progeny Applied to Climate and Air Quality Studies. IAEA Proceedings Series, IAEA, Vienna, pp. 77–96.

Zhang, K., Wan, H., Zhang, M., Wang, B., 2008. Evaluation of the atmospheric transport in a GCM using radon measurements: sensitivity to cumulus convection parameterization. Atmos. Chem. Phys. 8, 2811–2832.

Chapter 3

Mobility of Radionuclides in Tropical Soils and Groundwater

Timothy E. Payne[1] and Robert Edis[2]

[1]*Australian Nuclear Science and Technology Organisation, Kirrawee, New South Wales, Australia*

[2]*Melbourne School of Land and Environment, University of Melbourne, Victoria, Australia*

3.1. INTRODUCTION AND SCOPE

The broad objective of this chapter is to describe radionuclide behaviour in tropical soils and groundwater and, in particular, to highlight similarities and differences between the tropical and the comparatively well-studied temperate environments. The geographical area of interest comprises the vast region of the globe within approximately 25° to the north and south of the equator. Apart from the shared feature that the region generally enjoys a warmer climate than the temperate regions of the planet, there is considerable diversity of climate types—the local environments can include desert, rainforest, or mountains. This region contains many countries with high population density, increasing industrialisation, and a probable increasing dependence on nuclear power. These factors provide a strong imperative to understand the environmental behaviour of radionuclides that are associated with the operation of nuclear power reactors and associated facilities, including nuclear waste disposal sites.

In this chapter, we will attempt to outline the existing body of knowledge regarding the geochemical behaviour of key anthropogenic radionuclides and nuclear waste in the tropical environment. Reference will be made to significant research studies relevant to the behaviour of radionuclides in tropical soil and groundwater systems.

We will discuss case studies relevant to environmental contamination by radionuclides in various categories:

- Anthropogenic radionuclides released by deliberate explosions during testing of nuclear weapons and generally deposited by atmospheric pathways or moving through groundwater (from underground explosions)

Radioactivity in the Environment, Vol. 18. http://dx.doi.org/10.1016/B978-0-08-045016-2.00003-5

- Accidental releases of artificial radionuclides to the environment
- Natural radionuclides released by geological processes from undisturbed ore bodies and the Oklo natural reactors, which have been studied as analogous systems to nuclear waste repositories
- Experimental studies relating to radionuclides potentially released to tropical environments from nuclear waste sites or as a consequence of major accidents

The scope of this chapter excludes the releases of natural radionuclides from the uranium mining industry, as well as technologically enhanced naturally occurring radioactive materials (TENORM). As described in Chapter 1, this term refers to enhanced levels of natural radionuclides resulting from various human activities, such as mineral extraction and processing (Landa, 2007).

Given the relatively embryonic nature of the scientific literature on this subject, it is not anticipated that we will answer every question about radionuclide mobility in tropical environments. However, our aims will have been achieved if the reader is better able to pose the pertinent questions that are relevant in mapping out possible environmental pathways, particularly those that would not apply in a similar way to a temperate system.

3.2. GENERAL PRINCIPLES

To set the context for this chapter, it is useful to have a general understanding and definition of concepts relevant to radionuclide movement that apply to both temperate and tropical environments. We will illustrate these concepts by referring to previous studies that demonstrate the basic principles. These examples been selected on the basis of their relevance to tropical environments.

It is recommended that the reader refer to a standard text, such as Fetter (1993) or Domenico and Schwartz (1990), to supplement the brief discussion of some of the main principles presented in the following sections.

3.2.1 Hydrology and Water Movement

Groundwater modelling is considered to have originated in 1856 in Dijon, France, when Henry Darcy developed a mathematical relationship that related the change in hydraulic head over a given distance to the flow of water through a sand column (see Figure 3.1). Darcy defined the specific discharge (v) as the flow out (Q) divided by the cross-sectional area of the sand column. Darcy's law states that the specific discharge (v) is equal to the hydraulic conductivity (K) times the change in hydraulic head (dh) divided by the length (dl). Mathematically:

$$v = -K\frac{dh}{dl} \tag{3.1}$$

This equation is considerably simplified by only considering one-dimensional flow; more general equations are given in Fetter (1993). The proportionality

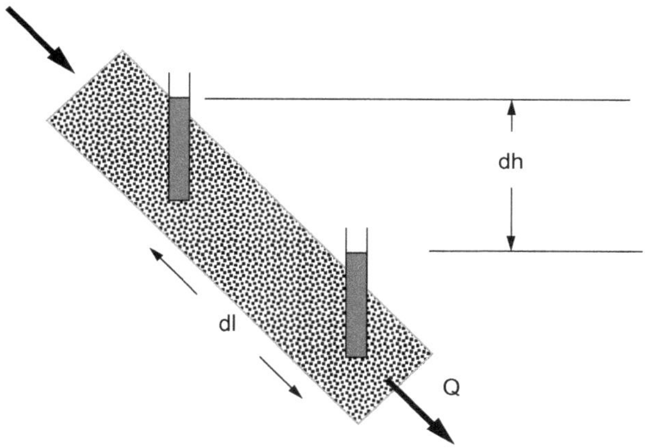

FIGURE 3.1 Simple apparatus for demonstrating Darcy's law.

constant (K) has the units of a velocity (m/s) and is a function of the aquifer material. It can vary over a large range, for example, from 10^{-12} m/s for a shale (or igneous rock) to 0.1 m/s for a gravel aquifer. Specific discharge is sometimes referred to as the "Darcy velocity," but because it does not account for the space occupied by the soil, it does not represent an actual velocity.

The average groundwater velocity ($\bar{v}$), or seepage velocity, is calculated by dividing the specific discharge by the porosity (n) of the medium:

$$\bar{v} = -\frac{K}{n}\frac{dh}{dl} \tag{3.2}$$

Porosity is defined as the percentage of the total volume of the rock that is void space, and can range from near-zero to greater than 50%. It is important to remember that the actual groundwater velocity can be much faster than the Darcy velocity because water only moves through the pore openings, rather than the entire rock (soil) volume (Domenico and Schwartz, 1990). Thus, the linear velocity ($\bar{v}$) will always be larger than the Darcy velocity and, for a given Darcy velocity, increases with decreasing effective porosity.

There are numerous geological environments in which water flow cannot be simply described by Darcy's equation. For example, many geological media are fractured rather than porous and, in these systems, water flow is less well understood (although considerable research has been undertaken in recent years). The knowledge of the vadose zone (the unsaturated zone above the water table) is also developing rapidly. This zone can be particularly important in radionuclide transport studies, because many releases of contaminants to the subsurface occur within or above the vadose zone. Atmospheric fallout (e.g., from nuclear weapons testing) occurs on the ground surface and the arrival of fallout radionuclides in the permanently saturated zone may be considerably delayed subsequent to the deposition event.

Some proposed nuclear waste sites (such as the controversial repository at Yucca Mountain in the United States) may be established in the vadose zone where the buried waste is not permanently saturated. The hydrological issues affecting this proposed nuclear waste repository (which appears unlikely to proceed) are discussed by Fabryka-Martin et al. (2006).

3.2.2 Solute Transport

The movement of a contaminant is affected by a number of processes that must be considered in addition to the rate of water movement.

Advection is the movement of solutes that are carried along with the flowing groundwater. The one-dimensional mass flux is estimated by multiplying the quantity of water per unit time by the concentration of dissolved solids.

Molecular diffusion is the movement of a solute from an area of greater concentration to lower concentration. This can occur even in the absence of groundwater movement and can be considered a consequence of random molecular motions. In one dimension, it is described by Fick's law:

$$F = -D_d(dC/dx) \tag{3.3}$$

where F is the mass flux of solute per unit time, D_d is the diffusion coefficient, C is the solute concentration, and dC/dx is the concentration gradient.

Mechanical dispersion is a process by which a solute in the groundwater becomes spread out through a larger volume. This is because groundwater actually travels at various rates in a real aquifer, both faster and slower than the average linear velocity. The flow is faster in the centre of pores compared to the edges, and also faster in larger pores (fractures). Some flow paths will be longer (more tortuous) than others. As a result, the solute becomes dispersed.

3.2.3 Chemical Principles

Solubility is an important factor in radionuclide mobility. The solubility of a particular mineral phase places an upper limit on its concentration in solution. When the solubility of a given solid phase is exceeded, it has the potential to precipitate. Computations of solubility require detailed calculations based on thermodynamic data. Provided calculations are based on relevant, correct, and complete thermodynamic data (and equilibrium is achieved), the solubility of a particular mineral places an upper limit on its concentration in solution. The importance of high-quality, critically reviewed thermodynamic data for these calculations has led to international efforts to develop comprehensive thermodynamic databases for specific radionuclides (NEA, 2002).

Solubility is dependent on redox state. For example, uranium is much more mobile in oxidising environments (where it exists in its hexavalent U(VI) state as the uranyl ion UO_2^{2+}) compared to reducing environments (as U(IV)). The role of redox state in radionuclide mobility was demonstrated by Yanagisawa et al. (2000), who reported that technetium applied as the relatively soluble pertechnetate (TcO_4^-) to flooded soil was transformed, in reducing conditions, to an insoluble and less mobile form, presumably through the influence of soil microorganisms.

A complex is a dissolved species that consists of an association between a cation (positively charged ion) and a ligand (an anion or neutral molecule). Many trace metals are transported in surface and groundwater in a complexed form, and the complex speciation is dependent on the nature and abundance of ions and ligand materials. Complexation is extremely important from a geochemical viewpoint for the following reasons (Langmuir, 1997):

- Complexation of a dissolved species tends to increase its solubility.
- Some elements are predominantly present in solution as complexes rather than free ions, as in the case for uranyl for pH values above 6 (Figure 3.2).
- Adsorption (see the following section) may be dramatically increased or decreased by complexation.
- Toxicity and bioavailability of metals depend on their aqueous speciation or complexation.

It is important to consider whether codisposed waste materials within radioactive waste could increase the mobility of radionuclides through complexation. Some organic ligands have been implicated in mobilising radionuclides, such as plutonium, at some low-level waste sites (Cleveland and Rees, 1981). Similarly, natural ligands, such as humic materials, may affect radionuclide mobility.

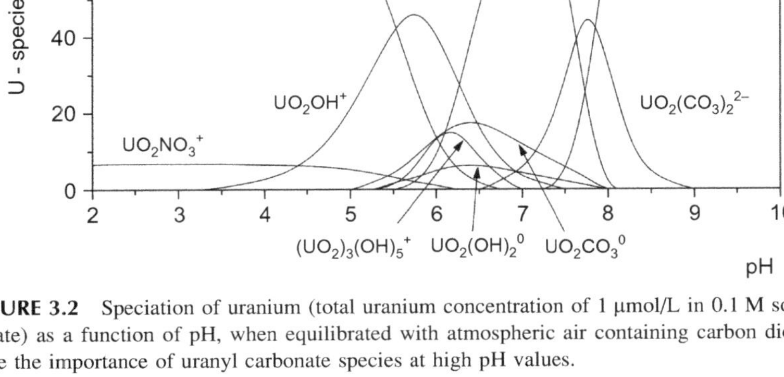

FIGURE 3.2 Speciation of uranium (total uranium concentration of 1 μmol/L in 0.1 M sodium nitrate) as a function of pH, when equilibrated with atmospheric air containing carbon dioxide. Note the importance of uranyl carbonate species at high pH values.

3.2.4 Adsorption

Adsorption is the uptake of a chemical species on the surface of a mineral. A simple case of adsorption is ion exchange, where a charged ion in solution is exchanged for an ion at the surface. It has been demonstrated that different ions have a greater or lesser affinity for a surface and competing ions will therefore tend to displace one another from ion exchange surface sites. This depends on their relative affinity for the surface, which is often referred to as 'selectivity' (Sposito, 1989). The number and properties of ion exchange sites are dependent on the type of material and this is a major point of difference between temperate and tropical systems.

The aqueous concentration of many trace elements is controlled by adsorption, which usually results in the dissolved concentration of a radionuclide being much lower than would be expected on the basis of solubility alone. Adsorption is often a particularly important process in the retardation of radionuclides, because these are often present in extremely low concentrations or below solubility limits.

The most simple model for adsorption is one in which the sorption density, Γ, is proportional to the dissolved sorbate concentration. This is the linear isotherm:

$$\Gamma = K_d[J] \tag{3.4}$$

where $[J]$ is the concentration of an adsorbing solute, J, and K_d is the distribution coefficient (see also Chapter 5). Distribution coefficients for radionuclides are commonly derived from adsorption experiments in which an aqueous phase containing a pollutant is equilibrated with a solid. One example is a study of the adsorption of ^{137}Cs and ^{90}Sr on soils from Bangladesh, where K_d values were obtained to assess the suitability of a proposed shallow, low-level waste disposal site (Mollah and Ullah, 1998). These authors also derived the retardation factor (R_f), which is a measure of the velocity of a contaminant relative to the rate of groundwater movement and can be calculated from the K_d value and other soil parameters.

A "constant K_d" model is often used in simplified models of radionuclide transport. In this approach, selected K_d values are incorporated in the transport codes used for migration predictions together with hydrological parameters, such as the rate of water movement. However, the K_d is usually dependent on a large number of factors, including the aqueous chemistry of the radionuclide in question and the nature of the geological material (Sheppard and Thibault, 1990). For example, Choi et al. (2005) demonstrated how the K_d values for ^{85}Sr and ^{137}Cs in a rice paddy could be influenced by the addition of lime and KCl. A study of the tropical northern Australian environment related K_d values for ^{134}Cs to their clay content and those for ^{85}Sr to the pH value (Twining et al., 2004). The pH was also shown to be a major factor influencing the adsorption of ^{60}Co, with the specific surface area of the soil also of importance (Payne et al., 2009). Therefore, experimental distribution coefficients are strictly applicable only to the experimental conditions under which they are measured and any change in a chemical parameter (such as pH or ligand concentration) may have a large effect on the K_d value.

Many models of adsorption that are more complex than the K_d concept have been developed (Kinniburgh, 1986). Sorption isotherms relate the amount of adsorbed contaminant to its dissolved concentration and may take into account some of the properties of the solid phase, such as the population of sorbing sites. The commonly used Langmuir isotherm is given by the following:

$$\Gamma = \Gamma_{MAX}\left(\frac{K_{ads}[J]}{1 + K_{ads}[J]}\right) \tag{3.5}$$

where K_{ads} and Γ_{MAX} are fitted parameters. This equation becomes linear at low $[J]$, corresponding to a linear isotherm with a K_d equal to $\Gamma_{MAX}.K_{ads}$. At high $[J]$, the value of Γ approaches a constant limiting value Γ_{MAX}.

The Freundlich isotherm has the equation:

$$\Gamma = A[J]^{\alpha} \tag{3.6}$$

with A and α adjustable constants. The Freundlich isotherm is linear on a log-log graph. Although the parameters associated with both the Langmuir and Freundlich isotherms are conditional constants and dependent on the chemical conditions under which they were measured, the isotherms are useful to describe experimental data in some instances. Cesium adsorption on several samples from arid-zone sites in central Australia was found to conform to the Freundlich isotherm (Payne et al., 2002).

More sophisticated chemistry-based sorption models, referred to as surface complexation models (SCMs), have become increasingly common for describing radionuclide adsorption, particularly data obtained in laboratory experiments. The SCM approach is based on developing a set of surface reactions describing adsorption, each with an associated equilibrium constant (Stumm, 1992). In many cases, the SCM also contains a mathematical description of the electrical double layer (EDL) at the interface (Davis and Kent, 1990). The EDL is a region near the surface where the electrical charge on the surface causes the distribution of cations and anions to differ from that in the bulk solution. The SCM takes into account the effect of surface charge on the sorption of any species by applying a correction factor derived from the EDL theory to the equilibrium constants for surface reactions. The advantage of the SCM approach is that it can be applied across a range of chemical conditions (i.e., pH, ionic strength, ligand concentrations) without adjustment of the associated parameter values.

The SCMs require considerable amounts of experimental data and invariably contain several fitted parameters, making them difficult to apply in real-world situations. Nevertheless, it has been shown that a simplified SCM approach can be used to model uranium adsorption on complex weathered rock samples from tropical northern Australia (Waite et al., 2000). In a recent paper, the pH dependence of K_d values for cobalt was modelled using the SCM approach (Payne et al., 2009). Whilst the SCM models were considerably simplified, the basic applicability of the SCM in simulating experimental radionuclide sorption data was demonstrated in both cases.

3.2.5 Other Phenomena

Numerous additional phenomena can affect radionuclide mobility, including processes related to radioactive decay such as nuclear recoil (Yanase et al., 1995; Janeczek, 1999), emanation of radioactive gases (particularly radon), and formation of colloidal particles in groundwater. In some environments, colloids have mobilised radionuclides, including plutonium, to a significant extent (Kersting et al., 1999). Microbiological and phytological processes can also have an impact. Some of these processes are relevant to the examples of radionuclide mobility discussed in Section 3.4. The transfer factors, which describe the uptake of radionuclides by plants, are explained in Chapter 5.

3.2.6 Summary

The preceding discussion demonstrates that the migration of radionuclides in tropical environments is subject to similar constraints to those that govern the distribution of radionuclides (and also nonradioactive contaminants) in temperate environments. Many of the cited studies of radionuclide mobility from tropical countries demonstrate these generally applicable principles. It is important when interpreting phenomena related to tropical environments that these common processes and mechanisms are understood in order to identify those factors that are distinctive to tropical environments and to ascertain how they impact radionuclide migration in these environments.

3.3. PROPERTIES OF TROPICAL SOILS

The tropical climate is characterised by high temperatures without significant seasonal variation and a relatively constant day length, which remains close to 12 hours throughout the year. Typically, rainfall is concentrated in summer months in a monsoonal pattern. As described by van Wambeke (1992), tropical soils are considered the products of interactions between this unique climate and the parent geological materials of the Earth's crust.

A full range of parent materials occurs in the tropics, including very young basalt and serpentine landscapes. Landscapes in the tropics can be placed in four broad categories that are useful when considering soil types and radionuclide behaviour: (1) old and tectonically stable regions, (2) young and tectonically active regions, (3) tropical karst, and (4) tropical coasts (Reading et al.,

1995). The residual landscapes of the tectonically stable regions are very highly weathered, often with marked duricrusts such as ferricrete, and are dominated by wide undulating plains with some isolated resistant residual hills. These landscapes occur throughout the mid-plate or shield zones in much of Africa, northern Australia, and some parts of Asia and South America. This contrasts to the tectonically active areas in which much stronger relief occurs through uplift and volcanism, with younger and therefore less weathered materials, even including reactive clays such as montmorillonite, in some cases. Karst landscapes, dominated by dissolution, also occur through the tropics and are mostly associated with limestone parent material. Tropical coasts are frequently fringed by extensive fluviomarine features, such as sediment plains, formed by the deposition of material eroded due to high-intensity rainfall. Other coastal geomorphologies that lead to distinctive soil properties are the coral reefs and atolls, algal mats, and mangrove mudflat systems.

Weathering within the equatorial climatic regime is the major process contributing to the properties of tropical soils. Due to relatively high temperatures and high-intensity rainfall in parts of the region, weathering is rapid and the strong leaching leads to an abundance of highly stable minerals such as kaolinite and sesquioxides (Naidu et al., 1998). The soil types of the tropics are described, with some variation in categorisation, in several texts. In the classification scheme used by van Wambeke (1992), the soils in the tropical areas consist of approximately 25% oxisols, 20% aridisols, 20% alfisols, 10% of ultisols, and 10% entisols, with several other soil types (as well as uncategorised soils) comprising the remainder. More information on the nature and distribution of tropical soil types using different classification schemes is given in Chapter 5.

The oxisols and other highly weathered soil types (including alfisols and ultisols), which predominate in the humid tropics, are important for agriculture and occupy extensive areas of land used for crop and food production. These soils are dominated by low activity sesquioxide minerals (i.e., oxides and hydroxides of iron and aluminium) and 1:1 layer silicates (kaolin), both of which have a variable surface charge (Naidu et al., 1998). The aridisols have an aridic soil moisture regime (as the name implies) and tend to be much less important in agricultural production. However, their properties may be of interest for potential repository sites.

In recent years, models of the behaviour of potentially toxic elements in soils have been developed for temperate soils, which tend to have quite different characteristics to those of tropical soils. Typically, temperate soils have much higher concentrations of more active clay minerals, including the 2:1 layer silicates, which possess permanent charge and therefore differ in ion exchange properties from tropical soils. As noted by Rieuwerts (2007), the major tropical soil types are distinguished by low activity clays, low organic matter contents, low pH values, and high levels of Fe-oxides. In tropical soils, the combination of relatively low pH, low organic matter, and the dominance of kaolinite (rather than the more strongly sorbing 2:1 layer silicate clays) tends to mitigate against the retention of heavy metals. In contrast, the Fe-oxide content might have a proportionally greater importance in heavy metal retention. For example, it has been reported that iron oxides are important in adsorbing uranium in tropical soils (Payne et al., 1994). The general conclusion of Rieuwerts (2007) is that heavy metals will tend to be comparatively mobile in tropical soils and appear to be relatively bioavailable.

Velasco et al. (2008) identified the two main factors distinguishing tropical soils and affecting radionuclide transfer factors as follows:

- In tropical environments, almost all organic material that reaches the soil surface decomposes rapidly, so the surface accumulation of soil organic matter is minimal. Consequently, there is rapid recycling of nutrients and contaminants into the vegetation. In temperate zones, the decomposition of organic debris is slower and the accumulation of soil organic matter is usually greater than the rate of decomposition, resulting in highly organic surface soil.
- In the tropics, due to the relatively highly aged soils and high mineral weathering rates, clays of low exchange activity, such as kaolinite, are more common than in temperate zones. This leads to soils that, despite having high clay content, have a low exchange capacity.

A particularly important soil environment that is widespread in tropical environments is the paddy field ecosystem. Because rice plants grow in flooded conditions, the behaviour of redox-sensitive elements, particularly technetium, is different in paddy fields than in comparable dryland agriculture (Yanagisawa et al., 2000; Uchida, 2007). Whilst most soils in the tropics are acidic, some agriculturally important areas have alkaline soils particularly associated with illuvial deposits, karst, and coral residuum.

Understanding the mobility of radionuclides under tropical conditions is a high priority because these soils provide food resources for large populations (see Chapter 5).

3.4. CASE STUDIES OF RADIONUCLIDE MOBILITY IN TROPICAL ENVIRONMENTS

In this section we will briefly review selected experimental and field investigations relevant to radionuclide mobility in tropical systems. These are grouped between radionuclide releases (to both atmosphere and groundwater) from nuclear weapons; releases from accidents; migration studies of natural radioactivity (uranium and thorium deposits); and finally, experimental studies relating to the siting and operation of nuclear facilities.

3.4.1 Radionuclides Derived from Nuclear Weapons Testing

There has been significant environmental dispersion of anthropogenic radionuclides as a consequence of nuclear weapons testing. UNSCEAR (1982) reported that 76% of ^{137}Cs fallout from nuclear weapons explosions occurred in the Northern Hemisphere and 24% in the Southern Hemisphere. Historically, the deposition rates of global fallout radionuclides in equatorial areas have not been as high as in the more temperate regions, where the highest deposition rates have been between 40°N and 50°N. However, there have been numerous nuclear tests with associated fallout in tropical areas. For example, the Bikini and Enewetak atolls in the Marshall Islands were the site of United States nuclear weapons testing from 1946 to 1958.

The coral soils at Bikini Atoll have an unusual composition and are made up almost entirely of calcium carbonate ($CaCO_3$) with some $MgCO_3$ and essentially no silicate clays (Robison et al., 2000). The source materials of the soils are coral, various other organisms (e.g., foraminifera), and shell-type materials. The soils are low in exchangeable potassium. The pH of soil water slurries is relatively high (7.7–8.8), and the organic matter content is 5% to 14% in surface soils. The cycling of radionuclides at the Bikini test site was reported by Robison et al. (2000). A significant finding was that ^{137}Cs was far more available for plant uptake, with much higher concentration factors than in temperate soils. Due to the unique composition and properties of the coral soil, ^{137}Cs uptake in food crops was much higher than would be expected for continental silica-based soils where cesium is quite strongly bound by clay minerals, particularly illite. The situation for plant uptake of ^{90}Sr in coral soils is the reverse of cesium. In this case, some of the ^{90}Sr is bound structurally into the $CaCO_3$ matrix and the large quantities of plant-available Ca in the coral soils reduce the uptake of strontium. Robison et al. (2000) reported that actinide behaviours are not significantly different in the studied soils when compared to temperate environments. Whilst the study focused on plant uptake, rather than the mobility of the radionuclides in the environment, it has significant implications for mobility. For example, the absence of silicate clays presumably makes the role of other minerals and organic matter in retaining cesium more significant. Indeed, Robison et al. (2000) reported that the organic matter contains 90% of the cesium in the Bikini Atoll soils. Thus, the cycling of Cs and Sr in these tropical atolls is completely different to that in temperate environments. However, it needs to be recognized that the coral soils are unusual soil types and other tropical environments may differ significantly.

Between July 1966 and January 1996, France conducted 193 nuclear tests above and below ground in French Polynesia. A total of 147 underground nuclear tests: were conducted at the Mururoa and Fangataufa Atolls. There were two types of tests: 178 were nuclear explosions with large releases of fission energy and in 15 tests the nuclear devices were exploded by conventional explosives. The latter were described as 'safety trials,' which simulated accident conditions. In these tests, the nuclear weapon cores were destroyed, with no or, on few occasions, very small release of fission energy.

Residual radioactivity in the terrestrial environment of the Mururoa and Fangataufa Atolls' nuclear weapon test sites has been investigated by

Danesi et al. (2002). Air filter samples, utilized to collect aerosols or resuspended radioactive particles, were analysed to evaluate the potential exposure due to inhalation. Wherever possible, samples of topsoil and depth profiles were also collected (in many locations there was no soil) to estimate the total surface contamination and to evaluate the migration of radionuclides through soil. At some sites, the soil essentially consisted of coral debris and sand. In areas where no soil was present, loose corals and sand samples were collected. Cores of coral rocks were taken to investigate the downward migration of radionuclides originally deposited on the surface. The vertical concentration profile was also an important parameter for the calculation of the overall radionuclide inventory from *in situ* gamma spectrometry. It was reported that, although the terrestrial environment of the Mururoa and Fangataufa Atolls still contained residual radioactive material attributable to the nuclear tests, the concentration levels are generally of no radiological significance and did not require remediation. It was concluded that particles containing plutonium and small amounts of americium resulting from safety trials are present in parts of the Mururoa Atoll and several kilograms of plutonium resulting from the atmospheric nuclear tests and safety trials are present in the sediments under the lagoons around the atolls. The total $^{238+239+240}$Pu inventory in the lagoons was estimated to be about 30 TBq. The inventories of ^{241}Am, ^{137}Cs, ^{155}Eu, and ^{60}Co were reported to be below 1 TBq each.

Whilst nuclear weapons testing in French Polynesia ceased in 1996, the underground test sites remain a potential source of leakage of radioactivity into the environment. Sensitive analyses of a number of water samples from underground monitoring wells and cavities created by the explosions have been undertaken by accelerator mass spectrometry (AMS; Jacobsen et al., 2000). The monitoring wells were in the carbonate strata, which surrounded the test zones, and the cavity chimneys were in the basalt where the tests were carried out. Three radionuclides (^{129}I, ^{36}Cl, and ^{14}C) were measured by AMS and the results were compared with measurements made for ^{3}H, ^{90}Sr, and ^{137}Cs using conventional methods. Some of the results were used to estimate the K_d values of the radionuclides to characterise sorption and to compare the mobility of the radionuclides in the carbonate rock surrounding the test zones and predicted levels in the lagoons. It was concluded that the elements I, Cl, and ^{3}H were all (essentially) conservative tracers and that the presence of tritium in the lagoons, as compared to the levels in the open ocean, provided some evidence that leakage was occurring from underground sources. However, the expected levels of ^{129}I and ^{36}Cl in the lagoons were predicted to be below the detection limits.

3.4.2 Radionuclides Released by Accidents

This section discusses the effects of releasing artificial radionuclides on the surrounding environment. Such releases have resulted from major reactor accidents including the Chernobyl disaster in 1986 (IAEA, 2006), the Fukushima disaster in 2011 (ANS, 2012), and various less well-known accidents involving radiation sources and nuclear facilities. Relatively few such accidents have occurred in tropical countries, but the expansion of nuclear energy and technologies into these countries implies that the possible impacts of similar accidents should be considered.

The Tohoku earthquake in March 2011 and the subsequent severe damage to several reactors at the Fukushima Daiichi nuclear plant resulted in significant releases of radionuclides into the environment, the consequences of which are still being assessed. An isotope of major concern has been ^{137}Cs (half-life of 30.1 years), which has potential deleterious human health impacts through various exposure routes, including agricultural crops and livestock farming. Both measured and estimated depositions of ^{137}Cs in Japan and surrounding regions, including the local seas and oceans, were reported by Yasunari et al. (2011). Their study considered how dispersion models could be calibrated using on-ground measurements and demonstrated the effect of mountain ranges in shielding some parts of Japan from the airborne plume.

Various radionuclides released from the Fukushima accident have been detected at significant distances from their source. For example, radionuclides including ^{131}I, ^{132}I, ^{132}Te, ^{134}Cs, and ^{137}Cs have been measured in Spain (Lozano et al., 2011), Italy (Ioannidou et al., 2012), Korea (Kim et al., 2011), and Seattle, Washington (Leon et al., 2011). The radioactive inert gas ^{133}Xe was also detected in Vancouver, Canada (Sinclair et al., 2011). The radionuclides ^{131}I, ^{137}Cs, and ^{134}Cs were detected in Taiwan, although at much lower levels than in Japan, North America, or Europe, and posed little risk to human health (Hsu et al., 2012). However, this showed that the impact from Fukushima will extend into the tropical environments, which is the concern of this book. In a few cases, the measurements of Fukushima-derived airborne radionuclides have been supplemented by studies of transfer of radionuclides into human foods, such as ^{129}I in milk in France (Parache et al., 2011). The studies reported thus far have been extremely valuable in establishing the global consequences of the disaster and the impact on various environments and human health. However, the early reports of long-distance radioactivity dispersion from the Fukushima event have predominantly focused on radionuclide detection in advanced temperate countries, and the impact on tropical countries is yet to be fully evaluated.

One of the most significant previous events specifically impacting tropical environments is the 1987 accident at Goiânia in Brazil. Lessons drawn from this incident, described as the worst accident involving a radioactive source that the world has seen (Hansen, 2008), continue to be relevant to radiation safety and security. Cesium chloride from a dumped radioactive source that had ended up in a scrap yard was distributed undetected for over two weeks. Some 250 people were contaminated and four died in the first month. The event focused international attention on the issue of safety standards for radioactive sources. Since the accident, the gradual replacement of sealed sources containing the soluble, powdery form of cesium chloride has been considered, although it is not easy to find the equivalent replacement for certain applications. The majority of papers regarding this incident relate to the incident, its causes, and mitigation. However the accident also has some implications for the mobility of radionuclides (specifically ^{137}Cs) released in tropical environments.

Wasserman et al. (2002) showed that soil-to-plant transfer factors for ^{137}Cs in the local contaminated oxisol soils (as well as another experimentally contaminated oxisol) were higher than previously observed in temperate climates. These differences were discussed in relation to the pedology and geochemical partitioning of this tropical environment. It was concluded that tropical soil characteristics, such as acidity, low availability of nutritive elements, and low content of 2:1 clays, may account for the high mobility of ^{137}Cs. A sequential soil extraction showed that most of the ^{137}Cs was bound to soil oxides, and that added ^{137}Cs remains potentially bioavailable, even a long time after contamination. Additionally, the low availability of nutrient elements (mainly K and Ca) and predominance of kaolinite and gibbsite in the clay fraction also led to higher transfer factors of ^{137}Cs from soil to plants. Plant elemental distribution studies at Goiânia suggested that there is a direct competition between Cs and K ions, indicating these elements could have a common accumulation mechanism and the deficiency of nutrients (such as K) would contribute to increasing plant uptake of ^{137}Cs (Carvalho et al., 2006).

Following the first rainfall after the contamination of the available soils and surfaces, ^{137}Cs quickly reached the Meia Ponte River, the main local waterway (Godoy et al., 1991). After the accident, ^{137}Cs was detectable in sediments up to 12 km downstream from the accident area. Because the accident occurred at the beginning of the wet season, downstream sediment transport was enhanced. This resulted in a decrease of ^{137}Cs concentrations in river sediments of the urban area of Goiânia by a factor of approximately 20 within 5 weeks of the accident. The main sedimentation area for suspended sediments in this portion of the river was a reservoir about 80 km downstream from the accident area.

Remedial action was undertaken after the accident to reduce dose levels at the site. These actions included removal of the most highly contaminated shallow surface layer and replacement with a shallow layer of soil or concrete (Facure et al., 2001). The long-term impacts of the accident have been the focus of ongoing studies. More than a decade after the accident, it was reported that a very high cesium concentration remained in the subsoil layer 10–40 cm below the surface (Anjos et al., 2002). It appeared that the contaminated soil layer may be coming slowly to the surface of the terrain because people were making holes in the soil to plant trees or to build small houses. In addition, the terrain has a downward slope in the direction of a local creek and the surface soil layer is slowly becoming exposed by rain and erosion. Thus, the site may require ongoing monitoring to ensure the safety of residents.

Goiânia´s legacy of a handful of cesium chloride is 3000 cubic metres of contaminated waste. Most of it is now buried in a near-surface repository on the outskirts of the city, where it must be isolated for the next 300 years.

3.4.3 Migration Studies Involving Natural Radionuclides

There have been several studies of radionuclide mobility in the vicinity of uranium or thorium ore bodies in tropical regions. Much of this research has been driven by the need to predict the behaviour of buried radioactive waste in the vicinity of nuclear repositories. This objective has been pursued vigorously in westernised countries in recent decades and the increased development of nuclear power in tropical nations will require consideration of safe disposal options for nuclear wastes.

To assist in this objective, natural systems exhibiting some of the properties of radioactive waste, such as uranium deposits, have been selected for the study of many of the relevant processes. This approach (known as the natural analogue approach) allows processes occurring over long geological timescales, and of considerable spatial extent, to be studied. Natural analogues can demonstrate the stability of some geological or engineered systems over long timescales; therefore, the results have been utilised for public relations and public education purposes. The applications of natural analogues for nuclear waste repositories have been extensively reviewed (Miller et al., 1994).

Several of the major natural analogue studies have been in tropical regions, although this is often incidental to the objectives of the study. Therefore, the results are applicable to geological nuclear waste disposal regardless of the regional location of the study site. Among the most intensively studied sites are the unique Oklo natural reactors in Gabon (western Africa), where naturally occurring fission reactions took place about two billion years ago. At that time, the ^{235}U content of normal uranium was about 3.2%, sufficient for criticality to occur. The reactor zones consisted of block-shaped accumulations of uranium ore about 20 m long and 1 m thick, which at the time were buried about 3500 m below the surface (Brookins, 1984). Over a period of approximately 400,000 years, the Oklo reactor produced and exposed the surrounding rocks to more than 30 nuclides, which are the products of neutron capture, nuclear fission, and radioactive decay. About 12 tonnes of ^{235}U were fissioned, producing 200 kg of plutonium, 730 kg of ^{99}Tc, and numerous other reaction products (Curtis et al., 1981). The evaluation of the stability of the uraninite matrix over very long periods of time provides information on the possible long-term stability of waste forms, such as spent fuel (Toulhoat et al., 1996). A comprehensive summary of the research at Oklo is provided by Janeczek (1999).

There have been numerous studies of the mobility of the radionuclides produced by the nuclear reactions at Oklo using sensitive analytical techniques such as AMS. In some cases, the mobility of a radionuclide, such as ^{99}Tc, has been inferred from the distribution of its progeny. Based on the zones of ^{99}Ru enrichment near the reactor zones, it was concluded that ^{99}Tc migrated a number of metres, but was then retained (Curtis et al., 1981; Janeczek, 1999). Actinide migration at Oklo appears to have been limited, with plutonium and neptunium being mostly retained near the reactor zone. The noble gases xenon and krypton escaped during the active period of the reactors, and iodine seems to have been mobile (Brookins, 1976).

The processes that have received the most attention at Oklo occurred in the vicinity of the original reactor zones, which were deeply buried at the time of criticality. However, there has also been interest in reactor zones possibly affected by surficial alteration processes, particularly the shallow Bangombé reactor (11 m depth), located 30 km south of Oklo. This study has included hydrogeology, groundwater chemistry, isotopic analyses, and measurement of ^{235}U/^{238}U isotope ratios in groundwater to evaluate the migration of uranium from reactor zones (Toulhoat et al., 1996). Comparison of Bangombé and a deeper (450 m depth) reactor zone at Okélobondo allowed the validation of modelling tools in two different sedimentary environments: one shallow, with a more homogeneous layering situated in an area of meteoric alteration, and the other, a deeper site, offering the opportunity to study radionuclide migration from the reaction zone through very heterogeneous sedimentary layers. The modelling results indicate that the chemical reactions retarding radionuclide transport are very different at the two sites. At Bangombé, the decomposition of organic material consumes oxygen and at Okélobondo, the oxygen is consumed by inorganic reactions resulting in uranium retardation in both cases (Gurban et al., 2003).

Oklo is the only known terrestrial example of a natural reactor and has provided a unique opportunity to study the transport of fission products and transuranic elements in a geologic environment. The very fact that evidence of criticality still exists attests to the possibility that radionuclides can be immobilised over timescales approximating half the age of the Earth, much longer than would be required of a repository (Curtis et al., 1981).

Another site at which radionuclide mobility has been studied is the Morro do Ferro, a deposit containing 30,000 tonnes of thorium, as well as light rare earth elements, located in Brazil. The deposit is near the surface in weathered rock and intersected by groundwater that has a short transit time to the accessible environment. The results from the site showed that thorium concentrations in local streams were extremely low (Eisenbud et al., 1984). However, some surface waters from the highest activity region of the deposit contained high thorium levels, exceeding its expected solubility by a factor of $\sim$200. The thorium was thought to be in the form of Th-humate complexes (Miekeley and Kuchler, 1987). This is another example of the important role of speciation in enhancing radionuclide mobility (see Section 3.2.3).

A recent study of groundwater in the Guarani aquifer in Brazil has been reported by Bonotto and Bueno (2008). These authors measured and investigated the potential impact of radionuclides in the aquifer. The main contributor to the human dose was concluded to be ^{226}Ra, although ^{222}Rn was considered to be of concern because of its possible presence inside spa facilities, which are common in the region. Whilst the mechanisms of exposure as determined in this study are not unique to a tropical region, the authors noted the relative lack of integrated studies of radioactivity in groundwater in South America. This highlights the need for such research to be undertaken in countries of the tropical regions of the world where geology, soils, land use, and patterns of human behaviour may differ from other, more studied and resourced regions.

Another major international study of radionuclide mobility in a tropical environment was the Alligator Rivers Analogue Project (Payne and Airey, 2006). This project was focused on the Koongarra uranium deposit (Figure 3.3) in the Alligator Rivers region of the Northern Territory of Australia (about 225 km east of Darwin). The primary mineralisation of the Koongarra No.1 ore body occurred adjacent to a graphitic schist unit, which acts as a reducing geochemical barrier. This ore body shows extensive evidence of radionuclide movement in shallow groundwater and one of its main geological features is the characteristic 'dispersion fan.' This is an area containing elevated concentrations of uranium and other radionuclides that have been transported from the zone of original uranium accumulation (the primary ore zone) into the surrounding weathered rocks (Figure 3.3). The research effort at Koongarra included extensive characterisation of the geological, hydrological, and geochemical conditions at the site. Patterns of the distribution of radionuclides (predominantly members of the ^{238}U decay chain, but also the rare nuclides ^{239}Pu, ^{99}Tc, and ^{129}I) were studied in both solid and groundwater phases. The project included detailed studies of uranium adsorption on mineral surfaces and of subsequent processes that may lead to long-term uranium immobilisation. Numerous models for uranium migration were developed during the project.

The scientific results of the study were reported in detail by Payne and Airey (2006). Numerous important processes were studied at the Koongarra site, including:

- The role of the downward movement of the weathering front triggering the mobilisation of uranium
- Mobility of uranium in intersecting groundwater leading to the development of the dispersion fan
- The immobilisation of uranium by adsorption on iron minerals and incorporation in iron nodules (Sato et al., 1997). As noted earlier, Fe-oxides appear to be extremely important in the retention of uranium in the soils at Koongarra and this is consistent with the kaolinitic clay mineralogy (Payne et al., 1994).

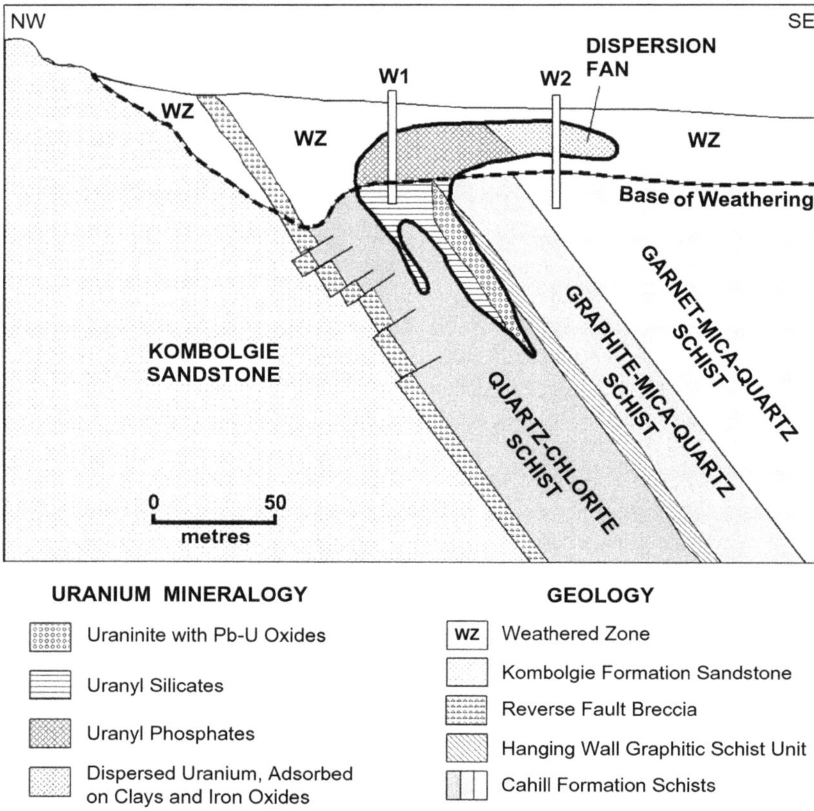

FIGURE 3.3 A cross section through the Koongarra uranium deposit. Boreholes such as W1 and W2 enabled water and solid core samples to be taken from various locations and depths in the ore body. *From Payne and Airey (2006), with permission.*

- The speciation of aqueous uranium (see also Figure 3.2) and the role of complexants, including phosphate, affecting its adsorption
- The mechanism of uranium sorption in complex materials, which was modelled using SCMs (see Section 3.2.4)
- The role of groundwater colloids in uranium and, particularly, thorium migration at the site
- Isotopic disequilibrium in the uranium decay series, particularly $^{234}U/^{238}U$ ratios. The unusually low ratio of ^{234}U to ^{238}U (<1.0) found in weathered zone groundwater at the site is one of the unusual isotopic effects attributed to water table fluctuations in the soil zone between wet and dry seasons (Yanase et al., 1995).

The previously mentioned studies of radionuclide mobility in the tropics demonstrate that, whilst many factors are similar to other locations, the unique properties of tropical environments, particularly in the intensely weathered near-surface soils, need to be considered when assessing the environmental behaviour of radionuclides. Issues related to the movement of water are particularly important, with the monsoonal climate leading to fluctuating water tables. Intense fluxes of water, alternating redox conditions, and the reactivity of iron minerals as radionuclide adsorbing phases (Sato et al., 1997) are amongst other factors that have the potential to affect radionuclide mobility more strongly in tropical and subtropical areas in comparison with more temperate climates.

3.4.4 Experimental Studies of Radionuclide Behaviour in Tropical Systems

These examples of radionuclide mobility involve the presence of radionuclides in the tropical environment from natural sources, bomb testing, or accidents. However, the number of case studies is relatively few and contrasts with the rather more numerous incidents and discharges from nuclear facilities in temperate environments. A well-known example is the discharge from Sellafield in the UK and associated contamination of the Irish Sea (Cook et al., 1997, Batlle et al., 2008). Similarly, the Chernobyl accident also impacted a significant area of predominantly temperate and sub-Arctic environments. As discussed earlier, the more recent Fukushima accident has led to extensive radionuclide dispersion, although the possible impacts on tropical environments remain to be elucidated.

As such, there have been relatively few opportunities to study radionuclide behaviour of anthropogenic radionuclides in tropical environments (apart from testing of nuclear weapons and smaller scale releases, such as the Goiânia accident). This is due to the relatively small number of nuclear facilities in tropical countries. As previously noted, part of the rationale for this book is the scarcity of relevant information on these issues.

Therefore, the final category of research discussed in this chapter covers experimental studies of the behaviour of radionuclides in laboratory or field experiments. These tend to be aimed at assessing the impacts of events that have not occurred, for example the performance of future nuclear waste repositories or releases from possible nuclear accidents. Whilst a large amount of migration and distribution data for radionuclides has been obtained for soils from cool temperate Northern Hemisphere regions, there has been limited research applicable to tropical regions, where the increasing use of nuclear power as an energy source is predicted. An accident at a nuclear power plant is a possibility deserving attention as nuclear power becomes more common in tropical and subtropical regions (Uchida, 2007).

The applications of controlled experiments for addressing these problems have been discussed by Payne et al. (2004). The advantages of using radionuclides for studies of environmental processes include the wide range of element concentrations that can be studied; the ability to simultaneously study several radionuclides in a single experiment; the direct applicability to the behaviour of radioactive waste or fallout; and the ability to study the mechanisms, reversibility, and kinetics of environmental reactions under controlled conditions. Various types of experiments have been undertaken, including field studies, pot studies, and batch adsorption experiments.

In an experimental (pot) study of the dynamics of Cs in tropical environments using soil from Bangladesh, Rahman et al. (2005) speculated that organic matter increased the availability of cesium. Their experimental results indicated that there is a weak, nonspecific interaction between Cs and soil organic matter and the organic matter decreases the affinity of the clay minerals for Cs, thus reducing their ability to immobilise Cs in soils. Therefore, the role of organic carbon was completely different to that found for the Bikini Atoll samples by Robison et al. (2000) (Section 3.4.1). In the latter case, the organic matter increased the affinity of the Cs for soils in which silicate clays were absent and the soils were dominated by carbonates.

An example of an experimental study of radionuclide behaviour is for K_d data obtained in laboratory experiments with samples of soils taken from field sites in the Northern Territory of Australia. A full account of this work is given by Twining et al. (2004). The experimental procedure utilized the tracers ^{65}Zn, ^{85}Sr, and ^{134}Cs, which were added as a mixed tracer solution. The pH dependence of the sorption of Cs, Sr, and Zn on the soil was experimentally studied. Measurements were obtained at the equilibrium pH of the soil–water system and at pH values one unit higher and lower than the equilibrium value. The pH dependence of the adsorption of Cs, Sr, and Zn is shown in Figure 3.4. The uptake of Cs is very strong, due to its very strong retention by clays, but shows little pH dependence. This is explained by the simple chemistry of the monovalent Cs^+ ion (Cornell, 1993). Strontium shows the greatest mobility, whilst Zn has an intermediate mobility and exhibits the strongest pH dependence. In general, cations exhibit increasing sorption (higher K_d) with higher pH, which is exemplified in this study by Zn and Sr.

Laboratory sorption experiments can be undertaken to determine the effects of various experimental parameters on radionuclide adsorption. As well as the effects of pH (which have already been discussed), sorption can be influenced by ionic strength, redox potential, the presence of complexing ligands, and other chemical variables (see also Chapter 5). Figure 3.5 illustrates the speciation of uranium as a function of pH in the presence of representative environmental ligands such as sulphate and citrate. The effect of the ligands on the adsorption of uranium by kaolinite, which is a common clay mineral in tropical environments, can be seen in Figure 3.5c. The results amply demonstrate the important role of pH in uranium mobility, as well as the potential impact of the presence of complexing ligands. In these systems that are equilibrated with atmospheric CO_2, the adsorption of uranium at high pH values is significantly decreased by the presence of stable aqueous phase uranyl carbonate complexes (Waite et al., 1994).

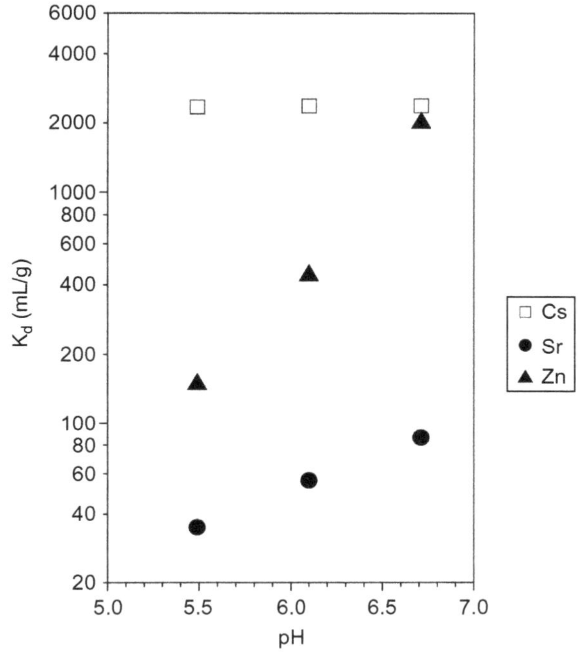

FIGURE 3.4 The effect of pH on the K_d for adsorption of cesium, strontium, and zinc on an Australian soil. *Data from Twining et al. (2004).*

3.5. FACTORS OF SIGNIFICANCE IN THE MIGRATION OF RADIONUCLIDES IN THE TROPICAL ENVIRONMENT

In this chapter, it has been shown that the migration of radionuclides in tropical environments is subject to the same scientific principles as in any other environment. Thus the basic migration mechanisms involve similar hydrological and geochemical parameters. However, the behaviour of radionuclides is also modified by many unique features of the tropical environment, as summarised in the following sections.

3.5.1 Climate and Rainfall

Tropical environments have a unique climate that impacts the properties of soils and the mobility of radionuclides in terms of the higher temperatures and, in particular, the patterns of rainfall. In many tropical environments there are a monsoonal climate, implying intense episodic rainfall events; a fluctuating water table; and cyclic saturation and unsaturation with associated redox changes and flooding. A major implication of the flooded surface environment (such as rice paddies) is that atmospheric radionuclides are often deposited onto

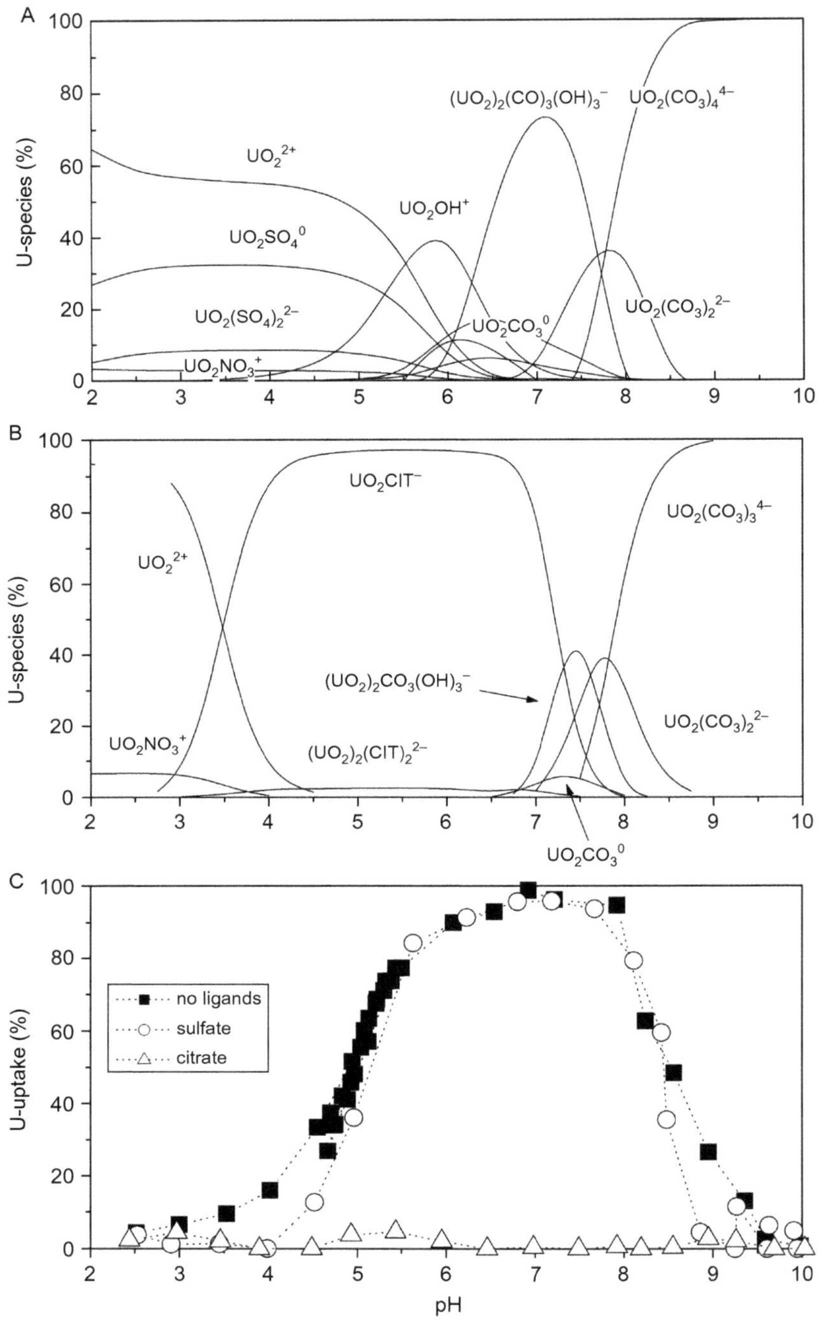

FIGURE 3.5 An example of an experimental study of the impact of uranium speciation on adsorption. The upper panels illustrate speciation of U(VI) at a total concentration of 1 μmol/L equilibrated with air and in the presence of (a) sulphate (0.01 M) and (b) citrate (100 μmol/L). Part (c) shows the adsorption of uranium on kaolinite (4 g/L) in the presence and absence of these ligands. Note the dramatic effect of citrate and minimal impact of sulphate on uranium adsorption. *Data from Payne (1999).*

standing water rather than onto soil (Choi et al., 2005). The intense effects on the tropical ground surface of sun baking and periodic fires also play a role.

3.5.2 Soil Mineralogy

There are many unique soil types in the tropics and factors such as soil mineralogy, porewater pH, the smaller cation-exchange capacity of tropical soils, the variable charge of common clay surfaces, and the possible associations of radionuclides with specific minerals (i.e., preferential sorbing phases) need to be considered (see also Chapter 5). In the relatively weathered soils of the tropics, clays of low exchange activity, such as kaolinite, are more common than in temperate zones. This leads to soils that, despite having high clay content, have a low exchange capacity. This can lead to greater mobility of radionuclides through the soil layers. This was demonstrated by the studies at Goiânia, where mobility of ^{137}Cs in the contaminated soils was higher than previously observed in temperate climates. These differences were attributed to tropical soil characteristics, such as acidity, low availability of nutritive elements, and low content of 2:1 clays (Wasserman et al., 2002).

3.5.3 Organic Matter

The majority of organic material in shallow soil horizons of tropical environments decomposes rapidly, and accumulation of soil organic matter is therefore minimal in many cases. In temperate zones, the decomposition of organic debris is slower, and soils often have a higher content of organic matter. This has implications for migration of radionuclides and, depending on the role of organic matter, may increase or decrease their mobility as demonstrated for the specific case of cesium discussed earlier.

3.5.4 Nutrient Status

The nutrient status of tropical soils can also be a consideration. Plant elemental distribution studies at Goiânia suggest there is a direct competition between Cs and K ions indicating a common accumulation mechanism. This implies that the deficiency of nutrients, such as K, could contribute to increasing plant uptake of ^{137}Cs (Carvalho et al., 2006). Similar results were reported by Twining et al. (2004), who suggested that trace nutrient deficiency of tropical soils may be implicated in increased transfer factors of trace metals, such as zinc, relative to temperate environments.

3.5.5 A Range of Unique Soil Types

Some tropical soil types, such as the coral atolls discussed by Robison et al. (2000), are dominated by carbonate minerals. In such soils, the cycling of radionuclides is highly unusual (as described in Section 3.4.1). In these soils, cesium is more mobile than expected in temperate environments, whereas strontium appears to be more strongly bound. The differences amongst tropical soils imply that it is important to consider diverse behaviour between soils from different tropical environments and not to erroneously conclude that the behaviour of one tropical soil will necessarily be similar to another soil from a different environment.

3.5.6 Environmental Changes

There is a high degree of environmental change underway in the tropics with ongoing modifications in land use (e.g., urbanisation and the conversion of tropical forest to pasture) as well as local impacts of the more general phenomenon of global climate change. Climate change will alter the distribution and delineation of tropical and temperate ecosystems, and it has been speculated that global warming will eventually affect the soil-to-plant transfer factors of radionuclides (Dowdall et al., 2008).

3.5.7 Unique Combinations of Conditions

Many of the significant factors noted in this chapter are not unique to the warm tropical regions. Flooding, for example, is a possibility in various ecosystems worldwide. Similarly, there is considerable diversity among tropical soils, as shown by the contrasting roles of organic matter in determining the behaviour of cesium between the tropical clay soils of Bangladesh and the carbonate rich soils of the Bikini Atoll. Recent initiatives on soil mapping, such as the GlobalSoilMap.net project (Sanchez et al., 2009), are necessary to improve soil management and predict the behaviour of radionuclides at regional scales. Whilst there is ample evidence of the unique nature of radionuclide migration in the tropics, it is also important to avoid generalisations on the basis of the work undertaken thus far. As noted previously, this chapter only provides a starting point of issues to consider when one is pondering which environmental factors may be of most significance when considering the impact of a potential radionuclide release in the tropics.

3.6. OVERALL CONCLUSION

Our overall conclusion is that there is an imperative to better understand the mobility of radionuclides in the tropics and only some tentative steps have been undertaken to address this issue. It would be expected that the outcomes of research in coming decades will considerably enhance knowledge on this topic.

REFERENCES

Anjos, R.M., Umisedo, N.K., Facure, A., Yoshimura, E.M., Gomes, P.R.S., Okuno, E., 2002. Goiânia: 12 years after the ^{137}Cs radiological accident. Radiat. Prot. Dosimetry 101, 201–204.

ANS, 2012. Fukushima Daiichi: ANS Committee Report. American Nuclear Society, LaGrange Park, Illinois.

Batlle, J.V.I., Bryan, S., McDonald, P., 2008. A process-based model for the partitioning of soluble, suspended particulate, and bed sediment fractions of plutonium and caesium in the eastern Irish Sea. J. Environ. Radioact. 99, 62–80.

Bonotto, D.M., Bueno, T.O., 2008. The natural radioactivity in Guarani aquifer groundwater, Brazil. Appl. Radiat. Isot. 66, 1507–1522.

Brookins, D.G., 1976. Shale as a repository for radioactive waste: The evidence from Oklo. Env. Geol. 1, 255–259.

Brookins, D.G., 1984. Geochemical Aspects of Radioactive Waste Disposal. Springer-Verlag, New York.

Carvalho, C., Anjos, R.M., Mosquera, B., Macario, K., Veiga, R., 2006. Radiocesium contamination behavior and its effect on potassium absorption in tropical or subtropical plants. J. Environ. Radioact. 86, 241–250.

Choi, Y.H., Lim, K.M., Choi, H.J., Choi, G.S., Lee, H.S., Lee, C.W., 2005. Plant uptake and downward migration of ^{85}Sr and ^{137}Cs after their deposition onto flooded rice fields: lysimeter experiments with and without the addition of KCl and lime. J. Environ. Radioact. 78, 35–49.

Cleveland, J.M., Rees, T.F., 1981. Characterisation of plutonium in Maxey Flats radioactive trench leachates. Science 212, 1506–1509.

Cook, G.T., MacKenzie, A.B., McDonald, P., Jones, S.R., 1997. Remobilization of Sellafield-derived radionuclides and transport from the north-east Irish Sea. J. Environ. Radioact. 35, 227–241.

Cornell, R.M., 1993. Adsorption of cesium on minerals: A review. J. Radioanal. Nucl. Chem. 171, 483–500.

Curtis, D.B., Benjamin, T.M., Gancarz, A.J., 1981. The Oklo reactors: Natural analogues to nuclear waste repositories. In: Hofmann, P.L. (Ed.), The Technology of High Level Nuclear Waste Disposal. United States Department of Energy, Washington, pp. 255–283.

Danesi, P.R., Moreno, J., Makarewicz, M., Radecki, Z., 2002. Residual radioactivity in the terrestrial environment of the Mururoa and Fangataufa Atolls nuclear weapon test sites. J. Radioanal. Nucl. Chem. 253, 53–65.

Davis, J.A., Kent, D.B., 1990. Surface complexation modeling in aqueous geochemistry. In: Hochella, M.F., White, A.F. (Eds.), Reviews in Mineralogy Vol. 23: Mineral-Water Interface Geochemistry. Mineralogical Society of America, Washington, pp. 177–260.

Domenico, P.A., Schwartz, F.W., 1990. Physical and Chemical Hydrogeology. John Wiley and Sons, New York.

Dowdall, M., Standring, W., Shaw, G., Strand, P., 2008. Will global warming affect soil-to-plant transfer of radionuclides? J. Environ. Radioact. 99, 1736–1745.

Eisenbud, M., Krauskopf, K., Penna Franca, E., Lei, W., Ballad, R., Linsalata, P., et al., 1984. Natural analogues for the transuranic actinide elements: An investigation in Minas Gerais, Brazil. Environmental Geology 6, 1–9.

Fabryka-Martin, J., Flint, A., Meijer, A., Bussod, G., 2006. Water and Radionuclide Transport in the Unsaturated Zone. In: McFarlane, A.M., Ewing, R.C. (Eds.), Uncertainty Underground. Yucca Mountain and the Nation's High Level Nuclear Waste. The MIT Press, Cambridge, Massachusetts, pp. 179–197.

Facure, A., Umisedo, N.K., Okuno, E., Yoshimura, E.M., Gomes, P.R.S., Anjos, R.M., 2001. Remains of Cs-137 contamination in the city of Goiania, Brazil. Radiat. Prot. Dosimetry 95, 165–171.

Fetter, C.W., 1993. Contaminant Hydrogeology. MacMillan, New York.

Godoy, J.M., Guimaraes, J.R.D., Pereira, J.C.A., Dorio, M.A.P., 1991. Cesium-137 in the Goiania waterways during and after the radiological accident. Health Phys. 60, 99–103.

Gurban, I., Laaksoharju, M., Madéc, B., Ledoux, E., 2003. Uranium transport around the reactor zone at Bangombé and Okélobondo (Oklo): Examples of hydrogeological and geochemical model integration and data evaluation. J. Contam. Hydrol. 61, 247–264.

Hansen, K., 2008. The global schoolhouse on Rua 6 – Goiania's lessons shape actions on radiation safety and security decades later. IAEA Bulletin 49 (2).

Hsu, S.-C., Huh, C.-A., Chan, C.-Y., Lin, S.-H., Lin, F.-J., Liu, S.C., 2012. Hemispheric dispersion of radioactive plume laced with fission nuclides from the Fukushima nuclear event. Geophysical Research Letters 39, L00G22. http://dx.doi.org/10.1029/2011GL049986.

IAEA, 2006. Chernobyl's Legacy: Health, Environmental and Socio-Economic Impacts (Report of the Chernobyl Forum). International Atomic Energy Agency, Austria.

Ioannidou, A., Manenti, S., Gini, L., Groppi, F., 2012. Fukushima fallout at Milano, Italy. In press. J. Environ. Radioact. http://dx.doi.org/10.1016/j.jenvrad.2012.01.006.

Jacobsen, G.E., Hotchkis, M.A.C., Fink, D., Child, D.P., Tuniz, C., Sacchi, E., et al., 2000. AMS measurement of I-129, Cl-36 and C-14 in underground waters from Mururoa and Fangataufa atolls. Nucl. Instrum. Methods Phys. Res. B. 172, 666–671.

Janeczek, J., 1999. Mineralogy and geochemistry of natural fission reactors in Gabon. In: Burns, P.C., Finch, R. (Eds.), Uranium: Mineralogy, Geochemistry, and the Environment, Reviews in Mineralogy 38, 321–392.

Kersting, A.B., Efurd, D.W., Finnegan, D.L., Rokop, D.J., Smith, D.K., Thompson, J.L., 1999. Migration of plutonium in ground water at the Nevada Test Site. Nature 397, 56–59.

Kim, C.-K., Byun, J.-I., Chae, J.-S., Choi, H.-Y., Choi, S.-W., Kim, D.-J., et al., 2011. Radiological impact in Korea following the Fukushima nuclear accident. J. Environ. Radioact. http://dx.doi.org/10.1016/j.jenvrad.2011.10.018.

Kinniburgh, D.G., 1986. General purpose adsorption isotherms. Environ. Sci. Technol. 20, 895–904.

Landa, E.R., 2007. Naturally occurring radionuclides from industrial sources: Characteristics and fate in the environment. In: Shaw, G. (Ed.), Radioactivity in the Terrestrial Environment. Elsevier, Amsterdam, pp. 211–237.

Langmuir, D., 1997. Aqueous Environmental Geochemistry. Prentice Hall, New Jersey.

Leon, J.D., Jaffe, D.A., Kaspar, J., Knecht, A., Miller, M.L., Robertson, R.G.H., et al., 2011. Arrival time and magnitude of airborne fission products from the Fukushima, Japan, reactor incident as measured in Seattle, WA, USA. J. Environ. Radioact. 102, 1032–1038.

Lozano, R.L., Hernández-Ceballos, M.A., Adame, J.A., Casas-Ruíz, M., Sorribas, M., San Miguel, E.G., et al., 2011. Radioactive impact of Fukushima accident on the Iberian Peninsula: Evolution and plume previous pathway. Environ. Int. 37, 1259–1264.

Miekeley, N., Kuchler, I.L., 1987. Interactions between thorium and humic compounds in surface waters. Inorganica Chimica Acta 140, 315–319.

Miller, W., Alexander, R., Chapman, N., McKinley, I., Smellie, J., 1994. Natural Analogue Studies in the Geological Disposal of Radioactive Wastes. Elsevier, Amsterdam.

Mollah, A.S., Ullah, S.M., 1998. Determination of distribution coefficient of ^{137}Cs and ^{90}Sr in soil from AERE, Savar. Waste Manag. 18, 287–291.

Naidu, R., Summer, M.E., Harter, R.D., 1998. Sorption of heavy metals in strongly weathered soils: an overview. Environ. Geochem. Health 20, 5–9.

NEA, 2002. The Use of Thermodynamic Databases in Performance Asessment. OECD Nuclear Energy Agency, Paris.

Parache, V., Pourcelot, L., Roussel-Debet, S., Orjollet, D., Leblanc, F., Soria, C., et al., 2011. Transfer of ^{131}I from Fukushima to the Vegetation and Milk in France. Environ. Sci. Technol. 45, 9998–10003.

Payne, T.E., 1999. Uranium (VI) Interactions with Mineral Surfaces – Controlling Factors and Surface Complexation Modelling. PhD Thesis. University of New South Wales, p. 310.

Payne, T.E., Airey, P.L., 2006. Radionuclide migration at the Koongarra uranium deposit, Northern Australia - lessons from the Alligator Rivers analogue project. Physics and Chemistry of the Earth 31, 572–586.

Payne, T.E., Davis, J.A., Waite, T.D., 1994. Uranium retention by weathered schists – the role of iron minerals. Radiochimica Acta 66 (67), 297–303.

Payne, T.E., Bertram, W.K., Itakura, T., Raven, M., 2002. Relationship of quantitative X-ray diffraction measurements of geologic materials to cesium sorption. Radiochimica Acta 90, 705–711.

Payne, T.E., Hatje, V., Itakura, T., McOrist, G.D., Russell, R., 2004. Radionuclide applications in laboratory studies of environmental surface reactions. J. Environ. Radioact. 76, 237–251.

Payne, T.E., Itakura, T., Comarmond, M.J., Harrison, J.J., 2009. Environmental mobility of cobalt as influenced by solid phase characteristics and groundwater chemistry. Appl. Radiat. Isot. 67, 1269–1276.

Rahman, M.M., Rahman, M.M., Koddus, A., Ahmad, G.U., Voigt, G., 2005. Soil-to-plant transfer of radiocaesium for selected tropical plant species in Bangladesh. J. Environ. Radioact. 83, 199–211.

Reading, A.J., Thompson, R.D., Millington, A.C., 1995. Humid Tropical Environments. Blackwell, Oxford, UK.

Rieuwerts, J.S., 2007. The mobility and bioavailability of trace metals in tropical soils: A review. Chemical Speciation and Bioavailability 19, 75–85.

Robison, W.L., Conrado, C.L., Hamilton, T.F., Stoker, A.C., 2000. The effect of carbonate soil on transport and dose estimates for long-lived radionuclides at a U.S. Pacific test site. J. Radioanal. Nucl. Chem. 243, 459–465.

Sanchez, P.A., Ahamed, S., Carré, F., Hartemink, A.E., Hempel, J., Huising, J., et al., 2009. Digital soil map of the world. Science 325, 680–681.

Sato, T., Murakami, T., Yanase, N., Isobe, H., Payne, T.E., Airey, P.L., 1997. Iron nodules scavenging uranium from groundwater. Environ. Sci. Technol. 31, 2854–2858.

Sheppard, M.I., Thibault, D.H., 1990. Default solid/liquid partition coefficients, Kds, for four major soil types: a compendium. Health Phys. 59, 471–482.

Sinclair, L.E., Seywerd, H.C.J., Fortin, R., Carson, J.M., Saull, P.R.B., Coyle, M.J., et al., 2011. Aerial measurement of radioxenon concentration off the west coast of Vancouver Island following the Fukushima reactor accident. J. Environ. Radioact. 102, 1018–1023.

Sposito, G., 1989. The Chemistry of Soils. Oxford University Press, New York.

Stumm, W., 1992. Chemistry of the Solid-water Interface. Wiley-Interscience, New York.

Toulhoat, P., Gallien, J.P., Louvat, D., Moulin, V., l'Henoret, P., Guérin, R., et al., 1996. Preliminary studies of groundwater flow and migration of uranium isotopes around the Oklo natural reactors (Gabon). J. Contam. Hydrol. 21, 3–17.

Twining, J.R., Payne, T.E., Itakura, T., 2004. Soil-water distribution coefficients and plant transfer factors for ^{134}Cs, ^{85}Sr and ^{65}Zn under field conditions in tropical Australia. J. Environ. Radioact. 71, 71–87.

Uchida, S., 2007. Radionuclides in tropical and subtropical ecosystems. In: Shaw, G. (Ed.), Radioactivity in the Terrestrial Environment. Elsevier, Amsterdam, pp. 193–209.

UNSCEAR (United Nations Scientific Committee on the Effects of Atomic Radiation), 1982. Source and Effects of Ionizing Radiation. United Nations, New York.
van Wambeke, A., 1992. Soils of the Tropics: Properties and Appraisal. McGraw-Hill, New York.
Velasco, H., Ayub, J., Sansone, U., 2008. Analysis of radionuclide transfer factors from soil to plant in tropical and subtropical environments. Appl. Radiat. Isot. 66, 1759–1763.
Waite, T.D., Davis, J.A., Payne, T.E., Waychunas, G.A., Xu, N., 1994. Uranium (VI) adsorption to ferrihydrite: Application of a surface complexation model. Geochimica et Cosmochimica Acta 58, 5465–5478.
Waite, T.D., Davis, J.A., Fenton, B.R., Payne, T.E., 2000. Approaches to modelling uranium(VI) adsorption on natural mineral assemblages. Radiochimica Acta 88, 687–693.
Wasserman, M.A., Perez, D.V., Bourg, A.C.M., 2002. Behavior of cesium-137 in some Brazilian oxisols. Commun. Soil. Sci. Plan. 33, 1335–1349.
Yanagisawa, K., Takeda, H., Miyamoto, K., Fuma, S., 2000. Transfer of technetium from paddy soil to rice seedling. J. Radioanal. Nucl. Chem. 243, 403–408.
Yanase, N., Payne, T.E., Sekine, K., 1995. Groundwater geochemistry in the Koongarra ore deposit, Australia (II): Activity ratios and migration mechanisms of uranium series radionuclides. Geochemical Journal 29, 31–54.
Yasunari, T.J., Stohl, A., Hayanoc, R.S., Burkhart, J.F., Eckhardt, S., Yasunari, T., 2011. Cesium-137 deposition and contamination of Japanese soils due to the Fukushima nuclear accident. Proc. Natl. Acad. Sci. U. S. A. 108, 19531–19534.

Chapter 4

Tropical Radiochemical Oceanography

Ron Szymczak
Nuclear and Oceanographic Consultant, Cronulla, New South Wales, Australia

4.1. INTRODUCTION

The purpose of this chapter is to provide radioecologists with a general understanding of the incidence and biogeochemical behaviour of radionuclides in the marine environment to assist the design, conduct, and interpretation of radioecology experiments. This chapter is mostly directed at tropical marine systems; however, information from other climatic regions is also presented. For the most part, reference is only made to the water column, that is, radiochemical oceanographic distributions and processes. For more detailed information on specific subjects, readers are advised to consult the references provided.

4.1.1 The Natural Incidence of Radioactivity

Radioactivity is natural and an omnipresent component of the universe. All matter around us was created by nuclear reactions and the concomitant radioactivity. The natural relative abundance of stable elements is a result of variations in the stability of nuclides. Of the more than 5000 atoms (nuclides) known, about 95% are radioactive — they are the norm rather than the exception. Practically all materials and environmental compartments on the Earth are either radioactive or naturally exposed to ionising radiation. Natural nuclear energy powers major geological processes on the Earth, such as internal convection cycles, earthquakes, volcanism, and tectonic drift. Natural radioactivity initiates and promotes the evolution of life. The Sun, a natural fusion reactor, provides energy in the form of light and heat, which regulates our climate and provides an environment conducive to life (Baxter, 1993).

Radioactivity in the Environment, Vol. 18. http://dx.doi.org/10.1016/B978-0-08-045016-2.00004-7

4.1.2 Radionuclides in Marine Systems

There are many radionuclides present in marine systems (for a comprehensive overview, see Burton, 1975 and Aarkrog, 2003). The major radionuclides contributing to the 13.7 mBq/L average total radioactivity in the oceans are shown in Table 4.1. Whilst by far the major contribution to ocean radioactivity is from naturally occurring radionuclides, several artificial radionuclides are also observed, at times with significant local enhancements (Burton, 1975; IAEA, 1999, 2005; Lu et al., 2007). This is discussed in greater detail in Section 4.3.

Sediments, seawater, and biota all play a role in the transport and cycling of contaminants. Ocean currents do not recognize territorial boundaries and some radionuclides can be carried between distant locations. For other radionuclides, particle scavenging and settling processes in the water column result in their accumulation in sediments (Baskaran and Santschi, 1993). Marine biota accumulate many radionuclides via direct absorption from the seawater and/or through ingestion of other organisms (Stewart et al., 2008).

Both vertical and horizontal distributions of physicochemical parameters (Table 4.2) vary considerably in marine systems from polar to equatorial

TABLE 4.1 Major Components of Radioactivity in the Marine Environment

Nuclide	Activity (mBq/L)	Origin
^{40}K	11800	Primordial
^{87}Ru	107	Primordial
^{234}U	48.1	Primordial
^{238}U	44.4	Primordial
^{210}Pb	11.4–118	Atmospheric
^{210}Po	28.1–85.1	Atmospheric and ^{210}Pb
^{3}H	11–22	Cosmogenic
^{14}C	7.4	Cosmogenic
^{3}H	1147—2738	Fallout
^{137}Cs	0.25–29.6	Fallout
^{90}Sr	0.74–18.5	Fallout
^{14}C	0.37–1.48	Fallout
$^{239\&240}Pu$	0.011–0.044	Fallout

$\Sigma = 13.7$ Bq/L.

TABLE 4.2 Physicochemical Parameters Influencing Biogeochemistry in Marine Systems

Parameter	Range
Temperature	−2 to 31°C
Pressure	100–100,000 hPa
Salinity	<1 to 37 (psu, mg/L, ‰)
Redox potential	−0.3 to +0.8 V
pH	7.3–8.4
Light	0–1600 μ mol m^{-2} s^{-1}

regions (for a detailed overview, see Anikouchine and Sternberg, 1981). During the transfer of radionuclides through various marine ecosystems and environmental compartments (seawater, particles, biota, etc.), they are influenced by biogeochemical processes similarly to stable elements and analogies in their behaviour can be drawn (see Section 4.2).

However, with most radionuclides, shorter half-lives, genetic relations to other nuclides, specialised modes of production, and their differing *in situ* behaviours result in variable fate and distributions. A direct consequence of these factors on members of uranium and thorium series decay families in the marine environment is that secular equilibria between parent and daughter pairs are significantly disrupted (Cochran, 1982). See Box 4.1 and refer to Chapter 1 for more details on radioactivity *per se* and decay series.

The decay of unsupported nuclides and regrowth of daughter products then provides many approaches to the measurement of kinetics (timescales) of environmental processes such as the chronology, transport and mixing of water masses, sediment accumulation, geochemistry, and air–sea gas exchange processes (Baxter, 1993). Detailed reviews of oceanographic applications of radionuclides have been undertaken by Chesselet and Lalou (1988), Cochran (1982), and Swarzenski et al. (2000).

4.1.3 Special Features of Tropical Marine Systems

Tropical marine systems experience high and consistent solar irradiation, temperatures, and precipitation, but a low Coriolis force and a large bouyancy flux, which influences the fate and dispersal of riverine discharges to coastal oceans. Terrestrial erosion is intensive, so tropical rivers carry high solute loads with high levels of dissolved organic carbon and particles enriched in iron, manganese, and aluminium oxide coatings. These factors lead to high

Box 4.1 Secular Equilibrium of Radionuclides

In a closed system of sufficient age, each daughter isotope in a given decay series would be in *secular equilibrium* (steady state) so that the radioactive decay rates of each would equal that of the primary parent and the total radioactivity would decrease only with the half-life of the parent nuclide. Under these conditions, each daughter product is produced at the same rate at which it decays. The parent nuclide regenerates the daughter at a rate that is governed by the half-life of the parent and secular equilibrium will be maintained if the system remains closed. On separation of a shorter lived nuclide from its immediate parent, the unsupported daughter nuclide decays with its own half-life. A period of about six daughter half-lives gives practically complete restoration (ingrowth) of secular equilibrium and this period also allows decay of an unsupported nuclide to only a few percent of its original concentration. Other situations may arise in which there are different relationships between parent and daughter half-lives, but the case of secular equilibrium is most important in this context.

rates of scavenging of dissolved oceanic components (e.g., metals and radionuclides) and subsequent sedimentation in coastal and shelf regions. With high levels of dissolved nutrients, coastal primary (phytoplankton) production rates can be extremely high, but seasonally alternate to very low rates with picoplankton representing an important component. High rates of organic matter production result in displacement of finfish stocks, which are common in higher latitudes with bacterial and invertebrate, often crustacean, dominated ecosystems (Nittrouer et al., 1994).

As tropical marine systems have significant sources of freshwater and terrestrial material (Nittrouer et al., 1994) and are an important sink for oceanic components (Sholkovitz and Szymczak, 2000; Swarzenski et al., 2004), their study is important to gain an understanding of river–ocean mixing and anthropogenic fluxes to the global oceans.

4.2. BIOGEOCHEMICAL BEHAVIOUR OF RADIONUCLIDES IN MARINE SYSTEMS

Riley and Chester (1989) provides an excellent grounding for understanding chemical oceanographic processes. In this section we examine the biogeochemical behaviours of both stable elements and radionuclides, which can be classified into four basic groups: (1) conservative, (2) nutrient-like, (3) scavenged, and (4) redox-controlled (Table 4.3).

4.2.1 Conservative Nuclides

Conservative elements, such as halogens, alkali, and alkaline earth elements, are present in seawater at relatively high concentration levels in constant proportion to salinity. They are homogenized in the ocean by water circulation on

TABLE 4.3 Oceanic Biogeochemical Behaviour of Some Stable Elements and Radionuclides

Geochemical Behaviour	Stable Elements	Radionuclides
Conservative	Chlorine, gold, potassium, rubidium, sodium, tungsten	Cesium, iodine, potassium, radium, radon, rubidium, tritium, uranium, strontium
Nutrient-like	Arsenic, cadmium, copper, iron, nitrogen, phosphorus, silver	Carbon, cobalt, beryllium, polonium, plutonium, silica, zinc
Scavenged	aluminium, arsenic, cobalt, copper, iron	americium, beryllium, bismuth, lead, iodine, protactinium, plutonium, thorium
Redox-controlled	arsenic, chromium, iron, manganese, selenium	neptunium, iodine, plutonium, uranium

Note: Some elements fall into more than one group (underlined).

the time scale of 10^3 years within their relatively long mean oceanic residence times ($> 10^5$ years) (Burton, 1975). Tritium (^3H) can be considered the most conservative radionuclide in this group as it presents as the tritiated water molecule. Distribution of tritium in the world's oceans is comprehensively reviewed by Broecker et al. (1986). As with its stable sibling isotopes present in seawater, ^{40}K, with its long half-life and conservative behaviour, is distributed uniformly throughout the oceans and in constant proportion to salinity (Burton, 1975). Although influenced by incident redox conditions, the conservative oceanic behaviour of long-lived uranium isotopes can also be considered as similar to that of the major conservative seawater component elements. Cesium also mostly behaves in a conservative manner and acts as a tracer of water mass movements. Iodine falls into this group; however, its behaviour is influenced by biological uptake and redox speciation (Burton, 1975; Choppin, 2006).

4.2.2 Nutrient-Like Nuclides

The nutrient elements, such as phosphorus, nitrogen, and silicon, are depleted in surface water due to biological uptake and are enriched in deep water by regeneration from particulate matter (Mann and Lazier, 1991). Stable elements (Ni, Zn, Cd, Ba, and Ge) and radionuclides such as ^{60}Co, ^{210}Po, ^{131}I, and plutonium isotopes also have this type of behaviour (Burton, 1975; Riley, 1989). An idealised ocean profile for $^{239+240}Pu$ is shown in Figure 4.1, indicating surface depletion and regeneration at depths of approximately 700–1000 metres.

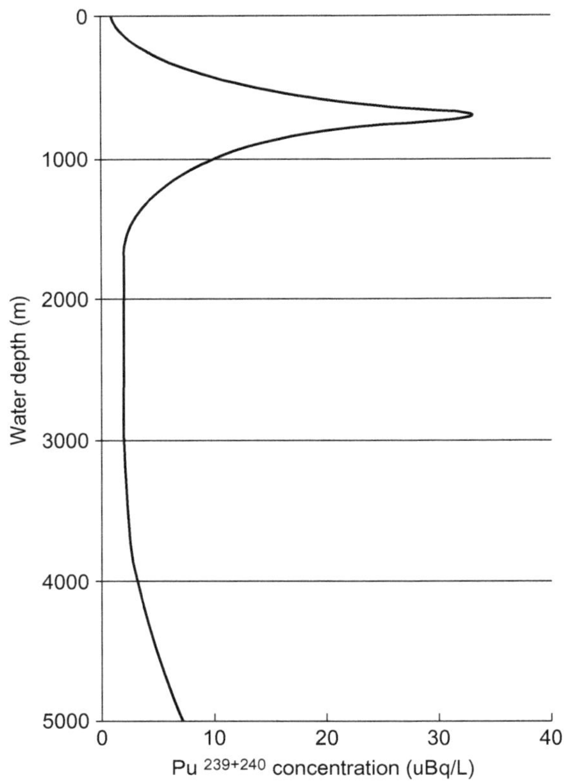

FIGURE 4.1 Idealised ocean profile for $^{239+240}Pu$ in the ocean. *Redrawn from Kinoshita et al. (2011).*

Some seaweeds (Box 4.2) have very high affinities to take up plutonium and iodine with typical concentration factors of up to 4,000 and 10,000, respectively (IAEA, 2004). However, different organisms concentrate different radioisotopes in varying rates and degrees (Stewart et al., 2008; see also Chapter 6 for more details on biological uptake). The seawater to organism concentration factors for marine biota are well detailed by the International Atomic Energy Agency (IAEA, 2004), although data are generally limited for tropical marine species. More site-specific seawater–organism concentration factor data are required for accurate radiological dose assessments (see Chapter 7) for tropical marine ecosystems. A detailed discussion of derivation and application of the International Commission for Radiological Protection (ICRP) reference animals and plants (RAP) approach to marine ecological radiological protection is given in Pentreath (2009).

Box 4.2 Transfer Factors for ICRP Reference Animals and Plants (RAPs)

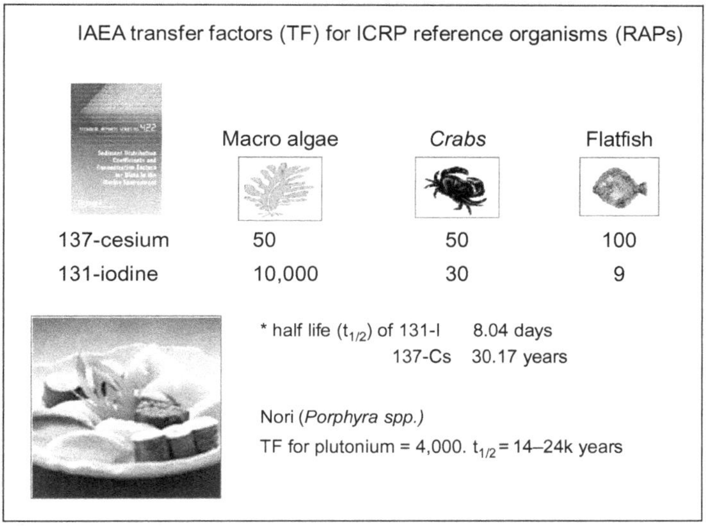

ICRP (2008) and data from IAEA (2004).

4.2.3 Scavenged Nuclides

Some stable elements, for example, aluminium, bismuth, cobalt, cerium, iron, and lead, are highly particle reactive in seawater and are consequently scavenged by various adsorption processes onto colloids and suspended particulate matter (Balistrieri et al., 1981; Baskaran and Santschi, 1993). Radionuclides with similar behaviours include ^{60}Co and ^{210}Pb, as well as americium, plutonium, and thorium isotopes (Baskaran and Santschi, 1993). They have short mean oceanic residence times (e.g., $<<10^2$–10^3 years). Despite the large terrestrial and atmospheric inputs to the oceans (Section 4.3), due to particle scavenging, surface waters generally have lower concentrations or activities of those elements and radionuclides than deep waters, with mid-depth maxima often observed (e.g., Figure 4.1), similar to nutrient-like profiles (Burton, 1975; Duran et al., 2004; IAEA, 2005; Hirose et al., 2011).

The adsorption affinity, or degree to which this group attaches to marine particles, is described by their sediment–water distribution coefficient (K_d; see Box 4.3).

Box 4.3 Distribution Coefficients (K_d)—Conditional and Operational Constraints

Distribution Coefficient (ratio) - K_d

$$K_d = \frac{C_{solid}\ (Bq/kg)}{C_{water}\ (Bq/L)}$$

- K_ds for highly particle reactive nuclides (eg. Th) are 10^5-10^6
- K_ds "do not" consider reaction rate kinetics
- K_ds are "conditional" and can vary greatly depending on the chemical environment (SPM, salinity, DOC, pH, redox, etc.)
- K_ds are "operational" as the distinction between dissolved and particulate species is defined by the filtration process
- K_ds are influenced by competitive chemical species

IAEA (2004) has also compiled considerable data on sediment–water K_ds for radionuclides. However, care should be taken utilising published data, as these vary according to the local nature of sediments as well as several other incident environmental and experimental factors (Abril and Fraga, 1996; Hatje et al., 2003). For example, considerable changes occur in the particle–water sorption ratios (R_s, which considers the weight of particles per unit volume of water) of beryllium and cesium according to salinity (Figure 4.2). Site-specific K_d studies are highly recommended.

4.2.4. Redox Controlled Nuclides

Some elements, like chromium, arsenic, iron, selenium, iodine, and plutonium, persist at more than one oxidation state in seawater (Table 4.4) and their oceanic behaviour and distribution patterns are variable (Burton, 1975; Stumm and Morgan, 1981; Choppin, 2006). The thermodynamically unstable reduced species of such elements are typically formed by biological mediation, light, or digenetic reactions, but the 'normal' oxidative oceanic environment greatly favours the higher valency species (Choppin, 2006). The behaviour of plutonium is especially complex since it can exist in four different oxidation states in the marine environment and the reduced forms, Pu(III) and Pu(IV), are more highly particle reactive than the oxidized forms Pu(V) and Pu(VI). Anoxic basins, such as the Black Sea, Cariaco Trench, and some fjords, are exceptions where reduced chemical species of both plutonium and uranium are stable (Burton, 1975; Hirose and Sugimura, 1993).

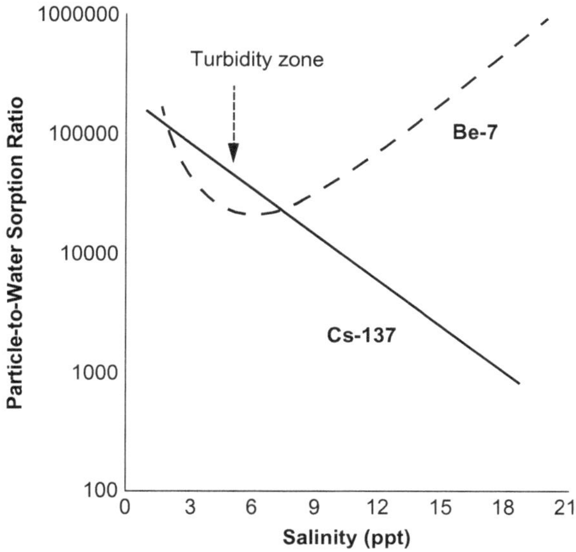

FIGURE 4.2 Changes in the particle-water sorption ration of beryllium and cesium according to salinity. *Redrawn from Olsen et al. (1989).*

TABLE 4.4 Oxidation States of Several Radionuclides in Seawater

Actinide	Oxidation State in Seawater	Other Oxidation States
Th	IV	
U	VI	IV
Np	V	IV,VI
Pu	IV	III,V,VI
Am	III	
Cm	III	

After Choppin (2006), with permission.

4.3. SOURCES AND SINKS OF RADIONUCLIDES TO THE TROPICAL OCEANS

This section outlines the types and incidence of radionuclides in marine systems. Generalised sources are depicted in Figure 4.3 and discussed in the following sections, whereas more details on the artificial radionuclides cesium and plutonium, as well as applications of some natural radionuclides, is discussed in Chesselet and Lalou, 1988. Chapter 1 provides additional useful information.

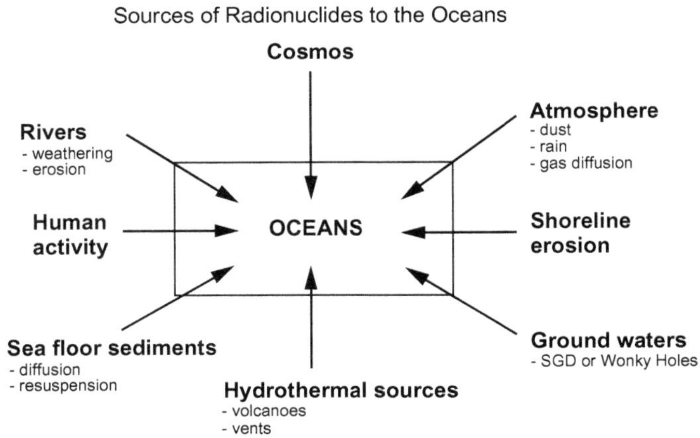

FIGURE 4.3 Sources of radionuclides to the oceans.

It is convenient to discuss ocean-occurring radionuclides in five groups:

1. Natural primordial radionuclides of elements that also have stable isotopes (e.g., ^{40}K)
2. Natural primordial parent and daughter radionuclides of the three natural decay series of ^{238}U, ^{235}U, and ^{232}Th
3. Natural cosmogenic radionuclides
4. Artificial radionuclides
5. Technologically enhanced naturally occurring radioactive materials (TENORMS)

4.3.1 Natural Primordial Radionuclides with Stable Siblings

The primordial radionuclides include those that have been present on the Earth since its formation approximately 4.5×10^9 years ago. For seawater of salinity 35 psu, which contains 0.40 g/L potassium, the radioactivity due to ^{40}K is 11.8 Bq/L. This accounts for over 90% of the total radioactivity in seawater. Natural potassium contains 0.0118% of ^{40}K (half-life 1.26×10^9 years), 93.1% ^{39}K, and 6.88% ^{41}K. Below ~100 m, where cosmic rays and photolysis of organic compounds diminish, the decay of ^{40}K is the dominant source of hydrated electrons, sustaining a steady-state concentration of about 6×10^{-26} M in these waters. ^{40}K has a riverine source to the marine environment, has a conservative geochemistry, and eventually decays to ^{40}Ar. The $^{40}K/^{40}Ar$ nuclide pair has applications in igneous and sedimentary rock dating (Burton, 1975).

Other nuclides in this group with half-lives ranging 10^{10} to 10^{15} years include ^{87}Rb, ^{50}V, ^{115}In, ^{138}La, ^{144}Nd, ^{147}Sm, ^{152}Gg, ^{176}Lu, ^{174}Hf, ^{187}Re, ^{190}Pt and ^{192}Pt. For many of these weakly radioactive nuclides, information on their occurrence in seawater is very limited and their long half-lives and general conservative behaviour would result in largely uniform distributions in the global oceans.

4.3.2 Natural Primordial U and Th Parent Nuclides and Daughters

The three radionuclides, ^{238}uranium, ^{235}uranium, and ^{232}thorium, are each parents to a family of daughter nuclides in a chain of radioactive decay that ends when a non-radioactive nuclide, typically lead, is formed. The uranium–radium decay chain begins with ^{238}U and ends via 18 intermediate states at the stable ^{206}Pb isotope (Table 4.5). ^{235}U is parent to the uranium–actinium decay chain leading to ^{207}Pb via 15 daughters. The thorium decay chain starting with ^{232}Th has ten intermediates and ends at ^{208}Pb (Burton, 1975). Figure 4.4 provides a generalised view of their oceanic behaviour.

TABLE 4.5 The Family of ^{238}U Series Decay Radioisotopes

Nuclide	Half-Life	Origin—Mode of Supply
^{238}U	4.47×10^9 years	Primordia–fluvial input
^{234}Th	24.1 days	*In situ* decay of ^{238}U
^{234}Pa	1.18 m	*In situ* decay of ^{234}Th
^{234}U	2.48×10^5 years	*In situ* decay of ^{234}Pa and fluvial input
^{230}Th	7.52×10^4 years	*In situ* decay of ^{234}U
^{226}Ra	1.62×10^3 years	Diffusion from ^{230}Th decay in sediments
^{222}Rn	3.83 days	*In situ* decay of ^{226}Ra
^{218}Po	3.05 m	*In situ* decay of ^{222}Rn
^{214}Pb	26.8 m	*In situ* decay of ^{218}Po
^{214}Bi	19.7 m	*In situ* decay of ^{214}Pb
^{214}Po	10^{-4} s	*In situ* decay of ^{214}Bi
^{210}Pb	22.3 years	*In situ* decay of ^{214}Po and atm deposition
^{210}Bi	5.0 days	*In situ* decay of ^{210}Pb
^{210}Po	138 days	*In situ* decay of ^{210}Bi
^{206}Pb	Stable	*In situ* decay of ^{210}Po

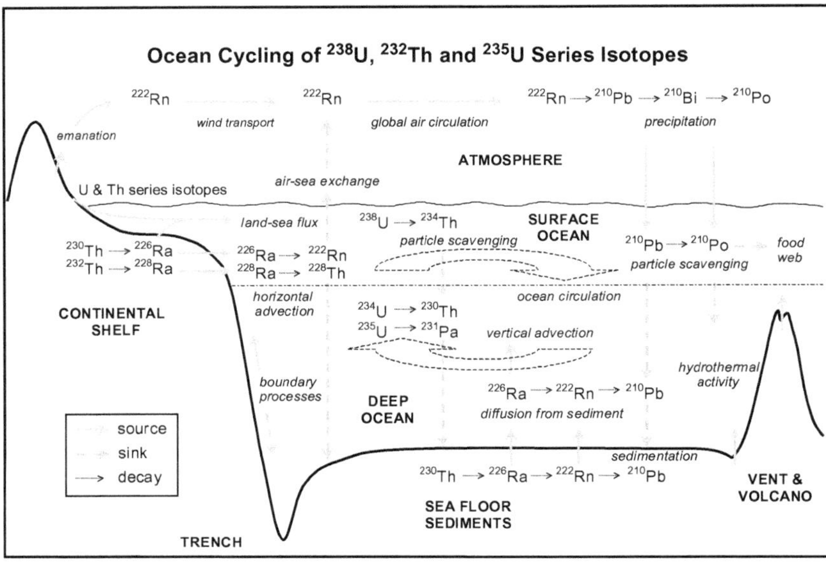

FIGURE 4.4 Ocean sources, cycling, and sinks of uranium and thorium series radionuclides.

4.3.2.1 Uranium

The primary source for uranium entering the oceans is rivers, following chemical weathering of continental crusts (Moore, 1967). A comprehensive recent review of estuarine behaviour of uranium and consequent ocean budgets was completed by McKee (2008). Other sources of uranium include pore water diffusion from pelagic sediments, low-temperature alteration of oceanic basalts, and hydrothermal vent emissions (Ku et al., 1977; MacDougall et al., 1979, Swarzenski et al., 2004). Uranium isotopes have a conservative biogeochemical behaviour and uniform distribution in open ocean waters due to formation of the stable $[UO_2(CO_3)_3]^{4-}$ complex. ^{238}U concentrations in oceanic waters have an average of 3.3 μg/L, or 44.4 mBq/L, and the $^{234}U/^{238}U$ activity ratio is extremely constant in the open ocean at 1.144 (Cochran, 1982; Chen et al., 1986). Particulate uranium concentrations are about 0.01 μg/L and represent an insignificant fraction relative to the dissolved component (Cochran, 1982). The conservative nature of this element allows uranium and its daughter products to be applied to tracing several *in situ* processes (Aller and Cochran, 1976; Cochran, 1982; Waples et al., 2006). The biogeochemical behaviours of radionuclides are discussed in more detail in Section 4.3.

4.3.2.2 Thorium

The basic biogeochemical characteristic of thorium in the marine environment is that it is highly particle reactive (the opposite of conservative uranium).

Primordial ^{232}Th enters the oceans primarily in detrital form, although there is some mobilisation during chemical weathering and soil formation. The naturally occurring thorium isotopes ^{234}Th, ^{230}Th, and ^{228}Th have *in situ* sources in seawater, resulting from the decay of U and Ra parents. Unlike uranium, which is dissolved during this process, thorium is largely retained on particle surfaces (Cochran, 1982).

Seawater activities of thorium isotopes are variable over several orders of magnitude and respond to the concentration and flux rate of suspended particles (Cochran 1982; Chen et al., 1986). Secular equilibrium with respective parent nuclides (^{234}Th/^{238}U, ^{230}Th/^{234}U, ^{228}Th/^{228}Ra) is dramatically disrupted by geochemical scavenging and particle flux rates are determined by analysis of the activity ratios (Cochran, 1982; Chesselet and Lalou, 1988). See more on ^{234}Th in Section 4.6.

4.3.2.3 Radium

Radium isotopes are derived from thorium parents in the U-Th decay series. Contrasting their parents, they are generally considered as biogeochemically conservative. With half-lives ranging from 3.66 days (^{224}Ra), 11.4 days for ^{223}Ra, to 5.7 years (^{228}Ra) and 1600 years (^{226}Ra), these isotopes have gained application as tracers of processes occurring in estuaries, salt marshes, coastal waters, and open oceans (Cochran, 1982; Moore, 2007). Measurements of high activities of radium in the coastal ocean, not explained by inputs from rivers or sediments coupled with measured high ^{226}Ra in salty coastal wells, led Moore (1996) to hypothesize that submarine groundwater discharge (SGD) was responsible for the elevated radium levels. In most open ocean regions, there is a marked and gradual vertical gradient in concentration of ^{226}Ra, with higher values in deep waters and an average concentration in the order of 10^{-13} g/L, amounting to about 10% of the equilibrium value with respect to the ^{238}U parent, but this is 2–3 orders of magnitude greater than could be supported by the immediate parent ^{230}Th in ocean water. As with conservative uranium isotopes, radium has been observed to associate with particulate matter, mineral deposits, and biological matrices (Cochran, 1982).

4.3.2.4 Lead

Although a number of lead isotopes exist, the only one of significance to our ocean studies is ^{210}Pb, which is in excess relative to its parent ^{222}Rn (from ^{226}Ra in the ^{238}U decay series) and which has a half-life of 22.3 years. The most significant source of excess ^{210}Pb to the surface oceans is via wet and dry deposition from the atmosphere, where it is produced from the decay of ^{222}Rn, although some exists from decay of the dissolved ^{226}Ra parent. The emanation of radon to the atmosphere is primarily from continental rocks and soils (see Chapter 2 for more details), with only about 2% resulting from flux from the surface oceans. The supply of excess ^{210}Pb delivered from continents by rivers, largely in particulate form, is considerably less than that derived from direct atmospheric input. The excess supply of ^{210}Pb to the surface oceans varies with distance from the continental sources. Latitudinal and longitudinal variation in the flux is controlled by: (1) air mass transit times, which affect the total production of ^{210}Pb as a function of distance away from continents; and (2) the mean aerosol residence time, which governs the deposition rate of ^{210}Pb. An additional source of ^{210}Pb in the oceans is *in situ* decay of dissolved ^{226}Ra via ^{222}Rn. This source is a function of the integrated ^{226}Ra activity in the water column and, although at depths of 4–5 km it may be comparable to the atmospheric flux (at that depth), it is not important in the coastal and surface ocean at depths of less than 100 m—the euphotic zone—which is important to most radioecological processes. The surface ocean distribution pattern observed for ^{210}Pb results from inputs from the atmosphere, *in situ* ^{226}Ra decay, upwelling, chemical and biologically mediated particle removal processes, and radioactive decay (Burton, 1975; Bacon et al., 1976; Cochran, 1982).

Similar to the thorium isotopes, lead has a high affinity for particle surfaces and is rapidly removed from solution by detrital matter and other sinking particles (Figure 4.5). Residence times for ^{210}Pb have been calculated to be in the order of 54 years for deep ocean waters and 0.63 years for the surface ocean above 100 m. Although ^{210}Pb (in combination with ^{210}Po) has been applied to studies of the fate of marine particulate matter and of biological productivity in the water column, its main application in coastal and marine studies has been as a geochronological tracer of sediment accumulation on the sea floor—this will be dealt with in later sections.

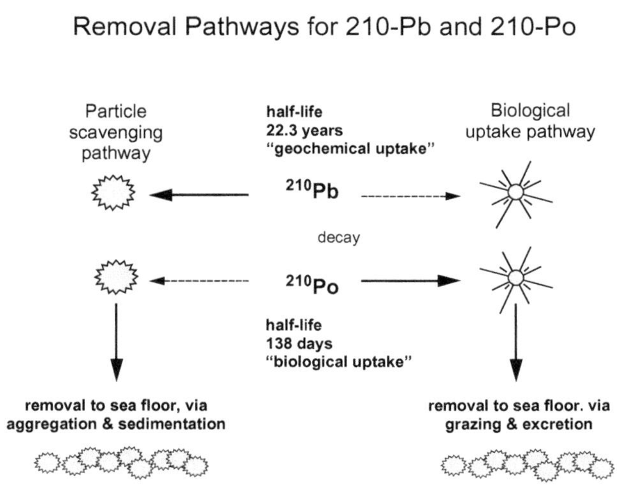

FIGURE 4.5 General schematic for removal pathways of ^{210}Pb and ^{210}Po in seawater.

4.2.3.5 Polonium

^{210}Po is primarily added to the oceans via decay of its ^{210}Pb parent. Despite the relatively short ^{210}Po half-life of 138 days, significant radioactive disequilibrium exists between ^{210}Po and its parent ^{210}Pb because of rapid biological removal of ^{210}Po in the oceans (Figure 4.5). Consequently, there exists a general deficiency of ^{210}Po relative to ^{210}Pb in surface waters and concomitant excesses in phytoplankton and zooplankton compartments (Bacon et al., 1976; Nozaki et al., 1976; Cochran, 1982). Polonium is readily accumulated by marine organisms and profiles in the ocean resemble those of nutrient-like elements (see Figure 4.1) whereby concentrations are low at the surface, increase to a mid-water maximum, and decrease again in deep waters (Ritchie and Shimmield, 1991). Fowler (2011) provided a comprehensive review of the occurrence and behaviour of polonium in the marine environment, and special focus on ^{210}Po in tropical marine systems is given in Section 4.4.

4.3.3 Cosmogenic Radionuclides

These are naturally occurring radionuclides that are continuously produced by nuclear processes and often have short half-lives relative to the rates of ocean processes (see Figure 4.6). These nuclides are produced by the interaction of cosmic radiation with atmospheric constituents (spallation) or extraterrestrial material accreted by the Earth; this group contains ^{7}Be,

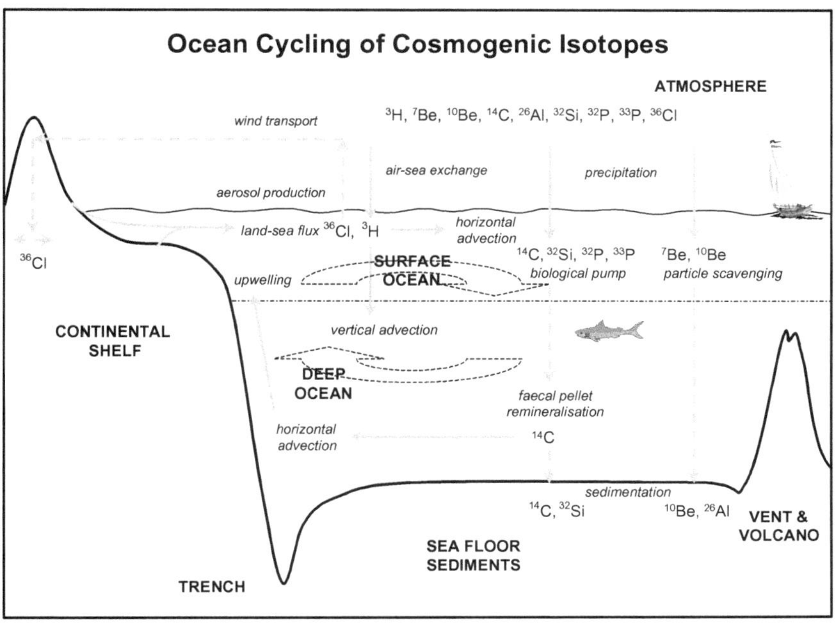

FIGURE 4.6 Ocean sources, cycling, and sinks of cosmogenic radionuclides.

^{10}Be, ^{14}C, ^{26}Al, ^{32,33}P, and ^{32}Si. Although they have gained importance in relation to tracing ocean processes, these nuclides generally do not reach levels in the environment to be of radiological concern (Burton, 1975; Chesselet and Lalou, 1988).

4.3.4 Artificial Radionuclides

Artificial radioisotopes arise from human activities/sources, such as nuclear power facilities (e.g., the 2011 Fukushima nuclear power plant accident, see Section 4.6.2), reprocessing plants, loss of radiation sources, re-entry of satellites, detonation and loss of nuclear weapons (e.g., tests in French Polynesia and the Marshall Islands, see Section 4.6.1), laboratory accidents, and discharges from nuclear-powered ships and submarines (IAEA, 1999, 2005). Radioisotopes included in this group include: ^{3}H, ^{14}C, ^{60}Co, ^{90}Sr, ^{91}Y, ^{99}Tc, ^{106}Ru, ^{144}Ce, ^{147}Pm, ^{151}Sm and various isotopes of Cs, I, Pu, and Am. There remains a legacy of radioactive contamination from earlier nuclear activities in many marine environments and anthropogenic radioactive substances have reached a variety of marine ecosystems from these different sources (Duran et al., 2004). Their distribution represents a cumulative result of radioactive decay, water movement, diffusion, uptake by biota, and particulate matter with subsequent vertical settling of and transport to ocean sediments. See Figure 4.7 for a generalized synopsis of the ocean cycling of artificial radionuclides.

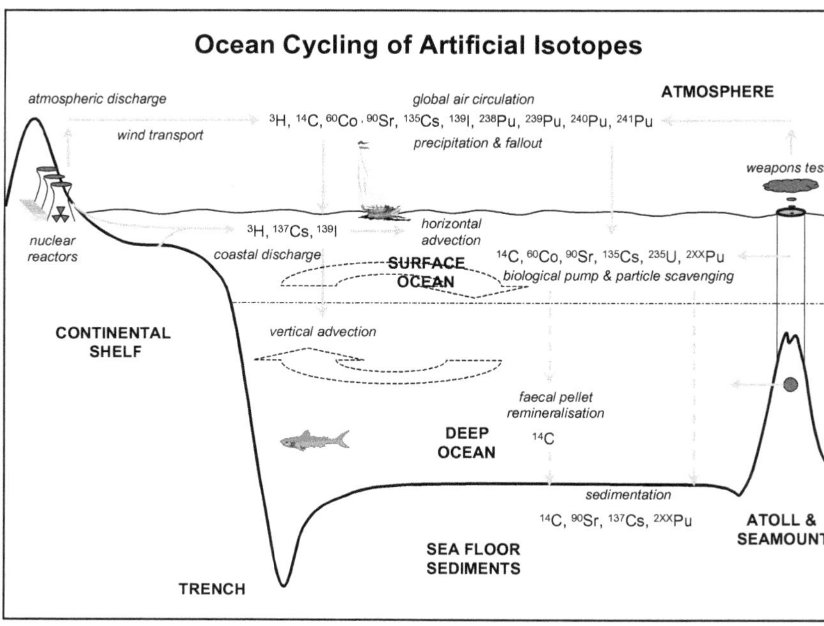

FIGURE 4.7 Ocean sources, cycling, and sinks of artificial radionuclides.

The dominant source of anthropogenic marine radioactivity is global fallout (Livingston and Povinec, 2002), mainly from the 542 atmospheric nuclear weapons tests carried out in 1950s and 1960s, releasing an estimated fission yield of 189 megatons (UNSCEAR, 2000), with some regions (e.g., Marshall Islands) receiving additional contribution from local tropospheric fallout (Nakano and Povinec, 2003). As one example, the total input for ^{137}Cs from global fallout was estimated to be 311 PBq for the Pacific Ocean, 201 PBq for the Atlantic Ocean, 84 PBq for the Indian Ocean, and 7.4 PBq for the Arctic Ocean (Aarkrog, 1971; Aarkrog, 2003). Other more localized sources of ^{137}Cs in the marine environment include fallout from the Chernobyl accident (UNSCEAR, 2000) and routine nuclear power plant operations (as a constituent of low-level radioactive liquid effluents). ^{239}Pu and ^{240}Pu are alpha-emitting nuclides with half-lives of 2.4×10^4 and 6.57×10^3 years, respectively, and are important indicators of radioactive contamination of the marine environment. The input functions of these radionuclides into the oceans and their latitudinal variations have been strongly affected by stratospheric–tropospheric mixing, resulting in the highest deposition rates at medium latitudes and lowest rates at poles and the equator. As the majority of large atmospheric tests were carried out in the Northern Hemisphere, and due to a slower Northern Hemisphere–Southern Hemisphere atmospheric mixing, the deposition into the Southern Hemisphere has been lower and delayed in comparison with the Northern Hemisphere (Livingston and Povinec, 2002, refer also to Chapter 2).

The benchmarking Geochemical Ocean Sections Study (GEOSECS) expedition in 1972/1973 showed the characteristic peaks for the general fallout distribution with maxima at 40°–50° N and S, as well as generally lower fallout in the Southern Hemisphere. However, the peak at 40°–50° N largely disappeared over the next 16 years, probably as a result of activity being transported into the Arctic Sea in addition to horizontal advection and mixing (Livingston et al., 1985).

Not so well studied are discharges of rare earth elements formed as fission products by thermal neutrons during nuclear power production. The yield is relatively large and several nuclides with long half-lives are produced (Kameda, 1962, Table 4.6). Natural concentrations of stable rare earth elements in the oceans are very low and analyses are difficult (Sholkovitz and Szymczak, 2000), consequently information on their seawater geochemical behaviour and concentration factors in marine organisms is limited.

4.3.5 Technologically Enhanced Natural Occurring Radioactive Materials (TENORMs)

Many industrial and domestic energy sources (like coal, oil, gas, wood, and peat) contain radioactive elements that are often concentrated as a result of combustion processes then released into the marine environment by various vectors. The naturally occurring radionuclides (primarily ^{40}K and the

TABLE 4.6 Fission Product Activity after 100 Days Cooling from 10^{11} MWh of Nuclear Power Production

Isotope	Half-Life	Tonnes	Curies (+100 days)
^{91}Y	57 days	111	2.8×10^{12}
^{106}Ru	290 days	35	1.5×10^{11}
^{144}Ce	275 days	490	1.6×10^{12}
^{147}Pm	94 years	7.3	3.3×10^{9}
^{151}Sm	73 years	0.7	2×10^{7}

Data from Kameda (1962).

uranium and thorium decay series) occur at significantly enhanced concentrations in both natural and man-made materials. Coal-powered electricity production and the oil and gas industries are significant sources of these discharges. Major estuarine or coastal contamination is often directly traceable to an industrial, power-generating, or mining source, although urban runoff is also often responsible for increases in radionuclides (Baxter, 1993).

As well as the association with nuclear and fossil-fuel power generation, radioactive enhancements are also, but not so publicly identified as, associated with industries producing pharmaceuticals, fertilisers, detergents, titanium oxides, rare earth element processing, mineral waters, paints, ceramics, dyes and colourings, clothing, lime burning and construction industries (Baxter, 1993).

4.4. RADIONUCLIDES IN TROPICAL MARINE ENVIRONMENTS

4.4.1 Cesium and Plutonium in the Indo-Pacific Region

The anthropogenic radionuclides ^{137}Cs and $^{239,240}Pu$ are important indicators of radioactive contamination of the marine environment, and are of primary interest because of existing inventories and possible health effects. ^{137}Cs, which has a half-life of 30.2 years, can also serve as a conservative tracer of the transport and accumulation patterns of contaminants in seawater.

Monitoring programs and studies, such as the IAEA's Marine Radioactivity Dose Assessment (MARDOS) project, have aimed at determining the concentrations of ^{137}Cs or $^{239,240}Pu$ in seawater, sediment, or biota and have produced a large amount of data and information on these radionuclides (IAEA, 1995). Based on MARDOS, it was estimated that the Collective Effective Dose Commitment (CEDC) due to ^{137}Cs in marine food was roughly 160 man-Sv for a world population of 5.3×10^{9} in 1990 (IAEA, 1995). The combined FAO areas 57, 61, and 71, which belong to a region of interest across Southeast Asia and Australia, contributed about 25% to this CEDC.

Atmospheric fallout is the main source of plutonium in the marine environment of the tropical Pacific Ocean with minor local sources of close-in fallout from the Bikini and Enewetak Atolls of the Marshall Islands (ca. 9° N, 168° E) and Mururoa Atolls (ca. 15° S, 140° W). The nuclear weapons tests carried out at those locations produced significant local fallout as radioactive clouds did not reach the stratosphere and, hence, radionuclides were not distributed globally. For example, in the case of $^{239,240}Pu$, about 10.9 PBq has been input into the ocean as global fallout and 6 PBq as local fallout in the Marshall Islands (Nakano and Povinec, 2003). Additionally, relatively small quantities of other alpha emitters were included in the wastes that were dumped into the seas. Of that material, about 96% comprised plutonium and americium isotopes (IAEA, 1999).

Global data on key anthropogenic radionuclides (IAEA, 1999), such as ^{137}Cs, show that physical radioactive decay contributes to changes in radionuclide concentrations over time at any given sampling station. Transport processes, in turn, account for observed spatial changes. For long-lived radionuclides, such as $^{239,240}Pu$, observed changes are mainly due to transport processes in the absence of new, significant inputs (Chiappini et al., 1999). ^{137}Cs levels in surface water in the Asia-Pacific regional seas, extending from 50° N to 60° S latitude and 60° E to 180° E longitude, have been declining to their present median value of 2.8 Bq/m^3, mainly due to radioactive decay, transport processes, and the absence of new significant inputs. $^{239,240}Pu$ levels in surface seawater are much lower, with a median of 5.9 mBq/m^3. $^{239,240}Pu$ appears to be preferentially scavenged by particles, and is therefore more readily transported down the water column (Chiappini et al., 1999). In surface sediment, the median ^{137}Cs concentration is 1.4 Bq/kg dry, which is higher than the $^{239,240}Pu$ median of 0.2 Bq/kg dry. The highest concentrations of ^{137}Cs and $^{239,240}Pu$ in surface sediment were observed between 35° N and 40° N, and in particular, in the East Sea. Coastal areas with higher sedimentation rates and with higher radionuclide concentrations, due to the additional contribution of runoffs from terrestrial catchment areas, show a different vertical profile of the sediment column than the deep open seas. For biota, the median ^{137}Cs activity concentration in fish (0.2 Bq/kg wet) is higher than in crustaceans (0.1 Bq/kg wet) and molluscs (0.1 Bq/kg wet). As in seawater, observed ^{137}Cs activity concentrations in fish are considerably higher than $^{239,240}Pu$ levels. The median activity concentration of $^{239,240}Pu$ in crustaceans (6 mBq/kg wet) is higher than in fish (1.5 mBq/kg wet) or molluscs (0.8 mBq/kg wet), but this is largely due to higher activity concentrations in non-edible parts.

Povinec et al. (2005) divided the Pacific and Indian Oceans into 17 latitudinal belts for which average ^3H, ^{90}Sr, ^{137}Cs, and $^{239,240}Pu$ activity concentrations, adjusted to the year 2000, were estimated. The highest concentrations were observed in the Sea of Japan/East Sea, the lowest in the Southern

Hemisphere, especially in the Southern Ocean. It has been interesting to learn that even 40 years after the main injection of anthropogenic radionuclides (produced during atmospheric nuclear weapons tests) into the global ocean, their distribution in surface waters is still dependent on latitude, although in some cases it is heavily influenced by circulation processes; in the case of Pu, it is influenced by scavenging processes in the water column.

Recently, several detailed studies have focused on the transport of cesium (Povinec et al., 2011; Sanchez-Cabeza et al., 2011; Tsumune et al., 2011; Aoyama et al., 2011a), plutonium (Gastaud et al., 2011; Hirose et al., 2011; Kinoshita et al., 2011), and other radionuclides (Aoyama et al., 2011b) in the Pacific Ocean. In addition, several groups are now undertaking further studies in relation to dispersal of the 2011 Fukushima Dai-ichi Nuclear Power Plant accident plume (see Section 4.6.2).

While certain areas, such as the Northwest Pacific Ocean and, in particular, East Sea/Sea of Japan and Yellow Sea, have been well studied and monitored for marine radioactivity, there appears to be limited data for other areas, notably in the South Pacific and Indian Ocean regions.

4.4.2 Uranium Series Isotopes in Tropical River–Ocean Systems

Considerable complexities and variations have been found in the behaviour of uranium isotopes in temperate and tropical river–ocean systems (McKee, 2008), and several studies have focussed specifically on the behaviour of uranium in tropical marine systems, such as the discharge areas of the Amazon and Fly Rivers (Swarzenski et al., 1992, 1995; Swarzenski and McKee, 1999; Swarzenski et al., 2004).

For example, Swarzenski et al. (2004) determined natural concentrations of ^{238}U and $\delta^{234}U$ in estuarine surface waters and pore waters of the Amazon (Brazil) and Fly (Papua New Guinea) Rivers to investigate uranium transport phenomena across river-dominated land–sea margins. Large, tropical river discharges are a major source of dissolved and solid materials transported to the oceans (see Section 4.1.3), and are important in defining oceanic mass budgets. They found that on the Amazon shelf, salinity plots of dissolved organic carbon, pH, and total suspended matter revealed two vastly contrasting water masses that were energetically mixed. In this mixing zone, the distribution of uranium was highly non-conservative and exhibited extensive removal from the water column, contrasting with temperate rivers (McKee, 2008). Uranium removal was most pronounced within a salinity range of 0–16.6 psu and likely the result of scavenging and flocculation reactions with inorganic (i.e., Fe/Mn oxides) and organic colloids/particles. The pore-water dissolved ^{238}U profile covaried closely with Mn(II).

Conversely, in the Fly River estuary, Swarzenski et al. (2004) found that ^{238}U appeared to exhibit a reasonably conservative distribution as a function of salinity. The absence of observed uranium removal did not necessarily imply non-reactivity, but instead an integration of concurrent uranium removal and release processes. It was likely that resuspension of bottom sediments, prolonged residence times in the lower reaches of the Fly River, and energetic particle–colloid interactions contributed to the observed estuarine uranium distribution.

The supply of uranium discharged from humid, tropical river systems to the sea appears to be foremost influenced by particle/water interactions that are ultimately governed by the particular physiographic and hydrologic characteristics of an estuary (Swarzenski et al., 2004).

4.4.3 ^{210}Lead and ^{210}Polonium in Tropical Oceanic Systems

Several studies have revealed the dramatic non-equilibrium behaviour of ^{210}Pb and ^{210}Po in oligotrophic and tropical oceans (Nozaki et al., 1976; Jeffree et al., 1997; Peck and Smith, 2000; Kim, 2001; Suriyanarayanan et al., 2008; Jeffree et al., 2012, in press).

Peck and Smith (2000) studied the distribution of ^{226}Ra, ^{210}Pb, and ^{210}Po in dissolved and particulate phases in near-surface waters (<300 m) in the Sepik River outflow area, Bismarck Sea, and western equatorial Pacific Ocean. ^{210}Pb/^{210}Po ratios of <1 were found in seawater and >1 in particulate matter. Using a 2-box model, they determined average residence times for ^{210}Pb and ^{210}Po in the dissolved phase were 8.0 years and 0.27 years, respectively, in the surface mixed (100 m) layer, and 170 years and 1.45 years in the deeper layer (100–300 m). The short residence time of ^{210}Po in both layers reflected its greater reactivity for association with sinking biological particulate matter. They also utilised ^{210}Pb/^{210}Po as a tracer to calculate the vertical flux of particulate organic carbon to be 104 ± 21 mg m^{-2} day and 180 ± 22 mg m^{-2} day in surface and deeper layers, respectively.

Jeffree et al. (1997) reported elevated ^{210}Po in both seawater and zooplankton from oligotrophic French Polynesian waters where activity concentrations were inversely and non-linearly related to zooplankton biomass and particle flux. Other studies found that this behaviour was also mimicked by certain nutrient-like trace element levels (Szymczak, 1997; Jeffree and Szymczak, 2000). Kim (2001) also found ^{210}Po deficient to ^{210}Pb in the oligotrophic Sargasso Sea (Bermuda) due to efficient uptake by the marine cyanobacteria *Trichodesmium* spp., which is a common bloom-forming species throughout many tropical marine systems. These studies postulated the enhancing effect of oligotrophy (low biological productivity) on the bioaccumulation in zooplankton of ^{210}Po, the most significant radionuclide with respect to the collective radiological dose to the global population from the consumption of sea foods.

Due to the bioaccumulation of ^{210}Po in marine biota (Stewart et al., 2008), its dose to those organisms (Cherry and Shannon, 1974), subsequent transfer and biomagnification through ocean food webs (Stewart et al., 2005), and dose to humans via seafood consumption (Pentreath et al., 1989), considerable interest still exists regarding ^{210}Po in tropical and oligotrophic marine systems. It is considered that increased oligotrophy due to postulated climate change and ocean warming (Jeffree and Szymczak, 2000) will result in further enhancement of ^{210}Po in tropical ocean systems. Fowler (2011) reviewed and highlighted the main gaps in understanding ^{210}Po biogeochemistry in marine systems and its application as an ocean process tracer.

4.5. MARINE RADIOACTIVITY DATABASES

4.5.1 Marine Radioactivity Information System

The IAEA's Marine Radioactivity Information System (MARiS) is a relational geographical information system (GIS)-based database (Figure 4.8) covering the distribution of radioactive and stable isotopes and, in the near future, other tracers (organic compounds, trace metals) in the marine environment. The MARiS database can be accessed online at http://maris.iaea.org. MARiS also includes oceanographic parameters, such as seawater temperature, salinity, and bathymetry. The database was developed in the Radiometrics Laboratory of the IAEA's Environment Laboratory (IAEA-EL) in Monaco. The map below shows the distribution of sites for which data on radionuclides in the world's oceans and seas were entered into the database. The reader will note the preponderance of Northern Hemisphere sites with particular emphasis around northern Europe and the North Pacific. This highlights the typical relative paucity of tropical and subtropical data.

The main objective of MARiS is to provide information on the radioactive contamination of the marine environment. Formerly this function was supported by the Global Marine Radioactivity Database (GLOMARD). MARiS contains data on the most important radionuclides in the world's oceans and

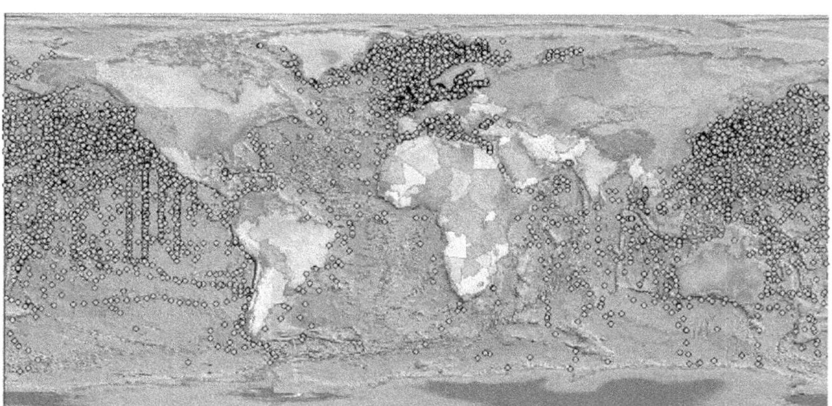

FIGURE 4.8 Geographical location of data held within the MARiS database (IAEA).

seas (in the open sea as well as in coastal zones), specifically in seawater, particulate matter, biota, and sediment. These data originate from published scientific papers, reports, and databases created within institutes or scientific programmes in IAEA Member States and can be used with appropriate quotation of sources and references. Quantification of contributions from the sources of radionuclides in the world's oceans and seas, computer modelling of the dispersion of radionuclides, and radiological assessment studies require that IAEA's Member States be provided with information on the past and present levels of radionuclides in the marine environment.

4.5.2 Asia-Pacific Marine Radioactivity Database

Within the past decade, information on artificial radionuclide levels in the oceans and seas of Asia (including the Indian subcontinent) and, to a limited extent, the South Pacific region were compiled into a database referred to as the Asia-Pacific Marine Radioactivity Database (ASPAMARD; see Duran et al., 2004). The development of ASPAMARD was funded jointly by the IAEA's Regional Cooperative Agreement (IAEA/RCA) and the United Nations Development Programme (UNDP) to: (1) provide reference or benchmark levels of key anthropogenic radionuclides in the regional seas against which the impact of future man-made contributions can be evaluated, (2) characterize the distribution and fate of key radioactive contaminants in the regional seas, (3) better understand transport processes and the fate and behaviour of radionuclides and analogue contaminants in the marine environment, and (4) assess dose associated with the ingestion pathway for seafood. ASPAMARD data is uploaded to the IAEA MARiS database and can be accessed using the MARiS hyperlink mentioned in the previous section.

4.6. CASE STUDIES

4.6.1 French Nuclear Testing at Mururoa and Fangataufa Atolls

A full report on the radiological situation at the atolls of Mururoa and Fangataufa was released by the IAEA (1998). Below are some edited excerpts from that report providing background information, radioactivity source terms, features of sampling and modelling procedures, and evaluations of the disposition and impact of the nuclear materials.

Mururoa and Fangataufa Atolls are situated at 21° 50′ S, 138° 54′ W and 22° 14′ S, 138° 45′ W, respectively, in the middle of the South Pacific Ocean, halfway between Australia and South America; they belong to the Pitcairn-Gambier island chain at the southeastern extremity of the Tuamotu archipelago in French Polynesia. Mururoa and Fangataufa Atolls have evolved from extinct submarine volcanoes, and each rests upon a massive volcanic substratum capped by a coral reef platform and surrounded by ocean water thousands of metres deep. Therefore, in vertical structure, each atoll consists essentially of two rock sequences: an igneous volcanic basalt basement and, capping it, sedimentary carbonate formations hundreds of metres thick. The carbonate formations are heterogeneous and fairly porous and the basalt basements are of lower permeability, but all are saturated with sea water. Driven by heat from the tectonic rocks, there is slow migration of relatively deep water from the ocean inwards, rising through the basalt basements and carbonate formations towards the lagoons.

The expériences nucléaires were of two types: 178 were nuclear tests, in which nuclear devices were exploded with large releases of fission energy, and 15 were safety trials, in which more or less fully developed nuclear devices were subjected to simulated accident conditions and the nuclear weapon cores were destroyed by means of conventional explosives, with no or — on a few occasions — very small releases of fission energy. Forty-one of the nuclear tests were atmospheric nuclear tests which took place — 37 at Mururoa Atoll and 4 at Fangataufa Atoll - between July 1966 and September 1974, and 137 were underground nuclear tests which took place - 127 at Mururoa Atoll and 10 at Fangataufa Atoll — between June 1975 and January 1996. Of the 15 safety trials, all of which were carried out at Mururoa Atoll, 5 were atmospheric safety trials and 10 were underground safety trials. Three underground safety trials had small releases of fission energy associated with them.

Environmental sampling and surveillance campaign

The Study involved an environmental sampling and surveillance campaign at Mururoa and Fangataufa Atolls, which provided the basis for evaluating a large amount of available data (mostly from French sources). The campaign was split into a terrestrial part, covering the surface soils and corals and the flora of the atolls, and an aquatic part, covering the atoll lagoons, the surrounding ocean, and their biota. The terrestrial part involved the collection of about 300 samples (vegetation, coconuts, sand, soil, corals, cores of coral bedrock, and aerosols), a large analytical effort (over 1000 radioanalytical determinations), and more than 100 in-situ gamma spectrometry measurements. The aquatic part involved gamma-spectrometric surveys of the seabed, in order to optimize sampling, and the collecting of over 300 samples of lagoon water, ocean water, sediment pore water, sediment, corals, and biota. Some 13,000 litres of water and about 1000 kilograms of solid samples were analysed, also representing a large analytical effort. No restrictions were placed by the French authorities on the sampling and surveillance campaign.

Dispersion of residual radioactive material throughout the ocean

The Study used regional and far field oceanographic models to estimate the concentrations in sea water at various locations and times of radionuclides released into the ocean from Mururoa and Fangataufa Atolls. The Study found that, except as a consequence of a hypothetical extreme disruptive event, the predicted long-term concentrations of radionuclides decrease to background oceanic levels beyond about 100 km from the atolls; thus, at Tureia Atoll the predicted concentrations will be around background levels and of no radiological significance.

Consequences of postulated disruptive events

The only disruptive event that was found by the Study to warrant a thorough assessment was the hypothetical major breakaway and slide of the carbonate formations in the northern zone of Mururoa Atoll, an area where the underground safety trials and some of the nuclear tests had produced cavity-chimneys, which penetrated into the carbonate formations. If such a hypothetical extreme event were to occur, ocean currents would carry the released radioactive material away from Mururoa Atoll and the highest potential annual dose would therefore be received by residents of nearby atolls. For the residents of Tureia Atoll, the dose in the first year following such a slide would not be more than a few thousandths of a millisievert — which is an extremely small fraction (a few parts in 1000) of the annual background radiation dose that the residents will unavoidably receive from natural radiation sources — even pessimistically assuming that all the plutonium involved in the slide went into solution.

Implications for biota

The Study assessed the dose rates to native biota resulting from the residual radioactive material at Mururoa and Fangataufa Atolls and, in the great majority of cases, found them to be similar to or lower than dose rates due to natural radiation sources. An exception is the potentially high dose rates that could be experienced by individual members of some species owing to plutonium contained in particulates — for example, from the sediment of the sandbank adjacent to the Colette motu in the northern part of Mururoa Atoll. Overall, the Study concluded that the expected radiation dose rates and modes of exposure are such that no effects on biota population groups could arise, although occasionally individual members of species might be harmed, but not to the extent of endangering the whole species or creating imbalances between species.

4.6.2 The 2011 Fukushima Nuclear Power Plant Accident, Japan

During March 2011, as a consequence of a high-magnitude earthquake and resultant tsunami impacting the eastern coast of Honshu Island, a series of incidents at the Fukushima Dai-ichi nuclear power plant (NPP) resulted in the release of radioactive materials into the environment.

From March 21 and in the days that followed, measurements of various radionuclides in the seawater in the vicinity of the power station soon after the accident revealed considerable contamination of the marine environment. A strong radioactive plume was observed near the NPP in Fukushima Daiichi. Radiometric analyses of the mainly gamma-emitting radioisotopes were undertaken on samples of seawater, sediments, and biota in the marine environment at coastal and offshore (30 km) monitoring stations. These data were published by the Japanese Ministry of Education, Culture, Sports, Science, and Technology (MEXT) and the Tokyo Electric Power Company (TEPCO) and updates were made available online each day. The radioisotopes measured included ^{131}I, $^{136/137/134}$Cs, and ^{132}Te.

^{131}Iodine (half-life of 8.04 days), a short-lived nutrient-like radionuclide, would be readily taken up by marine biota, but decay rapidly. Others, such as ^{106}Ru (half-life 368 days) and ^{134}Cs (half-life 2.06 years) and particularly ^{137}Cs, with a half-life 30.2 years, had the potential to remain in the marine environment for several years. These conservative radionuclides mostly remain in solution and would be carried over very long distances by the adjacent coastal Kuroshio Current to eventually join the Pacific Ocean circulation, with only some deposition to sediments. Plutonium was not reported to be present in the Fukushima NPP discharges.

In the marine environment, ^{131}I, ^{137}Cs, and to a lesser degree ^{134}Cs, were the main radionuclides monitored and reported by MEXT. Statistical analysis of the data reported from 23 March to 5 May 2011 for coastal and offshore (30 km) monitoring stations is presented in Table 4.7. Over 90% of the total discharge 27×10^{15} Bq of ^{137}Cs to the marine environment was released over this period.

Although significantly high levels of ^{137}Cs and ^{131}I were observed adjacent to the discharge source, median activity values for ^{137}Cs and ^{131}I (Table 4.7) at the sites 30 km offshore were less than double the total 13.7 Bq/L of ambient seawater. ^{131}Iodine decayed rapidly, and with dilution and dispersal it was no longer observable by late May.

Fukushima cesium isotopes will be able to be measured for some time due to very low minimum detection limits of radiochemical and radiometric techniques. $^{134/137}Cs$ isotope ratios will identify the contaminated water mass and provide transport kinetics as it transits the Pacific until diluted and dispersed into the background seawater activity of approximately 2 mBq/L for ^{137}Cs. Radiochemical oceanographers will be able to use the Fukushima identifying 'label' to validate North Pacific Ocean circulation models, and radioecologists will have an opportunity to study food web transfers of radionuclides.

4.6.3 $^{238}U/^{234}Th$ and Scavenging of Trace Elements by Marine Particles

^{234}Th is a naturally occurring, particle-reactive radionuclide produced in seawater from the decay of ^{238}U (Table 4.5). The nearly uniform concentration of ^{238}U in seawater leads to a theoretical generation of three atoms of ^{234}Th per minute per litre (Bhat et al., 1969). Its parent, ^{238}U, is dissolved in seawater under oxidizing conditions as the anionic uranyl carbonate complex $[UO_2(CO_3)_2{}^{-2}]$ and therefore behaves conservatively with salinity; whereas,

TABLE 4.7 Statistical Analysis of the MEXT/TEPCO Data Reported from 23 March to 5 May 2011 for Coastal and Offshore (30 km) Monitoring Stations near Fukushima Dai-ichi NPP

Nuclide	Unit	Max	Min	Average	Median	SD	Number
Offshore Sites							
^{137}Cs	Bq/L	90	0.08	12.6	8.64	15.8	65
^{131}I	Bq/L	88.5	0.22	20.7	12.3	20.9	84
Coastal Sites							
^{137}Cs	Bq/L	47,000	22	4390	330	9550	78
^{131}I	Bq/L	180,000	290	20,100	3210	37,800	77

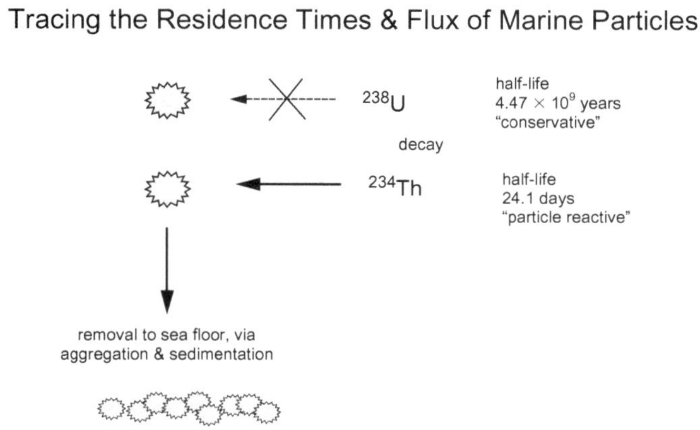

FIGURE 4.9 Simplified schematic of ^{234}Th particle scavenging from seawater.

^{234}This present as the particle-reactive hydrolysis product $Th(OH)_n^{(4-n)+}$ (Forster et al., 2009). ^{234}Th decays to ^{234} Pa (half-life 1.2 minutes), which then decays to ^{234}U, with the emission of a beta particle and a half-life of 24.1 days.

At equilibrium conditions (when there is no removal of ^{234}Th from seawater), the ^{234}Th activity is equal to the ^{238}U activity. The disequilibrium between ^{238}U and the measured total ^{234}Th activity reflects the net rate of particle export from the surface ocean. The timescale of the removal process, as measured by ^{234}Th, is in days to weeks, because the half-life of ^{234}Th is 24.1 days. Thus, ^{234}Th serves as an integrative tracer for particle flux that homogenizes short-term biological patchiness and temporal heterogeneity. The ratio of the activities, ^{238}U/^{234}Th, is taken as a measure of the state of disequilibrium, which is then related to residence time of the dissolved thorium in the particulate fraction. In the surface of the ocean, the activity of ^{234}Th is generally lower than that expected with respect to its production by its parent ^{238}U, which is generally interpreted as due to the uptake of ^{234}Th onto particle surfaces and subsequent sinking of particles into the water below (Cochran, 1982; see Figure 4.9).

There are numerous studies using ^{234}Th as a radiotracer for particle behaviour in marine systems and readers are directed to Aller and Cochran (1976); Cochran (1982); McKee et al. (1984); Rodriguez y Baena et al. (2006, 2007); Baskaran and Santschi (1993); Buesseler et al. (2006); Szymczak (1997); Chesselet and Lalou (1988); Cochran (1982); Swarzenski et al. (2000); and Waples et al. (2006).

4.7. SUMMARY AND CONCLUSIONS

The radiation environment of the oceans is predominantly a consequence of primordial natural processes, with origins in the formation of our universe. Based on their characteristic and different half-lives and biogeochemical behaviours, many of the natural radionuclides in marine ecosystems provide unique tools to interpret oceanic processes across multiple scales, both temporal and spatial. This is true for all systems, tropical or not.

As for nuclides derived or influenced by human activities, it may be concluded that the present activity concentrations of such radionuclides (e.g., ^{3}H, ^{90}Sr, and ^{137}Cs) in Pacific, Indian, and Southern Ocean surface waters are controlled by radioactive decay and by oceanographic processes (horizontal diffusion and advection, vertical mixing, and biogeochemical processes) following input processes due to global fallout. Due to radioactive decay, ^{3}H concentrations have decreased to one-fourth, and ^{90}Sr and ^{137}Cs to less than one-half the concentrations in the early 1960s, with corresponding decreases in spatial differences (IAEA, 2005). If we consider the main source for these artificial radionuclides as global fallout (Burton, 1975; IAEA, 2005) and average horizontal seawater diffusion coefficient of the order of 108 $cm^2\ s^{-1}$ (Nakano and Povinec, 2003), radionuclides can be dispersed about the peak at 35° latitudinally and longitudinally. However, advective ocean currents may impede such dispersal, resulting in further reduction in the spatial differences in radionuclide concentrations. Horizontal transport, vertical mixing, and biogeochemical processes combine to affect $^{239,240}Pu$ distributions, both at the surface and in deep waters, as observed in the Japan/East Sea and in the eastern equatorial Pacific. These particle-reactive and biologically mediated Pu isotopes arrived with global fallout, were scavenged, and settled to 700–1000 m and now upwell, along with nutrient rich subsurface waters, to produce surface enhancements (Gastaud et al., 2011). Because of its bioavailability and potential to transfer through marine food chains, Pu has been of particular radiological concern as well (Baxter et al., 1995).

Radionuclide concentrations in marine biota, although important from the viewpoint of dose assessment, are more difficult to characterize than those in seawater and sediment, given the limited data available and the intrinsic variations between species. Considering that ingestion is the main pathway of radiation exposure, that more than half of the world's population lives in Asia, and that seafood is a major dietary item in tropical regions, more concentration and transfer factor data, specifically for tropical biota, will be useful for future dose assessments (Stewart et al., 2005; Pentreath, 2009; Fowler, 2011).

Tropical marine systems have experienced lower levels of artificial radioactivity, which is generally sourced from global fallout (IAEA, 2005). However, tropical south and east Asia are the only regions in the world where electricity-generating capacity, and specifically nuclear power, is growing significantly (IAEA, 2012). With over 109 nuclear power reactors in operation, 21 out of the last 27 nuclear power plants connected to the world's energy grid are in Asia. Fourteen of the 17 new nuclear power plants being built globally are also in Asia (2010–2011), and plans exist to build more than 100 within the next 10 years. China and India have built 7 new nuclear power plants in the last 4 years, with 26 more under construction (IAEA, 2012). Africa is another region in the tropics that is likely to include a nuclear power option in any future development.

From this, an informed and strategic approach to the formulation and adoption of countermeasures and responses to nuclear discharges in tropical marine environments is required. This will need more region-specific measurements of artificial radioisotopes in the water column, marine biota, marine sediments, and suspended matter, in addition to more information about the impact of these radioisotopes on, and the risk to, endemic marine biota.

REFERENCES

Aarkrog, A., 1971. Radioecological investigations of plutonium in an arctic marine environment. Health Phys. 20, 31.

Aarkrog, A., 2003. Input of anthropogenic radionuclides into the World Ocean. Deep-Sea Research II 50, 2597–2606.

Abril, J.M., Fraga, E., 1996. Some physical and chemical features in the variability of K_d distribution coefficients for radionuclides. J. Environ. Radioact. 30 (3), 253–270.

Aller, R.C., Cochran, J.K., 1976. $^{234}Th/^{238}U$ disequilibria and diagenetic time scales. Earth and Planetary Science Letters 29, 37–50.

Anikouchine, W.A., Sternberg, R.W., 1981. The World Ocean, An Introduction to Oceanography, 2nd ed. Prentice Hall Inc., Englewood Cliffs, New Jersey, p. 513.

Aoyama, M., Povinec, P.P., Sanchez-Cabeza, J.A., 2011a. Preface - The Southern Hemisphere Ocean Tracer Studies (SHOTS) project. Progress in Oceanography 89, 1–6.

Aoyama, M., Fukasawa, M., Hirose, K., Hamajima, Y., Kawano, T., Povinec, P.P., et al., 2011b. Cross equator transport of ^{137}Cs from North Pacific Ocean to South Pacific Ocean (BEAGLE2003 cruises). Progress in Oceanography 89, 7–16.

Bacon, M.P., Spencer, D.W., Brewer, P.G., 1976. $^{210}Pb/^{226}Ra$ and $^{210}Po/^{210}Pb$ disequilibria in seawater and suspended particulate matter. Earth and Planetary Science Letters 32, 227–296.

Balistrieri, L., Brewer, P.G., Murray, J.W., 1981. Scavenging residence times of trace metals and surface chemistry in sinking particles in the deep ocean. Deep Sea Research 28A, 101–121.

Baskaran, M., Santschi, P.H., 1993. The role of particles and colloids in the transport of radionuclides in coastal environments of Texas. Marine Chemistry 43, 95–114.

Baxter, M.S., 1993. Environmental radioactivity: A perspective on industrial contributions. IAEA Bulletin 2, 33–38.

Baxter, M.S., Fowler, S.W., Povinec, P.P., 1995. Observations of plutonium in the oceans. Appl. Radiat. Isot. 46, 1213–1223.

Bhat, S.G., Krishnaswami, S., Lal, D., Rama, M., Moore, W.S., 1969. $^{234}Th/^{238}U$ ratios in the ocean. Earth and Panetary Science Letters 5, 483–491.

Broecker, W.S., Peng, T.H., Ostlund, G., 1986. The distribution of bomb tritium in the ocean. Journal of Geophysical Research 91, 14331–14344.

Burton, J.D., 1975. Radioactive nuclides in the marine environment. In: Riley, J.P., Skirrow, G. (Eds.), Chemical Oceanography, 2nd ed. vol. 3. Academic Press, London, pp. 91–139.

Buesseler, K.O., Benitez-Nelson, C.R., Moran, S.B., Burd, A., Charette, M., Cochran, J.K., et al., 2006. An assessment of particulate organic carbon to thorium-234 ratios in the ocean and their impact on the application of ^{234}Th as a POC flux proxy. Marine Chemistry 100 (2006), 213–233.

Chen, J.H., Edwards, L.R., Wasserburg, G.J., 1986. ^{238}U, ^{234}U and ^{232}Th in seawater. Earth and Planetary Science Letters 80, 241.

Cherry, R.D., Shannon, L.V., 1974. The alpha radioactivity of marine organisms. At. Energy Rev. 12, 3–45.

Chesselet, R., Lalou, C., 1988. The use of natural radionuclides in oceanography. In: Gury, J.C., Gueguenist, P., Pentreath, R.J. (Eds.), Radionuclides: A Tool for Oceanography. Elsevier Applied Science, London (UK), pp. 1–11.

Chiappini, R., Pointurier, F., Millies-Lacroix, J.C., Lepetit, G., Hemet, P., 1999. ^{240}Pu/ ^{239}Pu isotopic ratios and ^{239,240}Pu total measurements in surface and deep waters around Mururoa and Fangataufa atolls compared with Rangiroa atoll (French Polynesia). Sci. Total Environ 237/238, 269–276.

Choppin, G.R., 2006. Actinide speciation in aquatic systems. Marine Chemistry 99 (1–4), 83–92.

Cochran, J.K., 1982. The ocean chemistry of the U- and Th-series nuclides. In: Ivanovich, M., Harmon, R.S. (Eds.), Uranium Series Disequilibrium: Application to Environmental Problems. Clarendon Press, Oxford, pp. 385–429.

Duran, E.B., Povinec, P.P., Fowler, S.W., Airey, P.L., Hong, G.H., 2004. ^{137}Cs and $^{239/240}$Pu levels in the Asia-Pacific regional seas. J. Environ. Radioact. 76, 139–160.

Forster, S., Turnewitsch, R., Powilleit, M., Werk, S., Peine, F., Ziervogel, K., et al., 2009. Thorium-234 derived information on particle residence times and sediment deposition in shallow waters of the south-western Baltic Sea. Journal of Marine Systems 75, 360–370.

Fowler, S.W., 2011. ^{210}Po in the marine environment with emphasis on its behaviour within the biosphere. J. Environ. Radioact. 102 (2011), 448–461.

Gastaud, J., Povinec, P.P., Aoyama, M., Hirose, K., Sanchez-Cabeza, J.A., Levy, I., et al., 2011. Transport and scavenging of Pu in surface waters of the Southern Hemisphere Oceans. Progress in Oceanography 89 (2011), 92–100.

Hatje, V., Payne, T.E., Hill, D.M., McOrist, G., Birch, G.F., Szymczak, R., 2003. Kinetics of trace element uptake and release by particles in estuarine waters: effects of pH, salinity, and particle loading. Environ. Int. 29, 619–629.

Hirose, K., Sugimura, Y., 1993. Chemical speciation of particulate ^{238}U, ^{239,240}Pu and Th isotopes in seawater. Sci. Total Environ. 130/131, 517–524.

Hirose, K., Kim, C.S., Yim, S.A., Aoyama, M., Fukasawa, M., Komura, K., et al., 2011. Vertical profiles of plutonium in the central South Pacific. Progress in Oceanography 89 (2011), 101–107.

IAEA, 1995. Sources of Radioactivity in the Marine Environment and Their Relative Contributions to Overall Dose Assessment from Marine Radioactivity (MARDOS). IAEA TECDOC Series No. 838, pp. 11–49.

IAEA, 1998. The Radiological Situation at the Atolls of Mururoa and Fangataufa: Main Report. Radiological Assessment Reports Series 1. STI/PUB/1028, p. 282.

IAEA, 1999. Inventory of Radioactive Waste Disposal at Sea. IAEA TECDOC Series No. 1105, p. 121.

IAEA, 2004. Sediment Distribution Coefficients and Concentration Factors for Biota in the Marine Environment. Technical Report Series No. 422, p. 95.

IAEA, 2005. Worldwide Marine Radioactivity Studies (WOMARS): Radionuclide Levels in Oceans and Seas. IAEA TECDOC Series No. 1429, p. 187.

IAEA, 2012. The Power Reactor Information System (PRIS) - The Database on Nuclear Power Reactors. http://pris.iaea.org/public.

ICRP, 2008. Environmental Protection – the Concept and Use of Reference Animals and Plants. ICRP Publication 108, Pergamon Press, Oxford and New York.

Jeffree, R.A., Szymczak, R., 2000. Enhancing effect of marine oligotrophy on environmental concentrations of particle-reactive trace elements. Environ. Sci. Technol. 34, 1966–1969.

Jeffree, R.J., Carvalho, F., Fowler, S.W., Farber-Lorda, J., 1997. Mechanism for uptake of radionuclides by zooplankton in French Polynesian waters. Environ. Sci. Technol. 31, 2584–2588.

Jeffree, R.A., Szymczak, R., Peck, G.A., Fowler, S., 2012. Role of zooplankton in the biogeochemistry of Po-210 and Pb-210 in the oligotrophic South-West Pacific, in press.

Kameda, K., 1962. Study on abundance of rare earth elements in marine organisms. J. Radiat. Res. (Tokyo) 3–2, 89–103.

Kim, G., 2001. Large deficiency of polonium in the oligotrophic ocean's interior. Earth and Planetary Science Letters 192, 15–21.

Kinoshita, N., Sumi, T., Takimoto, K., Nagaoka, M., Yokoyama, A., Nakanishi, T., 2011. Anthropogenic Pu distribution in tropical East Pacific. Sci. Total Environ. 409, 1889–1899.

Ku, T.K., Knauss, K.G., Matheau, G.G., 1977. Uranium in the open ocean: Concentration and isotopic composition. Deep Sea Research 24, 1005–1017.

Livingston, H.D., Povinec, P.P., 2002. A millennium perspective on the contribution of global fallout radionuclides to ocean science. Health Phys. 82, 656–668.

Livingston, H.D., Bowen, V.T., Casso, S.A., Volchok, H.L., Noshkin, V.E., Wong, K.M., et al., 1985. Fallout Nuclides in Atlantic and Pacific Water Columns: GEOSECS Data Tech. Rep Woods Hole Oceanographic Institution.

Lu, J., Huang, Y., Wang, L., Li, F., Li, S., Hsia, Y., 2007. ^{3}H & ^{90}Sr background in water around Taiwan NPP. China. Radiat. Meas. 42 (2007), 74–79.

MacDougall, J.D., Finkel, R.C., Carlson, J., Krishnaswami, K., 1979. Isotopic evidence for uranium exchange during low temperature alteration of oceanic basalt. Earth and Planetary Science Letters 42, 27–34.

Mann, K.H., Lazier, J.R.N., 1991. The dynamics of marine ecosystems. Biological-physical interactions in the oceans. Blackwell Scientific Publications, Oxford.

McKee, B.A., 2008. U- and Th-series nuclides in estuarine environments. In: Krishnaswami, S., Cochran, J.K. (Eds.), U-Th Series Nuclides in Aquatic Systems. Elsevier Applied Science, London (UK), pp. 194–225.

McKee, B.A., DeMaster, D.H., Nittrouer, C.A., 1984. The use of ^{234}Th/^{238}U disequilibrium to examine the fate of particle-reactive species on the Yangtze continental shelf. Earth and Planetary Science Letters 68, 431–442.

Moore, W.S., 1967. Amazon and Mississippi river concentrations of uranium, thorium, and radium isotopes. Earth and Planetary Science Letters 2, 231–234.

Moore, W.S., 1996. Large groundwater inputs to coastal waters revealed by ^{226}Ra enrichments. Nature 380, 612–614.

Moore, W.S., 2007. Radium isotopes as tracers of submarine groundwater discharge. Nuclear and Isotopic Techniques for the Characterization of Submarine Groundwater Discharge in Coastal Zones. IAEA TECDOC Series No. 1595, pp. 139–148.

Nakano, M., Povinec, P.P., 2003. Modeling the distribution of plutonium in the Pacific Ocean. J. Environ. Radioact. 69, 85–106.

Nittrouer, C.A., Brunskill, G., Aller, R.C., Alongi, D., Bird, M., Burrage, D., et al., 1994. The gateway for terrestrial material entering the ocean. EOS 75 (16), 191–192.

Nozaki, Y., Thompson, J., Turekian, K.K., 1976. The distribution of ^{210}Pb and ^{210}Po in the surface waters of the Pacific Ocean. Earth & Planetary Science Letters 32, 304–312.

Olsen, C.R., Larsen, I.L., Lowry, P.D., McLean, R.I., Domotor, S.L., 1989. Radionuclide distributions and sorption behaviour in the Susquehanna-Chesapeake Bay system. Publication No. 2907. Environmental Sciences Division, Oak Ridge National Laboratory, p. 106.

Peck, G.A., Smith, D.S., 2000. Distribution of dissolved and particulate ^{226}Ra, ^{210}Pb and ^{210}Po in the Bismarck Sea and western equatorial Pacific Ocean. Marine and Freshwater Research 51, 647–658.

Pentreath, R.J., 2009. Radioecology, radiobiology, and radiological protection: Frameworks and fractures. J. Environ. Radioact. 100, 1019–1026.

Pentreath, R.J., Camplin, W.C., Allington, D.J., 1989. Individual and collective dose rates from naturally occurring radionuclides in seafoods. Radiological protection: Theory and practice - Proceedings of the 4th International Symposium of the Society of Radiological Protection. pp. 297–300.

Povinec, P.P., Aarkrog, A., Buesseler, K.O., Delfanti, R., Hirose, K., Hong, G.H., et al., 2005. ^{90}Sr, ^{137}Cs and $^{239,240}Pu$ concentration surface water time series in the Pacific and Indian Oceans, WOMARS results, J. Environ. Radioact. 81, 63–87.

Povinec, P.P., Aoyama, M., Fukasawa, M., Hirose, K., Komura, K., Sanchez-Cabeza, J.A., et al., 2011. ^{137}Cs water profiles in the South Indian Ocean – An evidence for accumulation of pollutants in the subtropical gyre. Progress in Oceanography 89, 17–30.

Riley, J.P., Chester, R. (Eds.), 1989. Chemical Oceanography. Academic Press, London.

Ritchie, G.D., Shimmield, G.B., 1991. The use of $^{210}Po/^{210}Pb$ disequilibria in the study of the fate of marine particulate matter. In: Kershaw, P.J., Woodhead, D.S. (Eds.), Radionuclides in the Study of Marine Processes. Elsevier Applied Science Ltd, London, pp. 142–153.

Rodriguez y Baena, A.M., Metian, M., Teyssie, J.L., De Broyer, C., Warnau, M., 2006. Experimental evidence for ^{234}Th bioaccumulation in three Antarctic crustaceans: Potential implications for particle flux studies. Marine Chemistry 100, 354–365.

Rodriguez y Baena, A.M., Fowler, S.W., Miquel, J.C., 2007. Particulate organic carbon: Natural radionuclide ratios in zooplankton and their freshly produced fecal pellets from the NW Mediterranean (MedFlux 2005). Limnology and Oceanography 52 (3), 966–974.

Sanchez-Cabeza, J.A., Levy, I., Gastaud, J., Eriksson, M., Osvath, I., Aoyama, M., et al., 2011. Transport of North Pacific ^{137}Cs labeled waters to the south-eastern Atlantic Ocean. Progress in Oceanography 89 (2011), 31–37.

Sholkovitz, E.R., Szymczak, R., 2000. The estuarine geochemistry of rare earth elements: Comparison of Amazon, Fly, Sepik, and Gulf of Papua systems. Earth and Planetary Science Letters 179 (2), 299–309.

Stewart, G.M., Fowler, S.W., Fisher, N.S., 2008. The bioaccumulation of U-Th series radionuclides in marine organisms. In: Krishnaswami, S., Cochran, J.K. (Eds.), U-Th Series Nuclides in Aquatic Systems. Elsevier, Amsterdam, pp. 269–305.

Stewart, G.M., Fowler, S.W., Teyssie, J.L., Cotret, O., Cochran, J.K., Fisher, N.S., 2005. Contrasting transfer of polonium-210 and lead-210 across three trophic levels in marine plankton. Marine Ecology Progress Series 290, 27–33.

Stumm, W., Morgan, J.J., 1981. Aquatic Chemistry. Wiley Press, New York, NY.

Suriyanarayanan, S., Brahmanandhan, G.M., Malathi, J., Ravi Kumar, S., Masilamani, V., Shahul Hameed, V., et al., 2008. Studies on the distribution of ^{210}Po and ^{210}Pb in the ecosystem of Point Calimere Coast (Palk Strait), India. J. Environ. Radioact. 99, 766–771.

Swarzenski, P.W., McKee, B.A., 1999. Seasonal uranium distribution in the coastal waters off the Amazon and Mississippi Rivers. Estuaries 21 (3), 379–390.

Swarzenski, P.W., McKee, B.A., Booth, J.G., 1992. Uranium geochemistry on the Amazon shelf: Seasonal differences in source functions and dissolved/colloidal partitioning of uranium isotopes. EOS 72, 64.

Swarzenski, P.W., McKee, B.A., Booth, J.G., 1995. Uranium geochemistry on the Amazon shelf: Chemical phase partitioning and cycling across a salinity gradient. Geochimica et Cosmochimica Acta 59, 7–18.

Swarzenski, P.W., Corbett, D.R., Smoak, J.M., McKee, B.A., 2000. The use of U–Th series radionuclides and transient tracers in oceanography: an overview. In: Hester, R.E., Harrison, R.M. (Eds.), Chemistry of the Marine Environment. The Royal Society of Chemistry, pp. 33–53.

Swarzenski, P., Campbell, P., Porcelli, D., McKee, B., 2004. The estuarine chemistry and isotope systematics of $^{234,238}U$ in the Amazon and Fly Rivers. Continental Shelf Research 24, 2357–2372.

Szymczak, R., 1997. Radionuclides as tracers of dynamics biogeochemical processes in tropical marine systems. Radioprotection 32 (C2), 183–188.

Tsumune, D., Aoyama, M., Hirose, K., Bryan, F.O., Lindsay, K., Danabasoglu, G., 2011. Transport of ^{137}Cs to the Southern Hemisphere in an ocean general circulation model. Progress in Oceanography 89 (2011), 38–48.

UNSCEAR, 2000. Sources and effects of ionising radiation, Vol 1; Sources. UNSCEAR Report to the General Assembly UNSCEAR. Vienna.

Waples, J.T., Benitez-Nelson, C., Savoye, N., van der Loeff, M.R., Baskaran, M., Gustafsson, O., 2006. An introduction to the application and future use of ^{234}Th in aquatic systems. Marine Chemistry 100, 166–189.

Chapter 5

Terrestrial Radioecology in Tropical Systems

Keiko Tagami[1], John R. Twining[2], and M. Angelica V. Wasserman[3]
[1]*Radiation Safety Research Center, National Institute of Radiological Sciences, Chiba, Japan*
[2]*Austral Radioecology, Oyster Bay, Australia*
[3]*Instituto de Engenharia Nuclear, Rio de Janeiro, Brasil*

5.1. INTRODUCTION

In previous chapters of this book, the sources, dispersal, and transport mechanisms for radionuclides and the nature of radioactive materials have been considered. In this Chapter we consider radionuclide uptake and translocation in tropical crops and ecosystems. Although some material in this chapter has been sourced from the general literature, much of it has been drawn from two International Atomic Energy Agency/Food and Agriculture Organization/International Union of Radioecologists (IAEA/FAO/IUR) Coordinated Research Programs (CRPs) established to focus on acquiring tropical data and look at factors affecting bioaccumulation in tropical systems. However, readers should be aware that the general literature relating to the environmental behaviour of radionuclides is under substantial review. The widely used IAEA Technical Report on radionuclide transfer in temperate environments (TRS No. 364; IAEA, 1994) has been updated to include more recent data, including tropical information and information on freshwater environments (TECDOC 1616; IAEA, 2009). Much of this effort is summarized in a special issue of the *Journal of Environmental Radioactivity*, Volume 100 (2009). Readers are encouraged to read the whole volume, particularly the overview provided by the editors (Calmon et al., 2009).

While many of the crops in tropical systems are of the same general class that exists in temperate regions (e.g., wheat, rice, maize, etc.), radionuclide uptake by those crops may be dissimilar depending upon environmental differences such as soil type, plant physiological response to higher temperatures, light intensity, rain periodicity, and so on. In addition, there are crops, agricultural methods, and terrestrial ecosystems particular to tropical environments.

Radioactivity in the Environment, Vol. 18. http://dx.doi.org/10.1016/B978-0-08-045016-2.00005-9

The human factors, such as agricultural practices (e.g., flood irrigation, slash and burn), will be considered to some extent here, but also in Chapter 7, which covers radiological dose assessment.

There is much less information available relating to radioecology for terrestrial animals compared to plants, particularly when referring to tropical and subtropical environments. However, the limited tropical data that are available will be tabulated and discussed as well.

Because we expect terrestrial ecosystems and agricultural practices in tropical and subtropical areas to be substantially different to those in more temperate regions, we will initially look at the different soil types and farming systems. We will then consider more specific radioecological matters later in this chapter.

5.2. TROPICAL SOIL SYSTEMS

5.2.1 Soil Formation

According to Sanchez (2000), nineteenth century soil scientists travelling from temperate regions to the tropics developed a misguided view about the uniformity of tropical soils and the presence and relevance of hardened layers rich in iron oxides, called laterites. Hence, tropical soils were described as red, acid, infertile soils that, when cleared of vegetation, became a worthless brick pavement in a few years, unsuitable for agricultural purposes. That limited perspective has changed markedly in the intervening period. Sánchez and Buol (1975) recorded that such characteristics covered only about 7% of the tropics and also occur in the United States. In actuality, the proportion of fertile soils with sufficient rainfall and no permafrost are roughly equal in tropical and temperate regions of the planet (45% and 41%, respectively).

Based on the Köppen-Geiger climate classification scheme (Kottek et al., 2006), the equatorial climate (including tropical regimes) occurs mainly in northern regions of South America, central Africa, northern Australia, Indonesia, southern India and Sri Lanka, Thailand, Malaysia, Vietnam, Laos, Myanmar, Philippines, and most of the Pacific island nations. Using this scheme, Peel et al. (2007) stated that the *A* classification corresponds to tropical soils and can be subdivided into tropical rainforest climate (*Af*), tropical monsoon climate (*Am*), and tropical savannah climate (*Aw*). Some authors replace *Aw* by *As* when the dry season occurs during summer. Subtropical divisions of the classification scheme include all of *B* (arid climates), except for *BWk* and *Bsk* (which are the cold deserts and steppes), and much of *C* (warm temperate), especially those areas with hot or warm summers (subsets *a* and *b*). This expands the area of interest to include almost all of Africa and the Mediterranean, southern Asia, most of Australia, the southern United States, Mexico, and Central America.

A classical picture of soil formation under different climatic zones (Strakov, 1967) well illustrates the consequence of heavy rain and high temperature on soil profiles of tropical regions (Figure 5.1). Tropical soils are typically deep and the upper layer is depleted of more complex clay minerals such as illite, montmorillonite, and vermiculite. Kaolinite is present, but the clay fraction is dominated by goethite and gibbsite, secondary minerals with a very simple structure compared to kaolinite, which is evidence of the higher degree of weathering. The abundance of some minerals and the absence of others can be very important to the fate of radionuclides in tropical soils, as will be discussed in Section 5.5.

5.2.2 FAO Soil Types in Tropical and Subtropical Regions

The United Nations FAO and UNESCO have identified 28 dominant soil groups, as classified by the World Reference Base for Soil Resources (WRB) Reference Soil Groups, all of which are present in the tropics, but some as only minor inclusions. Maps of their distribution can be found at http://www.fao.org/ag/agl/agll/wrb/mapindex.stm and more details on how to classify all soil types are given in the International Union of Soil Sciences Working Group WRB (IUSS, 2007). Table 5.1 provides a list of the WRB soils that are common in tropical and subtropical regions of the planet along with general descriptions of their properties and an overview of their distributions and abundances within those zones. Approximate cross-references to the American classification scheme, with a rationale for those relationships, can be found in Deckers et al. (2002).

Some soil types are not well described by the FAO systems. Of particular note in tropical systems are the atoll soils of mid-oceanic islands, often with low nutrient levels, high calcium concentrations, and high porosity. These soils have been particularly associated with extremely high ^{137}Cs accumulation and low ^{90}Sr accumulation by terrestrial plants (see Section 5.5.2).

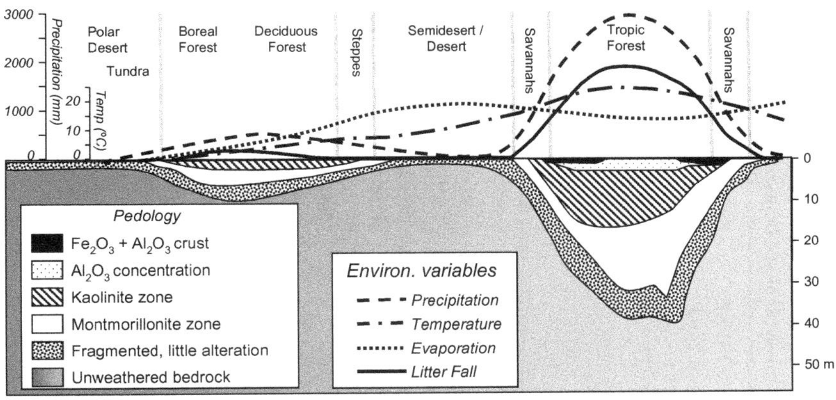

FIG. 5.1 Soil weathering under different climatic conditions (from Strakov, 1967).

TABLE 5.1 **FAO Soil Units Common in Tropical and Subtropical Areas, Including a Brief Description with Tropical Distributions and Typical Associated Soil Types and the Approximate Equivalent Taxonomy in the U.S. Classification**

FAO WRB Soil Type	Brief Description, Distribution, and Association in Tropical Areas	Approximate U.S. Taxonomy
Acrisols	Acrisols have more low-activity clay in the subsoil than in the topsoil as a result of long-term soil forming processes (especially clay migration). They are found in humid regions and are most extensive in southeast Asia, the southern fringes of the Amazon Basin, the southeastern United States, and in both east and west Africa. There are about 1000 Mha of Acrisols worldwide associated with Nitisols, Ferralsols, Plinthosols, Lixisols, Arenosols, Regosols, and Cambisols.	Ultisols
Alisols	Similar to Acrisols but with high-activity subsurface clays rich in exchangeable aluminium. Major occurrences of Alisols are found in Latin America, in the West Indies, in west Africa, the highlands of east Africa, Madagascar, and in southeast Asia and northern Australia, as well as in subtropical regions, and comprise at least 100 Mha. They are associated with Acrisols, Lixisols, and Ferralsols.	Ultisols
Andosols	Andosols are young soils associated with pyroclastic parent materials and therefore volcanic regions of the world. They are typically allophanes (amorphous aluminosilicate clays). They occur around the Pacific Rim, including North and central America, the Andean region, Indonesia, Philippines, Japan, and New Zealand, and along the African Rift Valley with scattered occurrence elsewhere. Altogether covering 110 Mha throughout the world, they are not a dominant soil but are regionally important. Associated soils are Cambisols, Luvisols, and Vertisols.	Andisols
Anthrosols	Anthrosols comprise soils that have been modified profoundly through human activities, such as by the addition of organic materials or household wastes, irrigation, and cultivation. Whilst they have a long history, they are not overly abundant as a worldwide soil type, comprising only about 0.5 Mha. Anthrosols are found in irrigation areas in dry regions (e.g., in Mesopotamia), near oases in desert regions, and in parts of India. Paddy soils occupy vast areas in China and in parts of south and southeast Asia. Anthrosols fertilized with household wastes and manure are found all over the world, where humans have existed.	Inceptisols, Plaggepts

Arenosols	Sandy soils featuring very weak or no soil development. Arenosols are one of the most extensive reference soil groups in the world; including shifting sands and active dunes, they cover about 1300 Mha, or 10% of the land surface. Vast expanses of deep Aeolian sands are found on the central African plateau between the equator and 30°S. Other areas of Arenosols occur in the Sahelian region of Africa, various parts of the Sahara, central and western Australia, the Near East, and China. Sandy coastal plains and coastal dune areas are of smaller geographic extent. They are frequently associated with Leptosols, Regosols, Calcisols, Solonchaks, and Podzols.	Entisols
Calcisols	Soils with substantial secondary accumulation of lime. Calcisols are widespread in arid and semi-arid environments and are often associated with highly calcareous parent materials. Nearly all of the estimated 1000 Mha is in the arid and semi-arid tropics and subtropics of both hemispheres. Associated soils are Regosols, Cambisols, Gypsisols, and Solonchaks.	Aridisols
Cambisols	Weakly to moderately developed soils with at least the beginnings of horizon differentiation in the subsoil evident from changes in structure, color, clay content, or carbonate content. Whilst much more common in temperate areas, the young alluvial plains and terraces of the Ganges-Brahmaputra system are probably the largest continuous surface of Cambisols in the tropics. Cambisols are also common in areas with active geologic erosion, where they may occur in association with mature tropical soils. Associated soils include Gleysols, Leptosols, Fluvisols, Acrisols, and Ferralsols.	Inceptisols
Durisols	Durisols are associated mainly with old surfaces in arid and semi-arid environments and accommodate very shallow to moderately deep, well-drained soils that contain cemented secondary silica (SiO_2) within 100 cm of the soil surface. These soils are extensive in Australia, South Africa, and America, but an estimate of their extent is not available. Durisols are normally associated with Gypsisols, Calcisols, Arenosols, Cambisols, and Vertisols.	Aridisols
Ferralsols	With a dominance of kaolinite and sesquioxides, Ferralsols are strongly weathered soils with chemically poor, but physically stable, subsoil. Restricted to tropical regions, mainly south and central America and central Africa, with scattered areas elsewhere, they are considered to be the 'classic' soils of humid tropics. There are an estimated 750 Mha of Ferralsols. Associated soils include Acrisols, Nitisols, Plinthosols, and Cambisols.	Oxisols

continued

TABLE 5.1 FAO Soil Units Common in Tropical and Subtropical Areas, Including a Brief Description with Tropical Distributions and Typical Associated Soil Types and the Approximate Equivalent Taxonomy in the U.S. Classification—Cont'd

FAO WRB Soil Type	Brief Description, Distribution, and Association in Tropical Areas	Approximate U.S. Taxonomy
Fluvisols	Young soils in alluvial deposits on floodplains, deltas, and tidal marshes. Widespread, but not a dominant type. The estimated total area is 350 Mha. Fluvisols are frequently associated with Histosols and Gleysols.	Entisols
Gleysols	Soils with permanent or temporary wetness near the surface. Of about 720 Mha worldwide, an estimated 200 Mha of Gleysols are found in the tropics, mainly in the Amazon region, equatorial Africa, and the coastal swamps of southeast Asia. Associated soils are Cryosols, Podzols, Histosols, Fluvisols, Calcisols, Gypsisols, Acrisols, Lixisols, Alisols, and Nitisols.	Inceptisols, Entisols
Gypsisols	Soils with an accumulation of secondary gypsum ($CaSO_4.2H_2O$). With only 90 Mha worldwide, often in the near and middle east, Gypsisols are uncommon. They have a distribution similar to that of Calcisols.	Aridisols
Histosols	Soils with thick organic layers typically developed on peats from mangrove or swamp forest in the tropics. Of about 275 Mha worldwide, about 40 Mha occur on coastal lowlands of the subtropics and tropics. Histosols are normally associated with Podzols, Gleysols, and Fluvisols.	Histosols
Leptosols	Very shallow soils over hard rock or in unconsolidated, very gravelly material, providing limited rooting capacity. With a total area of 1655 Mha worldwide, Leptosols are widespread and most common in mountainous areas (545 Mha; associated with Regosols and Cambisols) and deserts (420 Mha; with Arenosols, Calcisols, and Gypsisols).	Entisols, Lithic subgroups

Lixisols	Soils with subsurface accumulation of low activity clays and high base saturation. There are approximately 435 Mha of Lixisols worldwide, mainly in Brazil, west Africa, east Africa, and India.	Alfisols
Luvisols	Soils with subsurface accumulation of high activity clays. Of the 500–600 Mha worldwide, only a small proportion exists in subtropical and tropical regions; Luvisols occur mainly on young land surfaces.	Alfisols
Nitisols	Low-activity clays with good P fixation. Deep, strongly structured, dark red, brown, or yellow clay soils. Nitisols cover about 200 Mha on high plateau in eastern Africa, particularly in Kenya, Ethiopia, and Tanzania. Smaller areas occur at lower altitudes in India, the Philippines, Java, Central America, and Brazil.	Ultisols, Oxisols
Phaeozems	Soils with a thick, dark topsoil rich in organic matter and evidence of removal of carbonates. Phaeozems only cover an estimated 190 Mha worldwide. About 50 Mha are in the subtropical pampas of Argentina and Uruguay. Smaller areas are found in tropical montane regions.	Mollisols
Planosols	Soils with bleached, temporarily water-saturated topsoil on slowly permeable subsoil. Planosols are extensive in Brazil, northern Argentina, South Africa, and eastern Australia. Smaller areas occur in southeast Asia from Bangladesh to Vietnam and in the eastern United States. There are 130 Mha of Planosols worldwide, most of which occur in South America and Australia. Associated soils include Acrisols, Luvisols, and Vertisols.	Alfisols, Mollisols
Plinthosols	Wet soils with an irreversibly hardening mixture of iron, clay, and quartz in the subsoil. Active formation of plinthite is taking place in west Africa, parts of South America, India, and Western Australia. Ironstone, or hardened plinthite, is more widely spread and often associated with old, high-level peneplains of the southern continents. There are some 60 Mha worldwide; they are often associated with Ferralsols, Acrisols, and Alisols.	Oxisols,
Podzols	Acid soils with blackish/brownish/reddish subsoil with alluvial iron-aluminium-organic compounds. These are minor yet widespread soils, mainly associated with Acrisols and Arenosols in the humid tropics. Tropical Podzols occur on less than 10 Mha, mainly in uplifted coastal areas. Important occurrences are found along the Rio Negro and in French Guiana, Guyana, and Suriname in South	Spodosols

continued

TABLE 5.1 FAO Soil Units Common in Tropical and Subtropical Areas, Including a Brief Description with Tropical Distributions and Typical Associated Soil Types and the Approximate Equivalent Taxonomy in the U.S. Classification—Cont'd

FAO WRB Soil Type	Brief Description, Distribution, and Association in Tropical Areas	Approximate U.S. Taxonomy
	America; in the Malaysian region (Kalimantan, Sumatra, and Irian Jaya); and in northern Australia. They seem to be less common in Africa.	
Regosols	Soils with very limited or no soil profile development. Regosols only occupy an estimated 260 Mha worldwide, occurring mainly in arid areas, the dry tropics, and in mountainous regions. They are frequently associated with Leptosols and Arenosols.	Entisols
Solonchaks	Strongly saline soils. Solonchaks occur where evaporation exceeds rainfall and there is a seasonal or permanent water table close to the soil surface, or in coastal areas influenced by saline intrusions. Widespread but scattered distribution in Saharan Africa, east Africa, Namibia, central Asia, Australia, and South America. Estimates of their total area vary between 260 and 340 Mha. They are associated with Gleysols.	Aridisols
Vertisols	Dark-colored swelling and cracking clays affected by water There are 335 Mha of these soils worldwide. Vertisols in the tropics cover ~200 Mha occurring in central Sudan, east Africa, the Deccan Plateau of India, Texas, South America, and Australia. One-quarter of this is considered to be useful land. Most Vertisols occur in the semi-arid tropics, with an average annual rainfall of 500–1000 mm, but Vertisols are also found in the wet tropics, e.g., Trinidad (where the annual rainfall amounts to 3000 mm). They are associated with Luvisols, Cambisols, Gypsisols, and Solonchaks.	Vertisols

Mha, million hectares.
U.S. classifications from Deckers et al. (2002).
Adapted from IUSS Working Group WRB, (2007) and FAO AGL, (2003).

5.2.3 Agricultural Problems of Tropical Soils

Tropical soil systems suffer a range of constraints against crop development, which affect radionuclide transfer as discussed in Section 5.5. Table 5.2 was taken directly from Norman et al. (1995) and contains a list of the factors adversely affecting agricultural activities in the hot wet tropics, together with estimates of the area and relative proportions of soils affected by those factors in a number of key regions.

Low nutrient status is the predominant problem in all regions, affecting more than half of the available soils. In most tropical soils, the available nutrients are concentrated in the topmost, humic layers and are rapidly recycled back into the supported vegetation to promote growth and development. However, demand almost always greatly exceeds supply; hence nutrients are a

TABLE 5.2 Agricultural Constraints in Tropical Regions and the Areas Affected by Those Factors

Soil Constraint	Tropical America		Tropical Africa		Tropical Asia		Humid Tropics	
	Mha	*(%)*	*Mha*	*(%)*	*Mha*	*(%)*	*Mha*	*(%)*
Low nutrient reserves	543	(66)	285	(67)	101	(45)	928	(64)
Al toxicity	490	(61)	226	(53)	92	(41)	808	(56)
High P fixation	379	(47)	84	(20)	74	(33)	537	(37)
Acid, not Al toxic	88	(11)	92	(22)	74	(33)	255	(18)
Slopes >30%	145	(18)	22	(5)	73	(33)	241	(17)
Poor drainage	90	(11)	59	(14)	42	(19)	191	(13)
Low eCEC	68	(8)	87	(20)	10	(5)	165	(11)
Shallow depth	54	(7)	17	(4)	27	(12)	98	(7)
No major limitations	28	(3)	7	(2)	5	(2)	40	(3)
Shrink swell	11	(1)	2	(1)	3	(2)	17	(1)
Allophane	8	(1)	1	(<1)	4	(2)	13	(1)
Acid sulfate soils	2	(<1)	5	(1)	6	(3)	13	(1)
Gravel	2	(<1)	6	(1)	3	(1)	10	(1)
Salinity	3	(<1)	1	(<1)	4	(2)	8	(<1)
Sodic soils	5	(1)	3	(1)	1	(<1)	9	(<1)

eCEC, effective cation exchange capacity; Mha, million hectares.
From Norman et al. (1995).

limiting factor in these systems. Both this effect and the next most common problem, Al toxicity, can be explained by the higher temperatures and rainfall in tropical areas leading to more rapid leaching and the relative accumulation of oxides of Fe and Al. The enhanced Fe concentrations explain why tropical soils are often redder than those in temperate regions. The more intense rainfall and increased temperatures also lead to faster clay formation and desilication whereby the 2:1 clay minerals decay, releasing silicon, to 1:1 clay minerals (see also Section 5.5.3). During this degradation process the property of cation sorption in clays is reduced. The layers of 2:1 clay minerals are more negatively charged and 1:1 clay minerals are less negatively charged, therefore, they have less ability for cation sorption which, in turn, enhances leaching loss rates. The resulting low cationic exchange capacity (CEC) is another common problem in tropical soil systems. Greater details of the adverse effects of these factors are given in Section 5.5.

5.3. AGRICULTURAL SYSTEMS

5.3.1 Classes of Crops

A wide variety of crops is grown across the planet and each plant has its own identifiably characteristic form. However, despite these differences, broad classes of crops can readily be identified by their size, method of propagation, use by humans, and so forth. Because of this, useful plant types have always been assigned to more general categories of crops. This type of classification has been useful for generalized modelling of radioecological behavioural differences across and between these different crops (see Chapter 7 and Section 5.4).

The Indicative Crop Classification (ICC) (FAO, 2005), a new crop classification scheme drawing on the experience of ages, was developed for the 2010 round of agricultural censuses. The ICC classifies crops into categories based on three main elements:

1. **Product type**. The product type is provided in the structure of the Central Product Classification (CPC) (United Nations, 2005), especially at the group and class level. Thus, under the ICC, crops are first divided into groups (cereals, vegetables, etc.), and each group is subdivided by crop type (leafy/stem vegetables, fruit-bearing vegetables, etc.).
2. **Crop genus or species**. At the lowest level of the classification, each crop can be described by its scientific name. Thus, 'Lentils' (Class 75) is identified as the species *Lens culinaris*. However, it should be noted that the ICC is not a botanical classification, as the groupings are based more on the agricultural use of the crop rather than the botanical similarities between crops. Thus, 'Oilseed crops' (Group 4) is a grouping of crops of many different botanical types that produce the same type of product, in this case oil.

3. **Whether the crop is temporary or permanent**. The CPC does not always permit a temporary/permanent split, because this is not important in a product classification. However, this distinction is fundamental to a crop classification. Because of this, some CPC classes are divided into temporary and permanent subclasses. In the ICC, a separate code is provided to indicate whether the crop is temporary or permanent.

In this classification scheme, as a general principle, a particular crop is classified only once, regardless of how the crop is used. In the past, some crops appeared more than once (e.g., pumpkins for food and pumpkins for stock). The ICC provides only a broad level structure for the classification of crops. For Groups 1–8, the ICC is consistent with the CPC at the group level and generally consistent at the class level. At the subclass level, the two classifications are similar. Note that the different levels of the ICC — namely, groups, classes, subclasses, and orders — do not relate in any way to the same terms used in the botanical taxonomic hierarchy.

The nine major groupings under the ICC are (from 1–9, respectively): (1) cereals, (2) vegetables and melons, (3) fruits and nuts, (4) oilseed crops, (5) root/tuber crops with high starch or inulin content, (6) beverage and spice crops, (7) leguminous crops, (8) sugar crops, and (9) other crops. While this classification scheme is new, a very similar approach was used to assign plants that were studied in the radioecological programs described in Section 5.5 and recorded into bioaccumulation databases.

5.3.2 Regional Food Consumption

Typical consumption rates for each of the crop types vary in different parts of the planet (Table 5.3). For example, root crops are much more prevalent in Africa, fruit in Oceania, and vegetables in Asia. Given that each of these classes will behave somewhat differently with respect to bioaccumulation of radioactivity (see Section 5.5.1), that factor will give rise to differing radiological dose estimates for those communities (see Chapter 7).

Similarly, meat and milk consumption varies regionally (Table 5.4). Bioaccumulation into various animal products also varies depending upon the tissue type, the organism, and the radionuclide (see Section 5.7). This will also have a bearing on regional radiological dose assessments (Chapter 7) for different areas of the planet.

5.3.3 Farming Systems

Across developing regions of the planet, which are more common in tropical and subtropical environments, the FAO/World Bank (Dixon et al., 2001) has identified 72 farming systems that can be grouped into 8 major categories. The criteria used to arrive at the 72 systems were fairly broad, given the

TABLE 5.3 Consumption Quantity of Crops (kg/capita/year) in 2003

Region	Cereals	Vegetables	Starchy Roots	Roots and Tubers	Pulses	Fruits
Africa	146	54	130	35	8	52
Asia	162	141	45	10	5	48
Oceania	87	90	95	21	1	130
South America	118	47	64	16	11	93
North and Central America	126	100	53	11	7	112
Europe	128	123	79	15	3	104
World	151	117	64	15	6	62

From FAO database (http://faostat.fao.org/site/345/default.aspx), 13 January, 2009.

TABLE 5.4 Consumption Quantity of Livestock and Fish (kg/capita/year) in 2003

Region	Meat	Eggs	Milk	Fish, Seafood
Africa	14	2	38	7
Asia	28	8	45	17
Oceania	103	5	169	22
South America	65	6	109	8
North and Central America	94	13	200	17
Europe	85	12	238	22
World	39	8	80	16

From FAO database (http://faostat.fao.org/site/345/default.aspx), 13 January, 2009.

inherent variability in global farming practices. This method basically comprised an evaluation of the available natural resource base (water, land, grazing area, and forest; climate, including altitude; landscape and slope; farm size, tenure, and organization), considering the dominant pattern of farm activities and assessing household livelihoods. Table 5.5, summarizing the major categories of farming systems, has been adapted from Table 8.1 in

TABLE 5.5 Categories of Farming Systems Identified by the FAO and World Bank, Including Details of Total Land Areas within Developing Regions of the Planet, Proportion under Agriculture, Sustained Population, and Relative Productivity

Category Characteristic	Small Holder Irrigated Schemes[a]	Wetland Rice-Based[b]	Rain-fed Humid[c]	Rain-fed Highland[d]	Rain-fed Dry/Cold[e]	Dualistic (Large/Small)[f]	Coastal Artisanal Fishing[g]	Urban Based[h]
Number of systems	3	3	11	10	19	16	4	6
Total land (Mha)	219	330	2013	842	3478	3116	70	n.a.
Cultivated area (Mha)	15	155	160	150	231	414	11	n.a.
Irrigated area (Mha)	15	90	17	30	41	36	2	n.a.
Irrigated/ cultivated (%)	99	58	11	20	18	9	19	n.a.
Agricultural population (M)	30	860	400	520	490	190	60	40
Agric. persons/ cultiv. ha	2.1	5.5	2.5	3.5	2.1	0.4	5.5	n.a.
Market surplus	High	Medium	Medium	Low	Low	Medium	High	High

n.a. Data not available
[a]*Irrigated farming systems embrace a wide range of food and cash crop production schemes.*
[b]*Wetland rice-based systems are generally dependent upon monsoon rains supplemented by irrigation.*
[c]*Rain-fed farming systems in humid areas of high resource potential are characterized by the type of crop activity (notably root crops, cereals, industrial tree crops, e.g., palm oil, and commercial horticulture, either on a small scale or as plantations, including commercial horticulture) or by mixed crop–livestock systems.*
[d]*Rain-fed systems on high altitude or steep slopes are most often mixed crop–livestock systems and include traditional cut and burn agriculture.*
[e]*Rain-fed systems in dry or cold, low potential areas are typically mixed crop–livestock and pastoral systems that become more sparse and dispersed, with reducing productivity, as aridity or cold increases.*
[f]*Dualistic systems are mixed large commercial and smallholder schemes. They cover a variety of ecologies and diverse production patterns.*
[g]*Coastal artisanal fishing typically comprises mixed systems.*
[h]*Urban-based systems normally focus on horticultural and livestock production.*
From Dixon et al. (2001).

Dixon et al. (2001). Table 5.5 includes generalized descriptions of the broad characteristics of each category.

The largest single category in terms of agricultural population is that comprising the wetland rice-based systems, predominantly in south and east Asia, supporting some 860 million people. However, the rain-fed systems, in total, support more people (>1.4 billion) and cover a much larger land area (>6 billion ha). A large proportion of the regions covered are at subsistence level or slightly better. The three categories that produce a reasonable market surplus account for only about 130 million people.

From this information, it is clear that billions of people across the tropics and subtropics are supported by agricultural systems much different to those traditionally applied in more developed regions of the planet. Hence, the 'typical' radionuclide bioaccumulation data derived for crops grown using developed agricultural systems in temperate climates should be used with distinct caution in tropical and subtropical regions. For dose assessment purposes, preference should be given to data that are specific to the farming system under consideration (see Chapter 7).

5.4. RADIOECOLOGICAL CONCEPTS AND MODELLING

On the assumption that nonspecialists are referring to this book, we need to define some terms specific to radioecology. Additional, underpinning knowledge on radioactivity, radionuclides, and the biological effects of radiation is provided in Chapter 1.

5.4.1 Transfer Factors

In order to help predict the concentration of radioactivity in any particular organism when it is exposed to a radionuclide in the environment, the concept of a transfer factor (TF) has been developed. A TF is also referred to as an equilibrium concentration factor (CF), concentration ratio (CR or C_r), or bioaccumulation factor (BCF), amongst others. The International Commission on Radiation Units and Measurements in their Report 65 (ICRU, 2001) has defined the C_r as 'the ratio of the activity density of a radionuclide in the receptor compartment to that in the donor compartment', and gave the term 'transfer factor' as an acceptable alternative. Given that most of the literature to date referring to plant root uptake of radioactivity uses the TF terminology, we will retain that common usage in this chapter on the understanding that it is equivalent to the ICRU (2001) recommended term, C_r.

Because any of these factors are applied to assess radiological exposure pathways, it should be noted that they generally refer to the edible portion of the organism (e.g., grain, fruit, leaf of plants, or flesh in animals). The idea behind this concept is that, if equilibrium has been achieved (we will come back to this assumption), then the degree of radionuclide bioaccumulation in the food item is proportional to the concentration of the radioactivity in the supporting environment. A dry weight (DW) soil-to-plant TF is defined by the following equation:

$$Transfer\ Factor(TF) = \frac{Radioactivity\ in\ crop(Bq/kgDW)}{Radioactivity\ in\ soil(Bq/kgDW)} \quad (5.1)$$

The radioactivity in the soil is taken to be the average activity concentration in the top 20 cm, or 10 cm in the case of grasses given their shorter root development. This is an accepted international compromise (FAO/IAEA/IUR, 1998) arising from alternate measures that are often based on deposition per unit area (Bq m^{-2}), assuming atmospheric fallout. TF values in excess of unity imply bioaccumulation of activity. Values less than one imply strong binding of the radioactivity to the soil (that the plant is not actively accumulating that material) or that, if accumulated by the plant, the radioactivity is not being distributed to the crop part that is measured.

When determining a plant-to-animal TF, the same basic ratio is applied (alternate methods for this assessment are discussed in Section 5.7), but generally using fresh or wet weight data on the basis that it is usually fresh plant material that is consumed and that meat or milk is typically supplied to the consumer in that fashion. Exceptions to this may apply when dried material (such as hay or grain) is used as fodder or if dried meat or dried milk is consumed. Alternatively, when studying mineral metabolism, ash weight comparisons have been used on occasion. In any case, the basis of the comparison (i. e., fresh, dry, or ashed) should be stipulated when the data are reported, because the ratio can change substantially in relation to the state of the material measured. TFs in animals can be related back to the original soil activity concentrations instead of, or in addition to, the supporting vegetation. In some cases, the values may also be related to the areal radionuclide activity of the soil (e.g., Bq m^{-2}) assuming a fallout event onto the soil surface. Care should always be taken when reporting literature data such that the ratio used is clearly understood and stipulated.

However, the assumption of equilibrium between radioactivity concentrations in the organism and the supporting substrate having been achieved in any real-world situation is moot. If the release of any radioactivity is event based (also known as an acute exposure), rather than evenly consistent or routine (chronic exposure), then there will inevitably be both a geographical and a time-related variation in the environmental concentration of the radionuclide(s) in question, building up rapidly near the source of the material at the time of release, and then declining gradually as the material is removed by erosion and other natural dispersion processes. More will be written on these processes in Chapters 1, 3 and 6. Even with chronic exposures, or exposures that occur more than once (e.g., regular or irregular releases from nuclear facilities), the environmental concentrations do not often remain constant, because the amount of radioactivity released into the environment will change depending upon the different activities occurring within the facility and because processes for transport and mixing of released materials into the broader environment are many and varied.

Further, the concentration within organisms will also change through time, irrespective of the external concentrations. This is controlled by a complex range of different physiological processes. The main factors are the rates of bioaccumulation and excretion for any particular contaminant and the proportional sequestration and biodistribution amongst the various tissue types (see Chapter 6 for similarities in freshwater environments). Other influencing factors include overall metabolism (biomass changes affect dilution of the absorbed radioactivity) and physiological changes related to health, sex, and age (which affect relative bioaccumulation and excretion rates of major nutrients, and their radioactive analogues, in response to hormonal factors). All of these factors control the biological half-life ($B_{t½}$) of the radionuclide in question.

5.4.2 Bioavailability, Bioaccessibility, and K_ds

The concepts of bioavailability and bioaccessibility should also be considered. For uptake by plants, bioavailability refers to the proportion of total activity in the soil that is in a form amenable for uptake by the organism. Any radioactivity that is in an insoluble form, or that is strongly complexed with soil constituents (e.g., soil organic matter: SOM), will not be available for accumulation or will be difficult to accumulate. The greater the proportion of non-bioavailable radioactivity, the lower will be the apparent TF, given that it is reported as a ratio. The converse is true for bioavailable radioactivity. Hence, if at all possible, TF values should refer to the bioavailable concentration of radioactivity, as well as to the total concentrations.

One commonly applied approach to estimating the bioavailability of radionuclides in soils is to use, or estimate, the typical distribution of radioactivity between the soil and soil solution, the so-called 'distribution coefficient' (K_d). This term, also referred to as the partition coefficient (ICRU, 2001), is defined by the ratio of a radionuclide's sorbed activity concentration (Bq/kg) to the dissolved activity concentration (Bq/L) in the soil solution at equilibrium, as seen in the following:

$$K_d(L/kg) = \frac{Sorbed\ radioactivity(Bq/kg DW)}{Dissolved\ radioactivity(Bq/L)} \tag{5.2}$$

The K_d is strongly influenced by the way it is measured. For example, results of experiments undertaken to estimate the parameter are readily affected by factors such as OM content, pH, Eh, the mass-to-volume ratio of the test system, and the soil particle size distribution, to name but a few variables. Discussions of the influence of these factors, and others, on radionuclide availability in soil systems are expanded upon in Section 5.5, as well as in Abril and Fraga (1996). In addition, Chapter 3 discusses radionuclide behaviours in soils at some length, and Chapters 4 and 6 discuss K_ds in aquatic systems.

Despite its shortcomings, the K_d approach does have the benefit of providing a general indication of the relative bioavailability of radionuclides between different soil types, as well as between radionuclides in the same soil (see Section 5.5). Also, much of the earlier literature referred to K_d; hence it may be useful when comparing data from different sources. The most recent and relevant papers on K_ds are Gil-Garcia et al. (2009a) for Sr and Cs, Vandenhove et al. (2009) for NORMS, and Gil-Garcia et al. (2009b) for Cd, Co, Ni, Zn, I, Se, Sb, Pu, Am, and other radionuclides.

Bioaccessibility is a similar concept to bioavailability, but in this case it does not refer to the form of the radioactivity, but rather to whether or not the organism is able to make use of the available radioactivity. For example, if radioactivity is spread across the surface of the soil (e.g., following a fallout event), then it will be less accessible to deep rooted trees than to more shallow rooted plants, such as grasses. Similar issues exist if the radioactivity is removed from the system by transport processes (e.g., leaching or erosion) more quickly than the organism is able to accumulate it.

Some researchers question whether the concept of a TF, or at least the FAO/IAEA/IUR (1998) protocol to evaluate it, is valid in any case. In one example, Centofanti et al. (2005), studying radionuclide uptake in maize, derived very different values for transfer from an experimentally prepared soil to plants growing in pots compared with transfer from a less manipulated sample of the same soil to plants grown in the field. They found a significantly higher uptake of ^{57}Co and ^{134}Cs in the plants grown in the field soil, whereas no differences in the transfer of ^{54}Mn and ^{65}Zn between the two conditions were observed. Hence, the experimental method of TF assessment may be overly subjective. These findings also reinforce the need to acquire site-specific data wherever possible. Some of the other major issues regarding the applicability of the TF concept relate to the bioavailability issue discussed earlier and how that may vary through time after any contamination event. This so-called 'ageing effect' is discussed in more detail in Section 5.5.5, as are the effects of redox potential (Section 5.5.6), which will also vary between field and pot studies.

Despite these constraints, generalized assumptions such as chronic exposure at average concentrations and environmental equilibrium allow us to develop predictive models of radioecological behaviour that can be used to compare elements, species, or ecosystems. They can also be considered reasonable to some degree because individual and small-scale variability is typically overwhelmed when considering entire communities or populations of organisms across regions. As part of those assumptions, simple uptake and loss rate kinetic models can be used to simulate bioaccumulation, as discussed in the next section.

5.4.3 Biokinetic Models

Assuming constant environmental concentrations of a radionuclide, uptake kinetics in whole individuals are often described by a single-compartment, first-order exponential model:

$$CF_t = CF_{ss}(1 - e^{-kt}) \tag{5.3}$$

where CF_t and CF_{ss} represent the CFs at time t (in days, d) and at steady state, respectively, following the introduction of a constant concentration of radioactivity into the environment, and k is the biological loss rate constant (d^{-1}) (Whicker and Schultz, 1982). The CF_{ss} value is equivalent to the equilibrium TF value, defined above (Eq. 5.1).

Loss kinetics can be expressed in terms of percentage of remaining radioactivity, or the radioactivity at time t divided by the initial radioactivity measured in the organisms at the beginning of the depuration (loss) period, assuming that the organism is now growing in a radionuclide-free environment. The simplest loss kinetics are described by a single-component exponential model:

$$A_t = A_0\, e^{-kt} \tag{5.4}$$

where A_t and A_0 are the activities (%) remaining at time t (measured in days, d) and at time zero (i.e., 100%), respectively, and k is the biological loss rate constant (d^{-1}). The determination of k allows the calculation of the radionuclide's biological half-life ($T_{b1/2} = \ln 2/k$). Radioactive decay needs to be considered in these estimates when the isotopic half-life is sufficiently short to have an effect within the time frame of the biological process.

It should always be noted that these models are, at best, simplifications of the real situation. However, they do provide a basis for generic modelling of radionuclide biokinetics within quasi-natural systems and are often included as a component in more complex radiological dose assessment models (see Chapter 7).

A generalized model that accounts for multiple uptake pathways for any individual (e.g., from different prey organisms for a predator), in addition to distribution and subsequent loss from multiple tissue types within any organism, is described more fully in the Chapter 6. However, those more complex biokinetics models are equally pertinent to diverse terrestrial systems and may be amended and applied as required.

It should also be noted that the generalized models assume that the equilibrium concentration of a radionuclide in an organism is linearly proportional to the (bioavailable) concentration of that radionuclide in the supporting environment. That is, for twice the concentration in the environment the amount in the organism will be doubled. There is mounting evidence that this assumption is oversimplified. In many cases the proportionality will be less than a 1:1 relationship and may often be nonlinear (e.g., Yasuda and Uchida, 1995). In those cases, some have suggested the use of a power law function to describe the equilibrium plant concentration (e.g., Blanco Rodrıguez et al., 2006). Thus:

$$C_p = fC_s{}^g \tag{5.5}$$

where C_p and C_s are the equilibrium radionuclide concentrations in plant and soil, respectively, and g is the proportionality constant. For linear TFs, g would be set to 1. Converting Equation 5.5 into TF notation gives the following:

$$\mathrm{TF} = fC_s{}^{(g-1)} \tag{5.6}$$

Nonlinearity can, to some degree at least, be explained by changes in the proportion of bioavailable radioactivity decreasing with increasing concentration (e.g., DeForest et al., 2007). Another example is given in some recently published results for radium uptake by the bush passion fruit (*Passiflora foetida*) in areas affected by uranium mining in tropical Australia (Supervising Scientist, 2009). In these field samples, the TF decreased to a minimum of 0.004 ± 0.0002 based on total soil radium (0.3 ± 0.02 using estimates of bioavailable radium) at activity concentrations in excess of 10^3 Bq/kg. This agrees with the general pattern for ^{226}Ra uptake by plants described by Simon and Ibrahim (1987, 1990). As for K_ds, several other factors discussed in more detail in Section 5.5 may also influence TF linearity.

However, the concept of TF linearity has also been noted to hold true in many cases, across a range of species, radionuclides, and concentrations (e. g., Sheppard, 1980; Tracy et al., 1983; Vandenhove et al., 2001; Blanco Rodrıguez et al., 2002, 2006). Sometimes this is qualified for evaluations made above some critical range, particularly for essential elements or their radionuclide analogues. For example, linearity holds true for uranium soil concentrations above 20 μg U g^{-1} on an ash weight basis (Sheppard and Sheppard, 1985), but below that it will underestimate the degree of bioaccumulation. Nonetheless, for the purposes of modelling, TF linearity can be reasonably assumed unless specific information to the contrary is available.

The simplified models are also dependent upon the environmental concentrations of the radionuclides being so low they do not substantially influence biochemical adaptation in the affected organisms. Many organisms have bioaccumulation mechanisms for toxins and nutrients that work to regulate for constant elemental concentrations within their tissues or limit their extent. This is achieved by a variety of mechanisms, including enhanced uptake (of essential elements) and/or exclusion, excretion, and sequestration (of contaminants). When these mechanisms are overwhelmed, regulation no longer functions adequately and linear uptake can result. Further detailed generic discussion of these types of mechanisms (albeit for aquatic examples) can be found in Luoma and Rainbow (2005) and Rainbow (2002), as well as in Chapter 6.

More detailed literature reviews on soil-to-plant radionuclide TFs, although not restricted to tropical environments, can be found in Simon and Ibrahim (1987) and McGee et al. (1996).

5.4.4 Dynamic and Mechanistic Models

When referring to radioecological models, the reader should be aware that a number of more comprehensive models have been developed to describe the environmental behaviour of radioactivity. These models comprise a range of scenarios for a range of habitats and integrate most, if not all, of the simplistic functions discussed earlier at variable levels of refinement. That is not to say they are widely applicable; all have limitations on the range of radionuclides and habitats that they will model.

The IAEA provides a database of such models and additional, related information at the Radioecology Web page http://www.iaea.org/OurWork/ST/NA/NAAL/pci/che/radioecology/radioecology.php. One frequent feature of such models is that estimates are provided of the activity concentrations in environmental components over time, following a release of radioactivity. From this, time-variable doses are also often estimated. Chapter 7 gives details for this type of modelling.

Good examples of dynamic conceptual models (albeit for temperate systems) are provided in Absalom et al. (1999) using soil characteristics to estimate Cs uptake, and Absalom et al. (2001), which particularly considers organic soils. Variables included in those models comprise soil activity concentration, clay fraction, organic matter (OM) fraction, exchangeable K, solution ammonia concentration, cation exchange capacity for both organic and inorganic fractions, aerial interception of Cs, and TF. The models were applied to predict Cs uptake in a range of crops and the results obtained showed reasonable agreement with available data for wheat, barley, potato, and cabbage.

Tamponnet et al. (2008) provided discussion of two mechanistic models: CHEMFAST and BIORUR (see also Casadesus et al., 2008). These models are used specifically for modelling radionuclide sorption/desorption from soil matrices and radionuclide uptake by, and release from, plants. The role of abiotic factors (soil components and soil structure) and biological elements (organic compounds, plants, mycorrhiza, and microbes) on bioavailability of cesium, strontium, technetium, and copper were considered. However, it should be noted that the models were developed and validated using information and data derived from nuclear accidents. In other words, their specific pertinence to tropical systems should be considered before application. Nonetheless, they are good examples of how radioecological principles can be applied in a modelling approach.

5.4.5 Key Radionuclides

A number of key radionuclides make up the majority of available TF data. The reasons for this are their relative radiotoxicities, their abundance, and their relatively long half-lives, both radiological and biological. Some are reported mainly as a consequence of their ubiquitous presence in natural systems, and some simply because of their relative ease of detection.

Of the natural radionuclides, data are most often available for U- and Th-series radionuclides, particularly those with longer half-lives, and ^{40}K. Potassium-40 is not of any particular radiological significance, but its gamma emission at 1460 keV is easily detected in most environmental samples; hence, it can be readily reported. Potassium is also a chemical analogue of Cs, of which ^{137}Cs is a key radionuclide (see the following material). Uranium and thorium have sufficiently long half-lives that they can be readily detected by either alpha spectrometry or the fairly common inductively coupled plasma mass spectrometry (ICP-MS) methodologies. These elements are also known to be concentrated in tropical soils by the processes described in Section 5.2. Rosholt (1982) and Rabesiranana et al. (2008) reported that the laterization process concentrates thorium and uranium, resulting in a content in some tropical soils that is greater than that in the original unweathered bedrock.

Radium-226 and ^{210}Po are the most commonly evaluated U-series nuclides on the basis of their potential radiotoxicities. ^{226}Ra is a divalent cation taken up as a biological analogue of the major nutrient, Ca. It is also a strong alpha emitter, as are most of its short-lived progeny, hence it can generate relatively heavy radiological doses in tissues where Ca tends to concentrate, such as bones and blood marrow-forming tissues in animals. Polonium can occur in a wide number of chemical forms due to its potential to appear in valence states of -2, $+2$, $+4$, and $+6$. As such, Po can be highly mobile in biogeochemical systems. It is readily bioaccumulated and is also an alpha emitter.

Of the anthropogenic radionuclides, those generated or enhanced by human activities (see Chapter 1), the longer-lived fallout nuclides ^{137}Cs and ^{90}Sr (or their shorter lived gamma-emitting isotopes ^{134}Cs and ^{85}Sr) and, to a lesser extent $^{239/240}$Pu and ^{60}Co (^{57}Co), are the most frequently evaluated nuclides. Cesium and Sr are analogues of the major nutrient elements Na/K and Ca, respectively. Cesium in particular can be biomagnified up the food chain, reaching much higher levels in top predators than those that occur in the original inorganic substrate. Plutonium is an alpha-emitting radionuclide that is almost entirely due to human activities in nuclear science.

Technetium is also only produced by human activities. As an example of its potential environmental consequences, a model for a proposed radioactive waste storage facility in subtropical Australia identified ^{99}Tc as becoming the major (albeit small) potential contributor to human dose in the order of 1000 years after burial (Harries et al., 1998). However, ^{99}Tc is distributed in the environment at very low concentrations and, being a weak beta emitter, it is hard to detect. Hence, only very limited data are available on this radionuclide, although some studies using a nonradioactive chemical analogue, rhenium, have been undertaken (e.g., Tagami and Uchida 2004, 2005).

Additional radionuclides that have been studied to a greater extent than others include ^{65}Zn, ^{131}I, and ^{3}H. Tritium (^{3}H) is a low-energy beta emitter, hence, it is of generally low radiological concern, although organically bound ^{3}H is considered somewhat more harmful as a consequence of its substantially longer retention time within biological systems, compared to unbound ^{3}H. Tritium, whilst produced naturally by cosmic radiation in the upper atmosphere, was greatly enhanced and widely distributed in the environment as a consequence of atmospheric nuclear weapons testing in the 1960s and has been continuously output by nuclear reactors since that time. ^{131}Iodine, produced as a fission product, is of particular interest as an essential trace nutrient that concentrates in human thyroid glands. For ^{131}I, exposures are typically via the grass-cow-milk-human pathway or, in an aquatic example, by concentration in edible seaweeds. Zinc is also regularly reported. Despite being an essential trace element, the radioactive isotope ^{65}Zn is not widely distributed in the environment and its reporting may well be a consequence of its relatively easy production in nuclear reactors by neutron activation of stable Zn, and because it is easily detected by its abundant, strong gamma emission at 1115 keV.

5.5. COMPILATION AND EVALUATION OF TFs FOR TROPICAL AND SUBTROPICAL SYSTEMS

Quite a number of studies worldwide have been undertaken to derive TF values, often using the equations outlined in Section 5.4. These data were initially compiled in 1994 when the IAEA published the 'Handbook of Parameter Values for the Prediction of Radionuclide Transfer in Temperate Environments' (TRS No. 364) in collaboration with the IUR (IAEA, 1994). Environmental studies after the Chernobyl accident contributed to improved knowledge of the behaviour of artificial radionuclides in temperate areas, supplying additional relevant information for the management of contaminated agricultural areas.

Since that time, it has been recognized that most of the data acquired were related to northern, temperate zones and had less applicability in other regions of the planet. Hence, a number of activities have been undertaken to address those shortcomings, mostly facilitated by the IAEA in collaboration with partnership organizations such as the FAO and the IUR, including nationally supported research projects in a number of tropical and subtropical countries. These activities have included focused, region-specific studies through IAEA-led CRPs and data collation, modelling, and interpretation through programs such as the Environmental Modelling for Radiation Safety (EMRAS; http://www-ns.iaea.org/projects/emras/) project and the International Biosphere Modelling and Assessment (BIOMASS) project. The outcome of those efforts included new data, improved models, and a revision of the TRS No. 364 handbook (IAEA, 1994) to include more tropical data and data for freshwater environments (IAEA, 2009).

Following the amalgamation of the data, it became apparent that there was a high degree of variability within and between subsets of data. To try to explain at least some of that variability, diverse environmental factors pertaining to the sample were evaluated for their potential to influence the TF data. A comprehensive review of environmental processes affecting plant root uptake of radioactive trace elements and variability of TF data has been provided by Ehlken and Kirchner (2001). The following sections elaborate specific aspects that have substantial influence and how they may be affected by climatic region.

5.5.1 Effects of Plant and Soil Type on Bioaccumulation

One of the first observations made of TF values for most crop classes and soil types, across the board, was that they covered several orders of magnitude (IAEA, 1994). This suggested that the variables controlling uptake were numerous and would take some time to decipher. The pattern of widely differing TFs remained consistent even when only tropical and subtropical data were considered in isolation. Velasco et al. (2008), using the data compiled for the revision of TRS No. 364 (TECDOC 1616; IAEA, 2009), discerned a minimum value of two orders of magnitude difference across the range of TFs for Co and Zn and up to five orders of magnitude across the range of values for Ra and Cs. In Figure 5.2, reproduced from Uchida (2007), some of the data acquired through IAEA/FAO/IUR CRPs designed to assess TFs in tropical and subtropical regions of the planet are shown. These CRPs were led by Dr. Martin Frissel, previous European Secretary of the IUR. These data show TFs for Cs ranging from 0.0001 to 10, in other words, across approximately five orders of magnitude, agreeing with Velasco et al. (2008).

However, Figure 5.2 also displays another factor that has been observed as generally consistent within both temperate and tropical systems. That is, some classes of crop take up radionuclides to a greater or lesser extent than do others. Figure 5.2 shows that leafy vegetables generally accumulate more Cs than do beans and peas, and more again than grains and cereals. Other radionuclides showed similar differences between crop types. This observation led to the formation of a subsequent CRP entitled ‘Classification of Soil Systems on the Basis of Transfer Factors of Radionuclides from Soil to Reference Plants.’ This second CRP contributed further TF data to the FAO/IAEA/IUR database from which the IAEA (2009) revision partly derived.

Despite these observations, it needs to be recognized that using plant type as a potential classification scheme for TF evaluations involves very broad generalizations. Some classes of plants typically have a small range of potential TF values, whilst others have much wider ranges. Figure 5.3, taken directly from Velasco et al. (2008), shows this spectrum of results for cesium. In these data, the highest geometric means for Cs TFs are provided by non-leafy (fruits) and leafy vegetables, whilst those for grasses (seeds or stems)

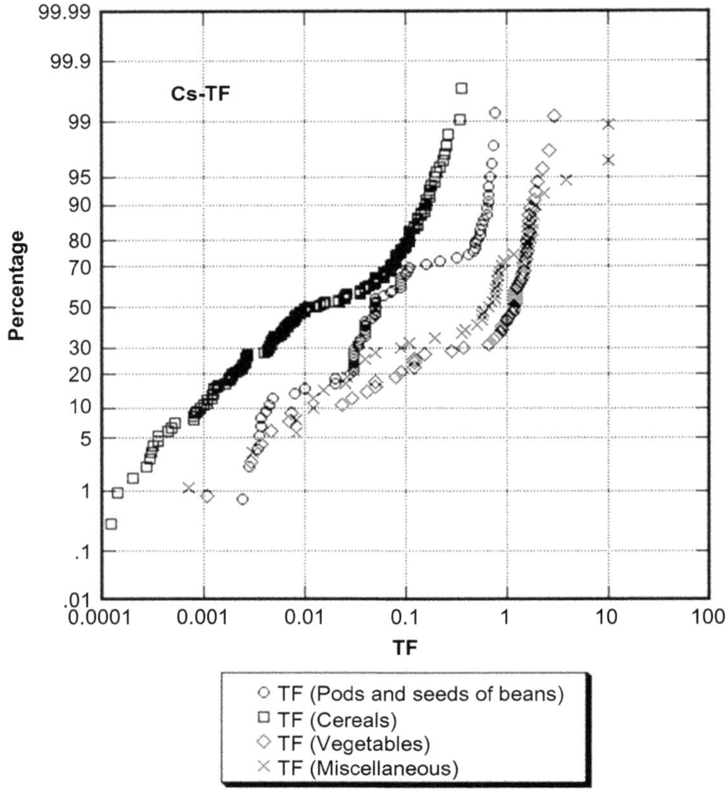

FIG. 5.2 Probability distributions of soil-to-plant TFs for Cs obtained for tropical and subtropical areas by the CRPs (taken directly from Uchida, 2007).

are amongst the lowest. In contrast, the grass seeds have also recorded the highest Cs TF values for any crop type.

The recent statistical analysis of the available TF data for tropical and subtropical environments by Velasco et al. (2008) also gave rise to the generic observation (amongst others) that soil type seemed not to have a decisive influence on TF values. This is at odds with the observations made of the data by the participants in the IAEA CRPs (Table 5.6) and also in relation to the evaluation of all radionuclide uptake data for fruit plants (as referred to in Section 5.6.2).

One of the specific reasons for the apparent discrepancy was that Velasco et al. (2008) applied the simplified codification of the soil type used by the CRP researchers when reporting to the database. In that case, the soil type was reported as sand, loam, clay, or organic, or a mixture thereof, as per the experimental protocol agreed at the start of the study (FAO/IAEA/IUR, 1998). The experimentalists providing the data eventually became aware that more refined descriptions of soil types were required. It is for this reason that

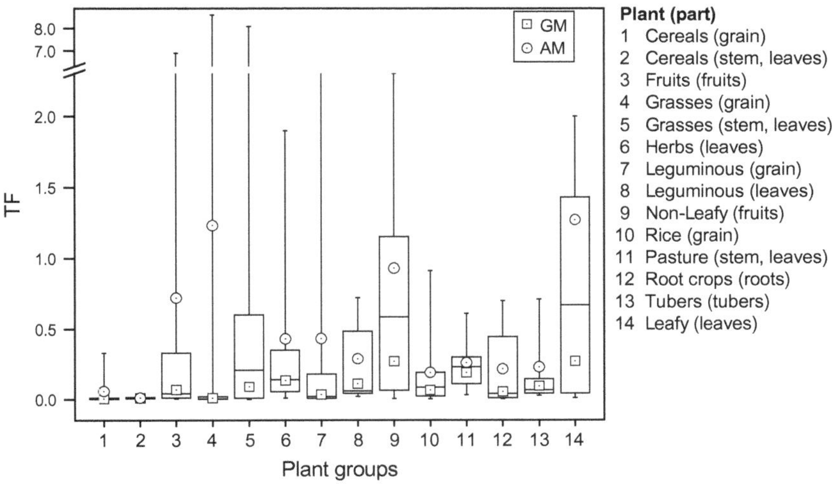

FIG. 5.3 Box-plot ranges of caesium TF values for different plant groups. The arithmetic (AM) and geometric means (GM) are also indicated (taken directly from Velasco et al., 2008).

TABLE 5.6 Generic Findings of the IAEA/FAO/IUR CRP 'Radionuclide Transfer from Air, Soil, and Freshwater to the Food Chain of Man in Tropical and Subtropical Environments'

1. Soil type is one of the dominant factors influencing soil-to-plant radionuclide movement.
2. There are no systematic differences of TFs between climatic zones. However, some extreme values have been reported in individual environments, particularly in the tropics.
3. Chemical factors, such as soil pH, affect uptake of Cs and Sr in tropical systems. Lower pH values generally increase plant bioavailability.
4. Increased potassium or calcium concentrations in soils generally decrease the plant uptake of Cs or Sr, presumably by competitive inhibition by analogous essential elements.
5. TFs are nuclide specific; therefore, anomalous plant uptake for a particular radionuclide does not necessarily indicate that a similar discrepancy will be observed for another radionuclide under the same conditions.
6. Some types of crop typically have higher TFs for a range of radioisotopes than do others (e.g., leafy vegetable compared to fruit).
7. If an unusually high or low TF for a nuclide is observed for one crop type on a particular soil, the same general trend is often observed for all crops studied in that environment.

the previous sections referring to WRB classification (Section 5.2.2) and the constraints on tropical soils in particular (Section 5.2.3) were included in this chapter.

Evidence in support of these divergent observations is given in Figure 5.4a and b, which compare data from the first tropical CRP with earlier data from temperate climates: in these examples, plotting Cs or Sr TFs against soil pH. It is apparent that the data from tropical systems, whilst on average the same as the earlier temperate data (agreeing with Velasco et al., 2008), tend to sit at the periphery of the other values and hence may be considered more extreme

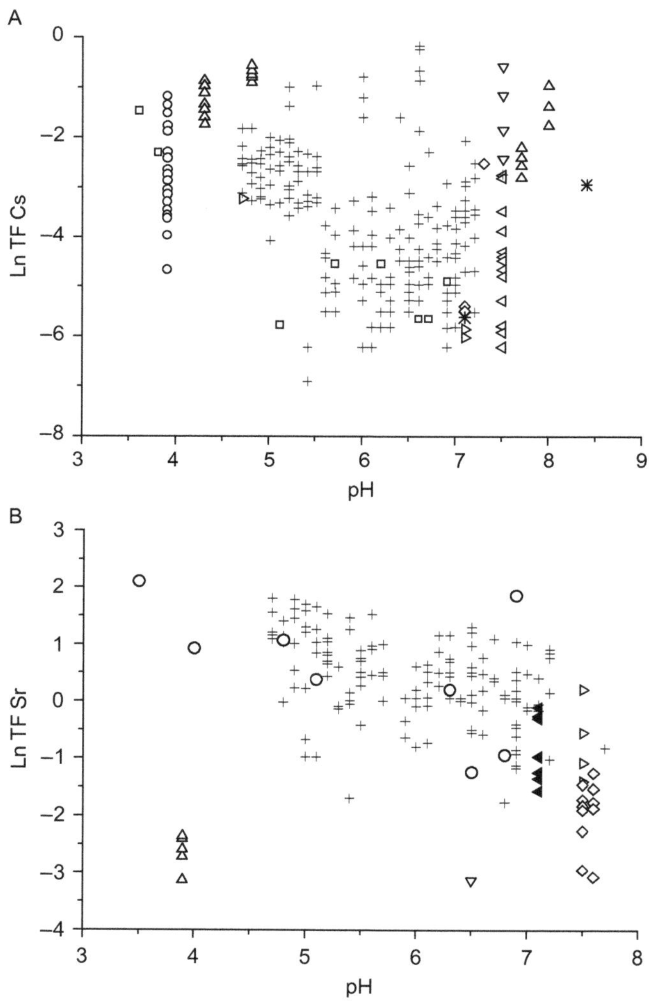

FIG. 5.4 Transfer factor (TF) values for A) Cs and B) Sr for a range of crops, plotted against pH. The data from temperate climates (+) are compared with values from the more recent tropical studies (all other symbols) (adapted from CRP notes by M. Frissel).

than the temperate data (as described by the members of the CRP). When TF comparisons are restricted to individual specific soil types, it is apparent that TF data are often more consistent within themselves.

A short compilation of TF data for ^{226}Ra and ^{137}Cs compiled from studies in tropical areas is compared with the IUR data for temperate regions in Table 5.7. From this limited number of direct comparisons, it is apparent that, in some cases at least, TFs in tropical areas can be markedly different for the same crop grown in temperate environments. In the example from Africa (Delmas et al., 1987), the difference is two orders of magnitude.

The following sections will evaluate the various factors pertaining to soil systems that may be affected by climatic conditions. These influences will result in changes in the rate and extent of elemental bioaccumulation from soil by plants and, in turn, affect the TF values subsequently determined.

5.5.2 Effects of Soil Chemistry on Bioaccumulation

As stated earlier, tropical soils are often substantially different from the temperate agricultural soils that comprise the bulk of the data reported in the literature. Some of the extreme tropical TF values shown in Figure 5.4a and b were obtained from environments such as coral soils (e.g., Robison et al., 2000; Simon et al., 2002); soils with high iron or aluminium oxide

TABLE 5.7 Soil to Plant TF in Tropical Areas Compared with Temperate Soils (IUR Data) for ^{137}Cs and ^{226}Ra (mean, *n*, range)

Crop	Radionuclide	TF	IUR Data	Reference
Beans	^{137}Cs	2.6 E-1(*n* = 16) [6.0E-2–4.1E-1]	2.9 E-2 (*n* =90) [1.6E-1–1.0E-3]	Wasserman, 1998
Beans	^{137}Cs	2.4 E-1 (*n* = 15)	2.9 E-2 (*n* = 90) [1.6E-1–1.0E-3]	Scardino and Helene, 1990
Beans	^{226}Ra	6.6 E-2(*n* = 4) [1.2E-2–9.1E-2]	6.0 E-3 (*n* = 6) [6.3E-3–1.4E-2]	Lima, 1988
Peanuts	^{137}Cs	1.0 E+0	3.0 E-2 (*n* = 84)	Delmas et al., 1987
Radish	^{137}Cs	1.6 E+0 (*n* = 5) [4.6E-1–3.5E+0]	5.5 E-2 (*n* = 8) [1.0E-3–1.3E-1]	Wasserman and Belém, 1996
Carrot	^{137}Cs	1.0 E+0 (*n* = 5) [8.5E-2–1.5E+0]	1.5 E-1 (*n* = 4) [2.6E-2–2.6E-1]	Wasserman and Belém, 1996
Carrot	^{226}Ra	9.2 E-2 (*n* = 4) [3.0E-2–1.5E-1]	4.1 E-3 (*n* = 4) [2.5E-3–5.7E-3]	Lima, 1988

content (e.g., Selagen, 1971; Wasserman et al., 2002); soils with extreme pH values (Mollah et al., 1998); and irrigated soil such as in rice paddies (e.g., Uchida et al., 2007; Tsukada, 2002b).

The atoll soils of mid-oceanic islands are an excellent example of the diversity available in tropical systems. These soils are basically coralline in origin and have a very high $CaCO_3$ content. The CEC of these soils is almost entirely correlated with the organic matter content (Walker et al., 1997). The high Ca concentrations affect the bioavailability of Sr to the plants by competing for biological uptake sites. This is not only due to saturation of the uptake mechanisms, but also to the fact that Ca, as the major nutrient, is preferentially suited to the uptake sites at the biological interface with the external environment, as a consequence of physicochemical attributes such as ionic radius (e.g., Choi et al., 2001). These factors combine in coralline soils to generate extremely low Sr TF values despite the low pH; in other words, against the general trend for increased Sr TF with increasing acidity (Figure 5.4b, Section 5.5.4).

Coral soils also have a very low K concentration. Simon et al. (2002) reported the mean ^{40}K concentration in Marshall Island soils as 10.7 ± 0.5 Bq/kg with a range over 200 samples from 2.3 to 55 Bq/kg. This compares with soils of volcanic origin that normally contain 40–1100 Bq/kg ^{40}K (Eisenbud, 1987). Cesium, as a monovalent cation, is a chemical analogue of the macronutrient K. Hence, the K deficiency in the atoll soils leads to a strong uptake of other available monovalent elements, such as Cs and thus to very high Cs TF values. Robison et al. (2009) showed a dramatic decrease in ^{137}Cs uptake into the fresh fruit of coconut trees growing in Marshall Island soils following the amendment of soil by the addition of K. The concentrations of ^{137}Cs in the tree trunk did not decrease, which reflects the generally low monovalent ion concentrations in the environment and the plant's need to retain K and its chemical analogues. However, the lower ^{137}Cs concentration in the newly produced fruit tissues following the amendment with K clearly indicates competitive inhibition of Cs uptake under these circumstances.

5.5.3 Low-activity Clays and Cationic Exchange

Cation exchange capacity (CEC) is a measure of the ability of the soil to adsorb positively charged ions. It arises as a consequence of the presence of materials such as organic matter and clays in the surface soils. Clays are aluminosilicates in which some of the Al or Si is replaced by elements with a different valence. For example, aluminium (Al^{3+}) may be replaced with iron (Fe^{2+}) or magnesium (Mg^{2+}) leading to a net negative charge on the clay. This charge will attract and bind cationic species such as Cs^{+} and Sr^{2+}. Organic ligands in the soil solution, which complex cations, provide a similar mechanism to reduce the bioavailability of radioactivity that has been introduced to the soil.

As was pointed out previously, the more intensive rain and higher temperatures of tropical and subtropical areas promote faster and more extensive degradation of 2:1 clay minerals into simpler structures such as kaolinite, goethite, hematite, and gibbsite, leading to the depletion of most 2:1 clay mineral types in the upper profile (Figure 5.5).

The clay degradation solubilizes cations that are subsequently leached from the superficial layers to deeper layers. This process simultaneously promotes the development of acid conditions in these soils and the loss of fertility (Section 5.2.3). Also, the specific surface area, as well as the quantity of permanently charged sites, diminishes in soils where oxides and 1:1 clay minerals like kaolinite are predominant, leading to reduced CEC. In soils where goethite, hematite, and gibbsite are the main minerals in the clay fraction, the charges are most dependent on pH that determines, for instance, a lower retention of cationic forms under low pH, as illustrated in Figure 5.6.

In the case of ^{137}Cs, the following impacts are expected to occur under these soil conditions:

- Absence of competition with K due to very low K concentration in these soils, enhancing Cs uptake by plants. Agricultural amendment with K diminishes this effect, and decreases the Cs TF (Boaretto et al., 2006).
- Nutritional stress due to the low fertility of soils, promoting low biomass production and significant contaminant concentration. However, due to the excellent physical properties of these soils, excellent biomass response

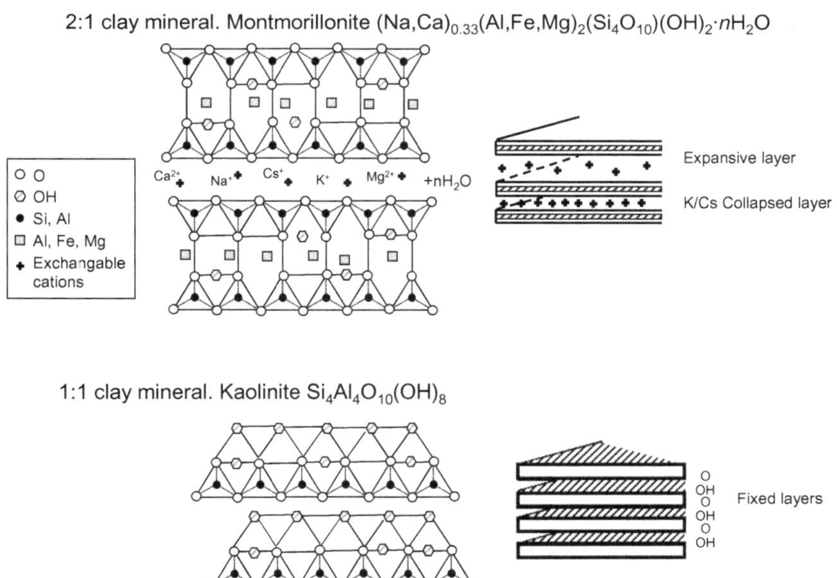

FIG. 5.5 Illustration of a 2:1 clay mineral type and a 1:1 clay mineral type.

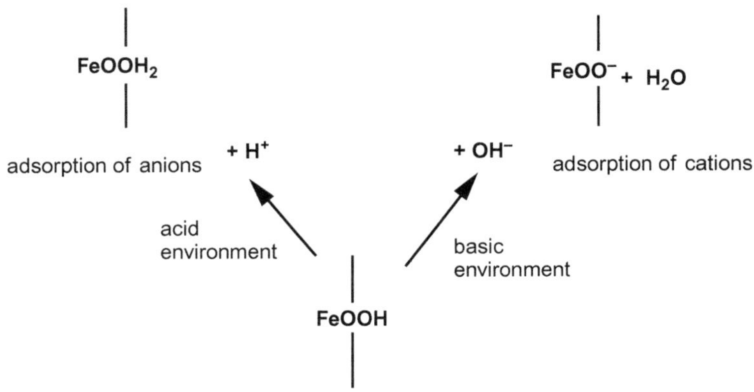

FIG. 5.6 Illustration of charge dependence on pH for goethite.

and radionuclide dilution can be achieved with normal agricultural practices (Wasserman et al., 2008a).

- Absence of irreversible fixation due to the deficiency of clay minerals with double, or expansive, layers.

The 2:1 clay mineral type, when saturated by K or elements with similar ionic radii such as ^{137}Cs, undergoes an irreversible union of the expansive layers, sometimes referred to as a collapse. By this mechanism, all elements that have penetrated into the between-layer space of the 2:1 clay mineral (Figure 5.5) are effectively trapped. Hird et al. (1996) pointed to the role of clay mineral type and the collapse of the expansive layer for irreversible fixation of ^{137}Cs in upland soils from Britain that were contaminated following the Chernobyl accident. This irreversible and progressive fixation of ^{137}Cs into 2:1 clay mineral types can occur in the majority of temperate soils, even if the process depends on K^+ ions (or elements with similar ionic radii to Cs) to induce the collapse. However, the process cannot occur in soils where 2:1 clay minerals are absent, as in the case of Ferralsols and Acrisols across the tropics. It can arise to a minor extent where these minerals occur only in trace quantities in subtropical soils, such as Nitisols, as observed by Wasserman et al. (2008b). These authors point out that in well-weathered soils, adsorption is reversible and dependent on pH and Eh, and are controlled mainly by iron and aluminium oxides, the main mineral compounds present, according to the model described in Figure 5.6.

5.5.4 pH Effects

Soil pH is primarily influenced by the chemistry of the parent rocks and minerals that have produced the soil, in addition to dissolved inorganic carbonates from rainwater and dissolved organic carbonates from decomposing biological material deposited on the soil surface. Soil pH often rises under flooded conditions due to the volatilization and denitrification of soil nitrogen. Plants can also directly influence pH in the vicinity of their roots (Hinsinger et al., 2003), just like the microflora of fungi, bacteria, and algae that are always present (see Section 5.5.7).

As stated earlier, the general effect of reducing soil pH values is to increase TF values. This is a consequence of the ability of increasing H^+ ion concentrations to, by and large, displace other cations from binding sites on organic ligands and mineral matrices within the soil. This makes those charged ions more bioavailable for uptake by the plant roots. However, it is apparent from Figure 5.4a and b that the effect is more pronounced for Sr than it is for Cs. This is because the degree of displacement is not consistent for each element and the effect varies for each different soil type. For example, the effect of pH on the distribution of Cs, Sr, and Zn between the particulate and dissolved phases in two similar agricultural soils (both arenic Acrisols) from tropical Australia (Twining et al., 2004) is shown in Figure 5.7.

It is apparent that, for the soils in Figure 5.7, there is little pH dependence on Cs adsorption, whereas for Sr, there is a change of almost an order of magnitude over the same range of pH values. The effect on zinc is even more marked with the adsorption of the radionuclide-to-soil particles growing by more than a factor of 100 as the pH of the solution increased from 4.3 to 6.8. The more strongly any radioactivity is adsorbed to particulates and other matrices within the soil solution, the smaller the proportion of bioavailable radioactivity and, consequently, the lower the apparent TF.

Wasserman et al. (2008b) studied geochemical partitioning of ^{137}Cs in tropical and subtropical soils and observed the important affinity of this radionuclide for Fe and for Al oxy-hydroxides (in the alkaline phase); these mineral inclusions constituting an important sink for Cs from soil solution. As the predominant minerals in the clay fraction of well-weathered soils are iron and aluminium oxides, it is expected that changes in pH will modify the charge of the soil. Consequently, these changes can lead to retention or desorption of ions depending on the direction of change. For example, an abundance of negative charges will be reached under basic soil conditions that, in turn, will retain more Cs^+ than under acidic soil conditions (Figure 5.6).

A good example of the pH effect on TF values under tropical conditions is given in Figure 5.8. The reduction of TF observed in Ferralsol was due to the increase of soil pH from 4.3 to 7.4 due to liming four years after contamination. Confirming this mechanism, contrary behaviour was observed for Goiânia soil that was initially basic (pH = 8.3) but then had its pH reduced to 7.7. This acidification slightly enhanced root uptake due to desorption of Cs (see the difference of TF values between 13 and 17 years after contamination in Figure 5.8). That behaviour can be explained by the effect of pH on the charges of iron oxide (see Figure 5.6) favoring stronger adsorption under more alkaline conditions or desorption when pH diminishes.

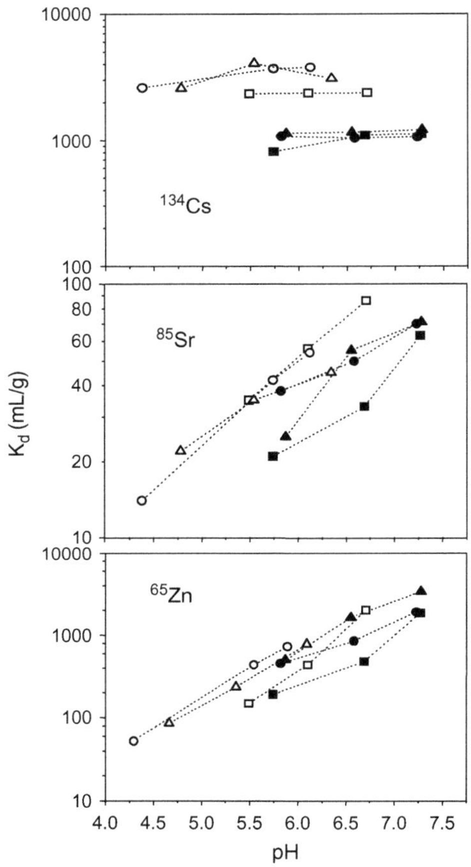

FIG. 5.7 Soil-solution distribution coefficient (K_d) values for adsorption of radionuclides on two Acrisols (solid and open symbols) from northern Australia. The central data points in each data-set are for experiments carried out at the equilibrium pH of the samples. The other two data points for each soil were obtained at pH values approximately one unit higher or lower.

As a consequence of wet and warm conditions at the experimental area, the Goiânia soil underwent a natural but progressive leaching process. Hence, a few years after decontamination, the pH started to change, as did the K content, leading to soil properties very similar to those of Ferralsols. The Ferralsol sample, in turn, received a phosphogypsum amendment to improve soil fertility that raised pH. From 2000 to 2008, the pH and K content in Ferralsol and in Goiânia soil remained very similar, as did ^{137}Cs soil to plant TFs for these two types of soils.

These results provide evidence that tropical soils deserve a special approach and that values from temperate climate areas should not be used in local dose assessments without due care. Based on 13 years of experimental studies with ^{137}Cs in Brazilian soils, Wasserman et al. (2008b) showed that

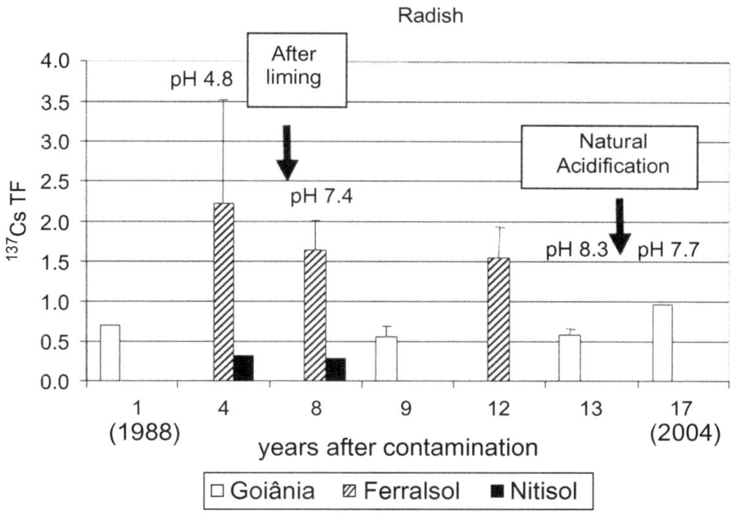

FIG. 5.8 Variation of ^{137}Cs soil to plant TF for radish with time as influenced by pH (adapted from Wasserman et al., 2009). Data for ^{137}Cs TF in the first year of contamination is from Lauria et al. (1994).

climate can have a role in the behaviour of radionuclides because soils from a tropical area (Ferralsols) presented high ^{137}Cs TF and no significant ageing effect (Section 5.5.5), whilst in soils from subtropical areas (Nitisols), ^{137}Cs showed a very similar behaviour to that observed in temperate soils affected by the Chernobyl accident.

5.5.5 Ageing Effects

The information in Figure 5.8 is also an example of the aging effect occurring in tropical and subtropical soils. In this case, the TF for Cs evolves over time in two types of tropical soil where 2:1 clay minerals are absent (urban soil from Goiânia and a Ferralsol) and one subtropical soil (Nitisol) with vermiculite in the clay fraction. More than 10 years after contamination, the tropical soils still present ^{137}Cs TF values one order of magnitude higher than the TF values recommended by the IUR for radish (0.05) to be used for dose assessment in the absence of regional values (IUR, 1989). On the other hand, the lower ^{137}Cs TF values that were measured in the subtropical Nitisol have not changed since contamination. This observation confirms a different mechanism controlling radionuclide behaviour in Nitisol, as compared with the other tropical soils. The effect is probably due to the presence of vermiculite which, although in trace quantities, traps Cs in the first years after contamination similarly to that generally observed in temperate soils less than three years after contamination (Riise et al., 1990) (see Section 5.5.3).

It is evident from the Brazilian studies just discussed that it is important to consider time since contamination as a factor when assessing TF data. Not accounting for this is one of the reasons that some individuals question the validity of the TF concept and experimental TF data, particularly those derived from pot (and to a lesser extent lysimeter) testing rather than field results (see also Sections 5.4.1 and 5.5.6).

Despite the previously mentioned result for Goiânia, it is known from many temperate climate bioaccumulation studies that TFs generally tend to decrease with time, even after accounting for radioactive decay and removal of radioactivity by processes such as erosion and crop harvesting (i.e., the radioactivity becoming less bioaccessible). This ageing effect can best be explained by reduction in the bioavailable fraction of the remaining radioactivity.

The implied reduction in bioavailability is the consequence of two processes. First, the more bioavailable fractions tend to be more geochemically mobile. Hence, they will be removed from the soil system more readily than the less bioavailable species. The second factor relates to the time that it takes for adsorption or complexation of applied radioactivity to soil components. Any chemical process takes some time to progress toward equilibrium; in some cases this period is substantial. Thus in the longer term, a higher proportion of the applied radioactivity will become unavailable due to a variety of adsorption processes. This effect on TFs is clearly evident in Figure 5.9, taken from a field study of Cs and Zn uptake in tropical Australia (Twining et al., 2003).

Frissel et al. (2002) recognized the need to account for the ageing effect when making dose assessments by generating reference TFs for a range of soil conditions following chronic long-term (i.e., steady state) releases as distinct from acute accidental releases. A selection of these values is given in Table 5.8.

5.5.6 Eh Effects

Some elements show a degree of redox sensitivity. Those of most interest from a radiological perspective are I, Tc, and Pu. Particularly in the monsoonal tropics, there may be a substantial shift in the aerobic state of the soil, and hence of the speciation of these elements, between the beginning and end of the growing season. These shifts depend upon soil porosity, the degree of soil water saturation, microbial activity, and other factors, including soil nutrition and chemistry. Given those circumstances, it is appropriate to consider the consequential possibility of seasonal changes in plant availability of the redox-sensitive elements in any radiological dose assessment studies. As chemical species often become more mobile when reduced, these radionuclides may become more bioavailable during the wet season, at which time most plants are most productive. Hence, redox reactions will tend to increase bioaccumulation of radioactivity, particularly in monsoonal systems.

This factor also has broader relevance to the types of experimental conditions that have been used to derive TF data, which range from pot studies to lysimeters

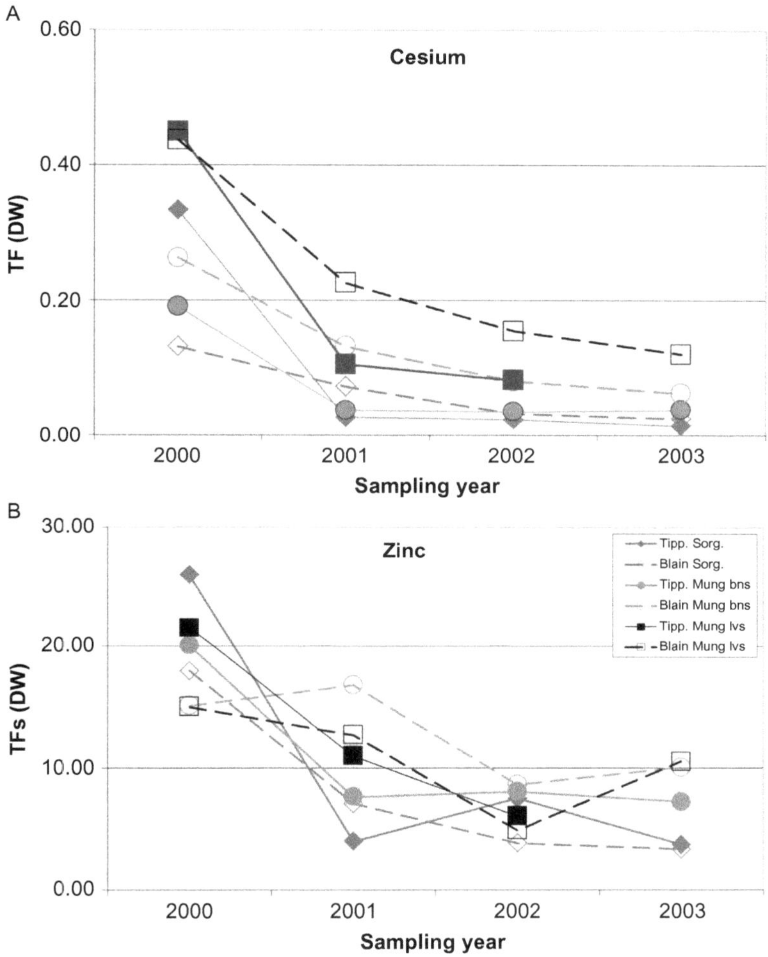

FIG. 5.9 Variability in TF for ^{134}Cs and ^{65}Zn in sorghum and mung plants over a period of three years following field application of radioactivity to two soils in the Northern Territory, Australia (Twining et al., 2003).

and up to field applications (see Section 5.4.1). Pots will be much more likely to remain well aerated and hence be less prone to redox changes than field experiments. Lysimeters, which typically contain a large soil mass, are more likely to compare favorably with field studies in relation to redox state.

It can be seen from the previous discussion that the bioavailability of elements and their radioactive isotopes can vary over orders of magnitude depending upon the chemical nature of the soils in question. Hence, these effects on the bioavailability of soil radioactivity are some of the largest factors determining the spread of TF factors that have been observed.

TABLE 5.8 Reference TFs of Cs for Cereals (DW)

Soil Condition for Clay and Loam Soils	Soil Type	TF for Routine Releases	TF for Accidental Scenario
High nutrient status, pH > 4.8	All soils	0.006 (0.002–0.01)	0.05 (0.02–0.1)
Medium nutrient status, pH > 4.8	Sand, peat, and other soils	0.05 (0.02–0.1)	0.2 (0.1–0.5)
Low nutrient status, pH > 4.8	Sand and other soils	0.3 (0.1–1)	0.7 (0.2–2)
Low nutrient status	Exchangeable K <0.05 cmol(+)/kg	2 (1–5)	5 (2–10)
Low nutrient status OR pH < 4.8	Wet, gleyic	3 (1–10)	8 (2–20)

Expected value (range); DW, dry weight.
Adapted from Frissel et al. (2002).

5.5.7 Effects of Soil Microbiology on Bioaccumulation

Healthy agricultural soils have a wide range of microorganisms as part of their ecosystem. Of particular interest in tropical systems are the plant-associated microflora of bacteria and fungi. These organisms are primarily responsible for the breakdown and recycling of organic matter and nutrients, which occurs very rapidly in tropical and subtropical systems. Several genera of bacteria also perform the important role of nitrogen fixation; a vital function in soils that regularly suffer from low nutrient status which is more common in tropical and subtropical systems (Table 5.2). In order to perform adequately, most bacteria require both moisture and a carbon source as a food substrate. In legumes, these resources are provided by the host plants, but for most other plant species organic matter becomes a limiting factor. Hence, any material that falls from the supporting vegetation is rapidly recycled. As radioactivity may be introduced into an ecosystem via foliar uptake, the recycling of fallen foliage is a mechanism for retention of radioactivity by plants.

Fungi in temperate systems are known to have relatively high Cs TF values (IAEA, 1994; Duff and Ramsey, 2008). Strontium TFs in fungi tend toward lower values than observed for higher plants, commensurate with reduced Ca uptake. However, fungi are very dependent on moisture and organic carbon. To that end, there are several genera that are symbiotic with plants, across all environments, but perhaps more critically so in the wet/dry tropics. Vesicular arbuscular mycorrhiza (VAM) and some other fungi associated with plant roots facilitate nutrient uptake by greatly enhancing the effective surface area of the root system (Garrett, 1981). It may be that, under such circumstances, radionuclides with chemical similarities to macro- and micronutrients will also be assimilated in greater quantity. Examples of this potential effect include radioactive Sr, Co and Ra for the macronutrients Ca, Cs for K, and Tc for the micronutrient manganese. A potential corollary is that increased plant performance related to fungal symbiosis may lead to greater above-ground biomass and hence lower plant TFs (by biomass dilution) despite greater overall uptake of elements. This latter effect is accounted for in dose assessment models (see Chapter 7). As with most biological systems, the relationship between plants and fungi in relation to bioaccumulation is not straightforward. An excellent review of the available information concerning Cs uptake by these systems is provided by Dupre´ de Boulois et al. (2008).

Fungi are effective bioaccumulators of a wide range of elements as a consequence of their ability to substantially influence the chemistry in the microenvironment surrounding their hyphae. They do so by pumping protons from their apical growth zone, which assist in the breakdown of organic matter and the solubilisation of mineral nutrients. These dissolved materials are then resorbed by the fungi in the adsorption zone of the hyphae (Garrett, 1981). The sensitivity of fungi to soil moisture variability in the wet/dry tropics was indicated in Twining et al. (2004) where fungal colony counts in field-collected samples of agricultural soils from the Northern Territory of Australia showed a 10 times increase over the wet season. Bacterial colony counts of the same samples showed no significant change.

The productivity of microbial systems together with plant activity in soils will also influence the chemistry of those systems. Plant exudates may significantly alter the local pH value or ligand concentrations and, as such, may result in release of radionuclides from the adjacent mineral surfaces and uptake to the plant. This has been reported in the case of radioiodine transfer of rice plants, where locally reducing conditions together with the presence of soil microorganisms was affecting radioiodine transfer (Muramatsu et al., 1990). The proton pumping of fungi will decrease pH and microbial respiration will also generally affect the aerobic conditions of the soil. Depending upon their porosity and soil moisture content, soils will tend to become more anaerobic, and hence chemically reduced, as a consequence of increased microbial activity. Redox state and pH are both known to influence TF (Sections 5.5.4 and 5.5.6).

For arid and tropical soils, Russell et al. (2004) studied the influence of microbes on ^{137}Cs availability and found that bacterial sulphate reduction substantially decreased adsorption, making Cs more available to plants, although extended incubation resulted in greater adsorption to the regolith. In a somewhat contrary finding, Parekh et al. (2008) found that the presence of soil microorganisms significantly enhanced the retention of Cs in organic systems (the soils studied by Russell et al. 2004, were not particularly organic). Sorption of ^{85}Sr was not significantly influenced by the presence of soil microorganisms. The complexity of radionuclide behaviour in organically rich systems is elaborated upon in the next section.

5.5.8 Effects of SOM on Bioaccumulation

Soil organic matter (SOM) is a complex substance originating from organic residues of diverse nature, continuously modified by dynamic processes such as biological activity; chemical, physical, and climatic factors; and the use and management of land by farmers (Fontana, 2006). Although they represent less than 5% of the composition of the majority of soils, organic compounds are key components in any terrestrial ecosystem. Variation in their distribution, content, and quality has an important effect on a range of processes occurring in soils.

Stability is one of the main criteria of the quality of SOM. Debris from fresh plants is gradually converted into more stable forms, which involves a variety of chemical, physical, and microbiological processes, as well as the efforts of other soil fauna. Furthermore, the interaction of SOM with minerals, principally clays, mediated by ions such as Ca^{++}, Al^{3+}, and Fe^{3+}, causes the formation of organomineral complexes of higher stability (Sokolik et al., 2001).

SOM is heterogeneous and its composition includes lipids, fulvic acids, humic acids, lignin, and humins. The less refractory organic materials present an extreme complexity and variety of compounds; however, it is the humic fraction (humic substances) that most substantially affects the behaviour of radionuclides (Gri, 2000). According to Cronan and Aiken (1985), when considering solubilisation, humic substances can be classified as:

Humin: Insoluble fraction at any pH value
Humic acids: Organic material soluble only in alkaline conditions
Fulvic acids: Organic material that remains soluble at any pH

According to Staunton (2002), in general, elements binding to fulvic acids are available for uptake by plants or become involved in transport processes, while more insoluble complexes tend to immobilize metals. The variable outcomes for SOM fractionation may explain why organic amendment can produce contradictory effects on the fate of radionuclides, as noted when discussing some Brazilian soils.

Organic amendment is a very common agricultural practice used by family farmers with the aim of improving, at low cost, the physical and chemical properties of soils. As pointed out in Section 5.3.3, these practices are more common in developing nations that are prevalent across the tropics and subtropics. This approach has been improved by the commercial expansion of organic farming practices, pointed out as one of the most suitable technologies for sustainable and healthy agriculture. However, the literature reports that an abundance of organic matter can favor ^{137}Cs transfer to plants (Staunton et al., 2002). According to Al-Oudat et al. (2006), ^{137}Cs root uptake increases, while ^{90}Sr plant uptake reduces, with increasing amounts of organic matter in soils.

In very weathered soils, the role of organic amendment in the potential mobility and bioavailability of ^{137}Cs and ^{60}Co was experimentally investigated by Wasserman et al. (2008a). The experiment was carried out in pots containing Histosol, Ferralsol, and Nitisol spiked with ^{137}Cs and ^{60}Co in one experiment and later with ^{241}Am in another. In order to promote changes in soil properties, Ferralsol and Nitisol, both with very low organic matter content and fertility, received different doses of organic amendment: 2 kg m^{-2} (the recommended dosage to cultivate radish), 4 kg m^{-2}, and without amendment. After contamination, radish (*Raphanus sativus*, L.) was cultivated to obtain soil-to-plant TF values for ^{137}Cs and ^{60}Co. The values obtained in these soils were compared with TFs obtained in Histosol, a soil with a naturally higher content of organic matter. The authors concluded that organic amendment can be used as a short-term functional countermeasure for ^{137}Cs contamination in Ferralsol and Nitisol for a couple of reasons. First, organic amendment increases productivity and thereby biomass dilution of accumulated activity. Second, it reduces the proportion of readily bioavailable radionuclides in soil. These factors combine to reduce the TF value by almost one order of magnitude over a short period of time (six months). In the same work, the geochemical partitioning of ^{137}Cs in these soils indicated a great affinity for labile organic matter (oxidisable phases), refractory organic matter, and Fe and Al oxyhydroxides (alkaline phases). The ^{137}Cs content associated with the alkaline phase showed a positive and significant correlation with the applied dosage of organic compost, confirming that the more refractory organic compounds added by the compost were responsible for the enhanced retention of ^{137}Cs.

Results for ^{60}Co obtained with sequential extraction (Wasserman et al., 2008a) were contradictory, since the application of organic compost decreased bioavailability of the radionuclide in the Nitisol, but not in the Ferralsol. Despite the variable ^{60}Co partitioning, the ^{60}Co transfers to plants in Ferralsol and Nitisol were substantially reduced after the addition of 4 kg m^{-2} of organic compost for cultivated radish, comparable with the low TF observed for Histosol.

Using the same experimental protocol as Wasserman et al. (2008a), the effect of SOM on the behaviour of ^{241}Am in highly weathered tropical soils was studied by Pereira (2009) in organic amended and unamended Ferralsol, Nitisol and Histosol spiked with radioactive solution sprayed onto the soil surface. This author observed that amended Nitisol (by addition of 4 kg m^{-2} of organic material) showed a decrease of ^{241}Am in the superficial layer (0–2.5 cm) followed by an enrichment of ^{241}Am in deeper layers, when compared with unamended soil (Nitisol control in Figure 5.10). The distribution of ^{241}Am with depth in the amended Ferralsol (Ferralsol 4 kg m^{-2} in Figure 5.11) also showed a decrease of ^{241}Am in the superficial layer (0–2.5 cm), with some enrichment of ^{241}Am lower in the soil profile. The ^{241}Am profile in the Histosol (the soil with naturally higher SOM content)

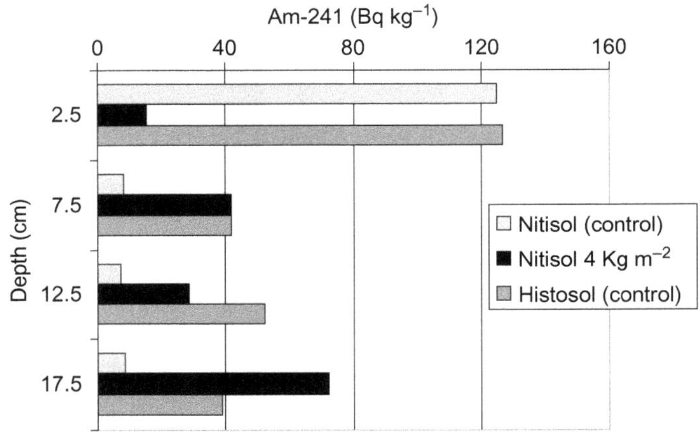

FIG. 5.10 Concentration of ^{241}Am with depth in unamended Nitisol (control), Nitisol with organic amendment (4 kg m^{-2}) and Histosol.

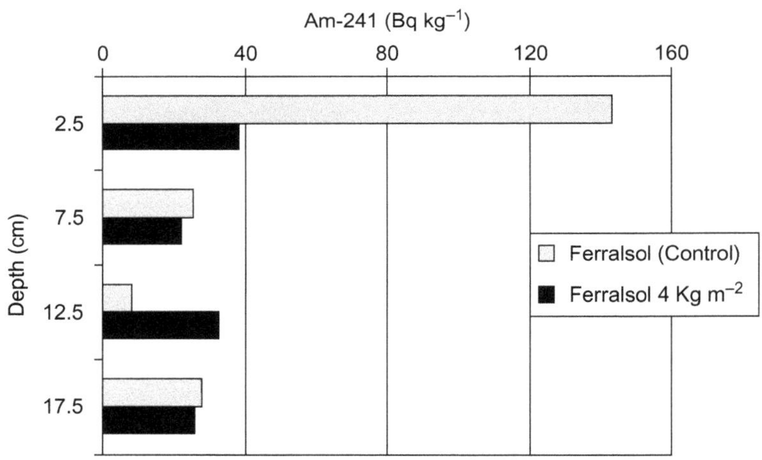

FIG. 5.11 Concentration of ^{241}Am with depth in Ferralsol and Ferralsol with organic amendment (4 kg m^{-2}).

also showed an enrichment of ^{241}Am in deeper layers, but the decrease in the superficial layer was not observed (Figure 5.10).

Organic matter fractionation can probably explain this behaviour. The great affinity of ^{241}Am for organic compounds allows association with labile organic matter and with refractory phases. With the ageing of organic amendment, the labile part is easily lost by decomposition processes and the ^{241}Am associated with this fraction is also lost to the solution because it is available for root uptake or for migration to deeper layers, while more stable organic compounds can retain the radionuclide associated with this fraction for longer periods, but as a less bioavailable quantity. As the organic matter turnover rate is very dependent on soil management (Syers, 1997), a variable proportion of radionuclides associated with organic compounds can be mobilized, depending upon the rate of decomposition. Indeed Pereira (2009) showed that a few years after the amendment, organic carbon content returned to its original values (Figure 5.12). Furthermore, no difference on the fractioning of SOM was observed between the Ferralsol and the Nitisol, while more mature organic soils showed significant reductions in the content of the soluble organic fraction (see fulvic acid in Figure 5.13).

The effect of organic amendment on the distribution coefficient (K_d) for ^{241}Am was also investigated by Pereira (2009). He showed that K_d values for tropical soils were extremely low (K_d Ferralsol $= 44.2 \pm 0.5$ L kg^{-1}, $n = 3$; K_d Nitisol $= 47.1 \pm 0.5$ L kg^{-1}, $n = 3$); compared to generic values for sandy soils (K_d = 2000 L kg^{-1}) or clay soils (K_d = 8100 L kg^{-1}) (IAEA, 1994). This indicates low retention of ^{241}Am in tropical soils compared with temperate soils. According to Pereira (2009), the K_d values were positively correlated with organic carbon and specifically with humic acids ($r = 0.943$; $n = 7$; $p = 99.9\%$) and humin fractions ($r = 0.933$; $n = 7$; $p = 99.9\%$), confirming the importance of these fractions for the immobilization of ^{241}Am.

All of these studies show how complex and dynamic the influence of organic matter can be on the fate of radionuclides, since it depends on the nature of existing organic compounds or those added by amendment. Interactions of organic matter with soil minerals form stable or soluble complexes with radionuclides that will enhance transportation processes. The fate of the radionuclides also depends on the use and management of land since there is a strong relationship with the practices that may engender organic matter decay, promoting associated radionuclide loss from the soil solution. Likewise, radionuclides can associate with more insoluble complexes that tend to be immobilized and retained for long periods, but also remain less bioavailable in that case.

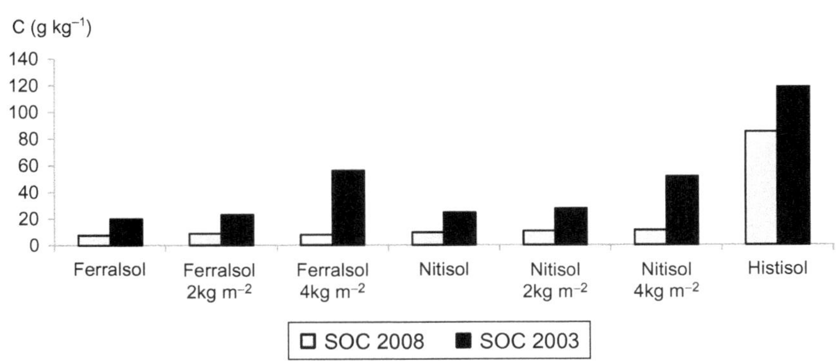

FIG. 5.12 Evolution of soil organic carbon (SOC) five years after organic amendment (after Pereira, 2009).

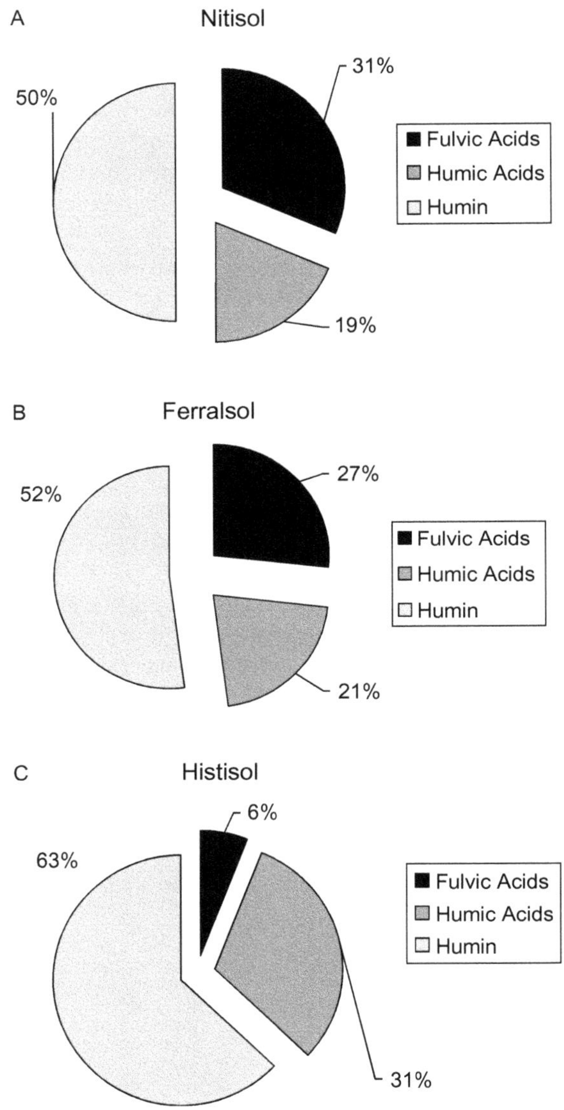

FIG. 5.13 Fractioning of soil organic matter in tropical soils (after Pereira, 2009).

5.6. TROPICAL CROPS

As stated earlier, the major crops consumed by humans or used as fodder by animals in tropical systems encompass the same groups as for temperate systems, based on the definitions prescribed in the FAO/IAEA/IUR Protocol (1998). The groups comprise cereals; fodder; grasses; leafy vegetables; root crops or tubers; pods, beans, and nuts; herbs; and miscellaneous (fruits, etc.). Even within these major classes of crops, as recognized by the IAEA (1994, 2009) in TRS No. 364 and No. 472 and the revisions thereof (EMRAS), some subclasses within these categories are grown in both regions. For example, grains in the tropics and temperate zones include rice, wheat, maize, sorghum, and others. That is, whilst the proportion of any particular crop grown and consumed may vary between climatic zones, the crops themselves are often replicated (although different strains may be grown in each region).

We have also already described that, whilst there may be regionally specific differences in TF between the various crop and soil types, on average there are no significant differences between temperate and tropical systems that cannot be generally explained by differences in soil type, nutrient levels, soil chemistry, microbial activity, and so on. However, there are some crop types that can be considered synonymously tropical, for example, rice, and tropical fruits. Because of this, we will be focusing on those major crop types here, where there is sufficient information for tropical systems to make comparative assessments with the temperate data.

5.6.1 Rice

Rice (*Oryza sativa,* L.) is one of the most dominant staple food crops in humid tropical countries across the globe (Table 5.5). Although the total planted area for rice is smaller than that for wheat, more than half of the world's population depends on rice as its main source of calories (i.e., rice supplies one third of the total caloric intake), and the demand for rice is increasing. Rice varieties are classified according to their area of cultivation: paddy or wet rice, which is suited to paddy fields; upland or dry rice, which is suited to dry fields; and floating rice, which is grown in flood-prone areas. Here, we have focused on rice grown in wet paddy fields, which is the most common type. As mentioned previously, cultivation methods have important effects on plant uptake of radionuclides from soil. However, concerning the other types of rice, little or no data are available; hence, further studies are needed for complete assessments to be made of them.

For paddy rice, no crop rotation is necessary since water-soluble toxic substances generated by the plant are removed from the soil with water flow, and thus rice has been planted in the same places for hundreds of years. Under flooded conditions, for example, a water layer of 5–15 cm deep, oxygen is depleted relatively quickly by the respiration of soil microorganisms and plant roots. Following that, variable degrees of anaerobiosis give rise to chemical reduction of mineral nutrients and pH increases within the soil. Subsequently, phosphate and iron species sorbed on the soil particles are gradually changed to soluble forms that are easily taken up by rice plants. In addition, under reducing conditions, nitrogen is present as the positive ammonium ion, which is efficiently used by rice plants. These soil conditions are substantially different in unsaturated (dry) fields; hence, elemental concentrations in rice are likely to be in some cases markedly different from those of other cereals grown under unsaturated field conditions. In Figure 5.14, concentrations of 30 stable elements in brown rice samples are compared with those for cereal samples collected in Japan, that is, in nine wheat and barley (WB) samples (Uchida et al., 2007; Tagami and Uchida, 2009). Most of the elements in rice are not significantly different to those in WB. Exceptions are K, Ca, Fe, Cu, As, Sr, Mo, and Ba (by t test, $p < 0.05$). Concentrations of rare earth elements Th and U are amongst the lowest concentrations of all elements in both brown rice and WB, and the concentrations tend to be lower in brown rice than in WB. For Ca, Sr, and Ba, the values are significantly lower for brown rice. It should be noted that Sr and other alkaline earth metals, for example, Ca and Ba, have almost the same concentrations in paddy fields as in upland field soils (Uchida et al., 2007). The results indicate that uptake of Ca, Sr, and Ba by WB is probably higher than by brown rice. On the other hand, As and Mo concentrations are significantly higher in brown rice than in WB. Molybdenum and Fe, essential trace elements to plants, are redox sensitive. Hence, the waterlogged condition might affect their mobility in paddy fields. Thus, there are similarities between WB and brown rice samples, but there are also differences between them.

When a rice grain is carefully investigated, several concentric layers are easily observed. Brown rice is produced by removing the outermost layer called the hull. Thus, brown rice is sometimes called hulled rice. White, or polished, rice is obtained when brown rice is milled to sequentially remove the bran, the germ layer, and the aleurone layer (a layer filled with

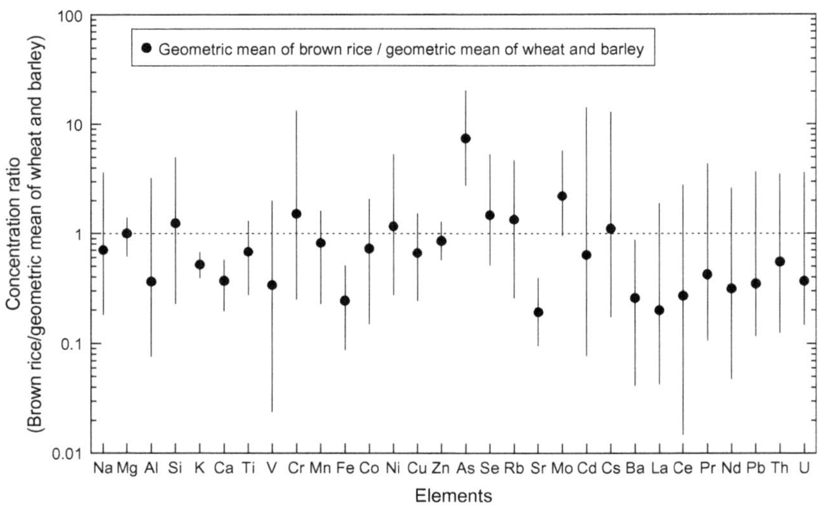

FIG. 5.14 Comparison of geometric means of elemental concentrations for brown rice and for wheat and barley. Bars show maximum and minimum concentration ratio for brown rice (modified from Uchida et al., 2007; Tagami and Uchida, 2009).

health-supportive and essential fats). Thus, white rice is simply a refined starch. As a consequence of retaining or removing these various layers, the chemical compounds and elements in brown rice and white rice are different, implying that TF values for brown rice and white rice would differ.

In order to summarize rice TFs, data were collected from papers in international journals, books, and proceedings with a peer review process. These papers appear in this chapter's reference list marked with an asterisk (*). Unpublished data of the same quality were also added for stable elements. To avoid double counting, summary data from reviews were excluded. If the grain treatment process (white or brown rice) was not clearly indicated, such data were considered as brown rice. Table 5.9 summarizes TF values for brown rice and white rice, respectively, based on dry weight.

It should be noted that there are three transfer pathways for radionuclides into rice: direct deposition onto foliage, water–plant pathways, and soil–plant pathways. The direct deposition pathway is important if the fallout concentration is high (Ichikawa et al., 1962). However, because the TF definition implies that soil is the major source, data for radionuclides in plants that might include the direct deposition pathway (e.g., field observation data before 1987) have been eliminated from the table for the sake of easy comparison. This is because the proportional contribution of each pathway to the total plant radioactivity is difficult to identify. Similarly, the data from laboratory experiments that clearly showed irrigation water was contaminated with radionuclides were excluded from this table. Nonetheless, the water–plant pathway is also important and the excluded data, which can be expressed as an areal TF, are reported elsewhere (Leung and Shang, 2003; Choi et al., 2005). Also, note that the physicochemical forms of radionuclides are important factors for their transfer to plants and indeed the forms change with time, which is known as the 'ageing effect' (Yamagata et al., 1965; Myttenaere et al., 1969; Mousny and Myttenaere, 1981; Noordijk et al. 1992; Tagami and Uchida 1996; Krouglov et al., 1997) (see Section 5.5.5).

The TFs were obtained for 14 radionuclides for brown rice and of 4 radionuclides for white rice comprising fallout nuclides (e.g., ^{90}Sr and ^{137}Cs), neutron activation products (e.g., ^{54}Mn, ^{60}Co, etc.), and naturally occurring radionuclides (e.g., ^{210}Pb, ^{226}Ra, ^{238}U, etc.). Because there are no available TF values for some radiologically important radionuclides, information on relevant stable elements is also indicated. As already pointed out in this chapter, types of soil and cultivation information are important factors because the biogeochemical behaviours of radionuclides depend on soil properties and conditions. However, it was also noted previously that comparisons between climatic zones show no substantial difference, on average. The TF data including soil information such as sand, loam, clay, and organic classifications can be found in IAEA (2009).

It should be noted that, under equilibrium conditions, that is, under long-term contact with the environment, radionuclides behave similarly to their

TABLE 5.9A TFs of Radionuclides and Stable Elements from Soil to Brown Rice Grain (DW)

Element	*N*	AM	SD	GM	GSD	Confidence Interval		Range	
						−95%	*95%*	*Min*	*Max*
^{54}Mn	5	2.9E-1	1.5E-1	2.6E-1	1.7	8.8E-2	7.5E-1	1.2E-1	5.2E-1
^{60}Co	5	6.7E-3	3.5E-3	5.9E-3	1.8	1.8E-3	1.9E-2	2.2E-3	1.2E-2
^{65}Zn	5	1.7E+0	9.3E-1	1.5E+0	2.0	3.8E-1	5.6E+0	5.8E-1	2.7E+0
^{90}Sr	149	2.3E-1	9.4E-1	4.6E-2	4.4	2.4E-3	8.7E-1	5.5E-3	6.0E+0
^{95m}Tc	2	<0.0002							
^{125}I	8	4.6E-3	2.5E-3	3.8E-3	2.1	9.0E-4	1.6E-2	1.1E-3	7.6E-3
^{137}Cs	398	3.6E-2	8.1E-2	8.0E-3	5.2	3.0E-4	2.1E-1	4.3E-4	6.1E-1
^{210}Pb	2	8.4E-3	5.2E-3	7.5E-3	1.9	2.0E-3	2.8E-2	4.7E-3	1.2E-2
^{210}Po	2	1.3E-2	5.2E-3	1.3E-2	1.5	5.6E-3	2.9E-2	9.4E-3	1.7E-2
^{226}Ra	166	1.6E-3	4.2E-3	6.3E-4	2.8	7.9E-5	5.1E-3	7.1E-5	2.8E-2
Th	67	7.3E-4	4.0E-3	1.1E-4	3.8	7.7E-6	1.5E-3	4.0E-6	3.0E-2
U	459	1.8E-3	1.0E-2	8.7E-5	6.5	2.1E-6	3.7E-3	5.5E-6	9.0E-2
$^{239+240}Pu$	3	1.4E-3	6.8E-4	1.3E-3	1.8	4.0E-4	4.1E-3	6.7E-4	2.0E-3
^{241}Am	3	2.5E-3	1.2E-3	2.3E-3	1.8	7.0E-4	7.5E-3	1.2E-3	3.3E-3
Na	63	1.2E-3	1.2E-3	9.2E-4	2.0	2.2E-4	3.8E-3	1.7E-4	6.9E-3

K	63	2.5E-1	1.7E-1	2.1E-1	1.7	7.1E-2	6.2E-1	7.4E-2	9.4E-1
Mn	63	5.1E-2	3.1E-2	4.4E-2	1.8	1.4E-2	1.4E-1	8.0E-3	1.9E-1
Fe	63	3.2E-4	1.9E-4	2.8E-4	1.7	1.0E-4	7.8E-4	7.5E-5	1.3E-3
Co	62	1.1E-3	8.8E-4	8.7E-4	1.9	2.4E-4	3.1E-3	2.1E-4	6.4E-3
Ni	63	1.9E-2	1.8E-2	1.4E-2	2.2	2.9E-3	6.4E-2	3.4E-3	9.1E-2
Cu	63	1.2E-1	6.9E-2	9.8E-2	1.8	3.0E-2	3.2E-1	2.2E-2	3.5E-1
Zn	63	2.6E-1	1.1E-1	2.4E-1	1.5	1.1E-1	5.4E-1	7.3E-2	6.8E-1
As	63	1.2E-2	7.6E-3	9.6E-3	1.9	2.7E-3	3.5E-2	2.7E-3	3.4E-2
Se	54	7.9E-2	5.3E-2	6.6E-2	1.9	1.9E-2	2.3E-1	9.5E-3	2.9E-1
Rb	63	2.4E-1	3.9E-1	1.2E-1	3.1	1.2E-2	1.2E+0	8.8E-3	2.2E+0
Sr	61	4.0E-3	3.0E-3	3.2E-3	1.9	8.8E-4	1.2E-2	6.6E-4	1.6E-2
Mo	63	9.5E-1	6.5E-1	7.8E-1	1.9	2.2E-1	2.8E+0	2.4E-1	3.4E+0
I	14	6.9E-3	5.1E-3	5.0E-3	2.6	7.1E-4	3.5E-2	1.5E-4	2.0E-2
Cs	63	1.9E-3	2.9E-3	9.7E-4	3.1	1.0E-4	9.3E-3	9.0E-5	1.6E-2
La	58	7.0E-5	8.4E-5	4.9E-5	2.2	1.0E-5	2.4E-4	7.4E-6	4.9E-4
Ce	46	6.0E-5	6.5E-5	3.7E-5	2.9	4.4E-6	3.1E-4	1.9E-6	3.4E-4
Pb	65	4.9E-4	8.3E-4	3.2E-4	2.2	6.9E-5	1.5E-3	8.8E-5	5.9E-3

AM, arithmetic mean; GM, geometric mean; GSD, geometric standard deviation; SD, standard deviation.
Data sources are identified in the reference list using an asterisk (*).

TABLE 5.9B TFs of Radionuclides and Stable Elements from Soil to White Rice (DW)

Element	N	AM	SD	GM	GSD	Confidence Interval		Range	
						−95%	*95%*	*Min*	*Max*
^{90}Sr	90	1.8E-2	2.4E-2	9.4E-3	3.0	1.0E-3	8.6E-2	2.1E-3	8.8E-2
^{137}Cs	227	2.6E-2	4.7E-2	4.7E-3	6.6	1.1E-4	2.0E-1	1.3E-4	1.9E-1
Th	74	1.5E-4	3.1E-4	6.1E-5	3.5	4.8E-6	7.7E-4	6.4E-6	2.2E-3
U	108	3.6E-5	2.9E-5	2.9E-5	1.8	8.5E-6	1.0E-4	6.2E-6	1.9E-4
Na	83	7.7E-4	1.0E+0	5.8E-4	2.0	1.4E-4	2.3E-3	9.5E-5	4.7E-3
K	97	8.8E-2	2.8E-1	7.5E-2	1.7	2.5E-2	2.3E-1	2.4E-2	3.2E-1
Mn	83	2.0E-2	5.5E-1	1.8E-2	1.7	6.1E-3	5.2E-2	3.7E-3	6.8E-2
Fe	83	8.4E-5	1.2E+0	7.4E-5	1.6	2.8E-5	1.9E-4	2.9E-5	2.3E-4
Co	79	5.3E-4	9.7E-1	4.5E-4	1.8	1.5E-4	1.4E-3	1.4E-4	2.4E-3
Ni	83	1.5E-2	6.9E-1	1.1E-2	2.3	2.0E-3	5.5E-2	2.2E-3	6.8E-2
Cu	83	9.4E-2	3.7E-1	7.9E-2	1.8	2.4E-2	2.6E-1	2.0E-2	4.4E-1
Zn	83	2.0E-1	2.5E-1	1.9E-1	1.5	9.1E-2	4.1E-1	6.4E-2	3.9E-1
As	83	8.3E-3	6.8E-1	6.4E-3	2.1	1.5E-3	2.8E-2	1.2E-3	2.9E-2
Se	83	5.5E-2	4.4E-1	4.7E-2	1.9	1.3E-2	1.7E-1	2.7E-3	1.4E-1
Rb	97	8.2E-2	4.6E-1	4.2E-2	3.1	4.4E-3	4.0E-1	2.1E-3	6.6E-1

Sr	100	1.1E-3	9.7E-4	8.4E-4	2.1	1.9E-4	3.7E-3	1.7E-4	4.9E-3
Mo	83	9.1E-1	6.7E-1	7.5E-1	1.9	2.1E-1	2.7E+0	2.4E-1	3.3E+0
I	28	3.1E-3	2.3E-3	2.0E-3	3.2	2.0E-4	2.0E-2	1.3E-4	7.8E-3
Cs	92	8.2E-4	1.1E+0	4.4E-4	3.0	5.1E-5	3.9E-3	4.3E-5	6.4E-3
La	58	7.7E-5	1.5E+0	3.6E-5	2.8	4.8E-6	2.7E-4	4.6E-6	1.4E-3
Ce	39	7.1E-5	1.9E+0	3.3E-5	3.1	3.5E-6	3.1E-4	6.3E-6	5.7E-4
Pb	82	2.2E-4	3.6E-4	1.3E-4	2.5	2.1E-5	8.3E-4	1.7E-5	1.8E-3

AM, arithmetic mean; GM, geometric mean; GSD, geometric standard deviation; SD, standard deviation.
Data sources are identified in the reference list using an asterisk (*).

naturally existing stable isotopes so that naturally existing elements can be suitable analogues. However, under freshly contaminated conditions, TF values for radionuclides would often be higher than for naturally occurring elements due to their different mobility in soil-to-plant systems. Indeed, the values obtained from pot experiments (short-term contamination) are usually higher than values obtained from field observations of fallout radionuclides (long-term contamination), due to the ageing effect (Section 5.5.5) that occurs with the water management used in paddy fields (Table 5.10). There are a number of reports on TF values for Sr and Cs isotopes. These elements, as well as many others, do not distribute uniformly in the rice plant. For Sr, only 0.7% and 1.1% of the total ^{90}Sr in the whole rice plant were found in polished rice and brown rice, respectively; whereas 5.3% and 7.4% of the total ^{137}Cs were observed in polished rice and brown rice, respectively (Wang et al., 1998). Tsukada et al. (2002a) reported that 7% and 10% of the total stable Cs were distributed in polished rice and bran, respectively.

Some TF values for Tc in rice and wheat plants have been obtained from laboratory radiotracer experiments (Yanagisawa and Muramatsu, 1992). The mean TF for rice plants is less than 0.2×10^{-3} for the hulled rice and 5.1 for the lower leaf blade. For wheat plants, higher TF values are found: 2.7×10^{-3} for the hulled wheat and 230 for the lower leaf blade. The Tc concentrations in the grains of both plant types are much lower than those in their leaves, and TF values of Tc for rice grain are lower than those for wheat grain. The level of Tc in soil solution collected from the flooded soil of rice pots decreased rapidly with time, whereas the Tc concentration in the soil solution of wheat pots decreased slowly. When a Tc tracer was added as $TcO_4{}^{1-}$ to a flooded paddy soil sample, Tc was easily transformed to immobile forms due to the reducing conditions generated by microbial activity (Tagami and Uchida, 1996). Thus, the low TF value observed for the rice plants could be explained by fixation of Tc in the soil.

The TF values of radioiodine for brown rice were obtained as 2.5×10^{-3} for Andosol and 5.9×10^{-3} for Gray lowland soil (Muramatsu et al., 1989, 1993). Measured TF values of radioiodine from the soils to polished rice are about 0.7×10^{-3} for the Andosol and 0.2×10^{-2} for the Gray lowland soil. Iodine sorbed on the soil solid phase is readily dissolved into the soil solution under flooded conditions, as it is more soluble in anoxic than oxic conditions; however, the TF values of iodine for brown rice and polished rice are extremely low. One of the reasons for this is that iodine is mainly distributed in the leaves, and hardly any translocates into the grains.

The naturally occurring elements, Ra, Th, and U have been studied as natural analogues. Their TFs for white and brown rice are compared with cereals or maize (grain) values listed in TRS No. 364 (Figure 5.15). The values summarized in this chapter for white and brown rice are close to the best-estimate value for maize. However, the uranium TF tends to be low in rice samples. Uranium is a redox sensitive element, so paddy field conditions might affect its mobility.

TABLE 5.10 Soil-to-Rice TFs of Sr and Cs Based on Dry Weight Crop

	Soil[a]–rice Grain	Radionuclide		Reference
		Strontium	*Cesium*	
Pot experiment				
Bangladesh	Sandy loam—not specified Alluvium—not specified	0.82 (7.8E-1–8.9E-1)	0.28 (2.5E-1–3.3E-1) 9.0E-3–2.6E-2	Mollah et al., 1998 Rahman et al., 2005
P. R. China	Loam–white rice Loam–white rice Clay–white rice Loam–brown rice Clay–brown rice	6.4E-3[c] (2.1E-3–2.0E-2) 1.8E-1 ± 3.5E-2[b,c] 6.7E-2 ± 9.4E-3[b,c]	4.0E-3[c] (2.8E-4–5.7E-2) 5.7E-3 ± 2.8E-3[b,c] 0.11–0.53 0.68–2.2	Li et al., 2006
South Korea	All–brown rice	1.2E-1	3.1E-2 (1.0E-2–1.0E-1)	Choi et al., 2001
Vietnam	All–not specified	3.1E-1 ± 2.8E-1[b]	1.7E-1 ± 3.4E-1[b]	Quang et al., 2003
	Al–not specified	1.1E-1 (5.1E-2–3.3E-1)	5.0E-2 (9E-4–1.4)	Binh et al., 2003
Field observation (fallout)				
India	All–not specified	2.2E-1[d] 2.0E-1[d]	1.4E-2[d] 5.6E-2[d]	Narayana et al., 1995; Radhakrishna et al., 1996

continued

TABLE 5.10 Soil-to-Rice TFs of Sr and Cs Based on Dry Weight Crop—Cont'd

	Soil[a]–rice Grain	Radionuclide		Reference
		Strontium	*Cesium*	
Japan	All–brown rice All–white rice	1.4E-2[d] 4.0E-3[d]	4.1E-3[d] 1.6E-3[d]	Komamura et al., 2006
	All–white rice	3.8E-3 ± 2.6E-3[b]	2.6E-3 ± 2.8E-3[b]	Komamura and Tsumura, 1994
	All–white rice	2.1E-3 ± 0.7E-3	1.6E-3 (2.1E-4–1.2E-4)	Tsukada et al., 2002b, 2005
	All–brown rice	—	2.8E-3 (4.3E-4–4.1E-2)	Uchida and Tagami, 2007a; Uchida et al., 2007; Tsukada et al., 2007
South Korea	All–brown rice	—	4.5E-3 (1.2E-3–1.1E-2)	Choi et al., 2001
Taiwan, China	All–white rice	3.7E-2 (0.5E-2–8.8E-2)	9.7E-2 (0.3E-1–1.88E-1)	Wang et al., 1998
	All–brown rice	8.6E-2 (2.7E-2–2.2E-1)	1.4E-1 (0.7E-1–2.7E-1)	Wang et al., 1998
	All–not specified	3.1E-2[b]	2.2E-3[b]	Lu et al., 2006

[a]*Soil types listed are in different classification systems but as shown in the literature. 'All' means typical soil in that country.*
[b]*Arithmetic mean.*
[c]*Values were converted from fresh weight basis TF to dry weight basis TF by applying dry/wet ratio of 0.85.*
[d]*Calculated from soil and rice concentration.*

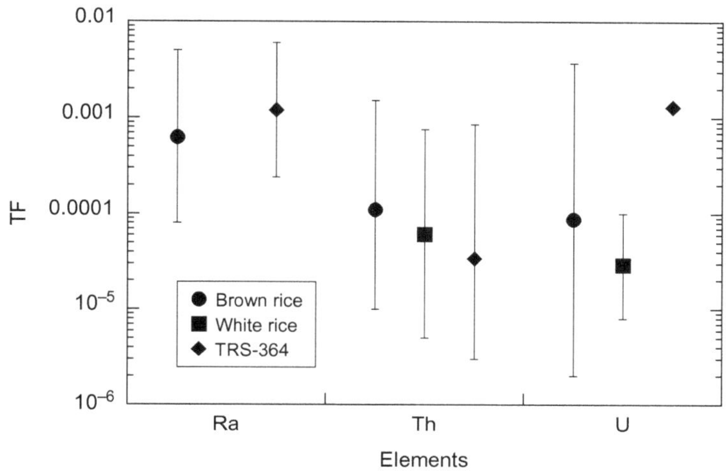

FIG. 5.15 Comparison of DW TFs of Ra, Th and U for white rice, brown rice and cereals or maize grain in TRS-364 (IAEA, 1994). Bars show upper and lower 95% confidence limits for TFs.

5.6.2 Tropical Fruit

Fruits form an obviously important component of human diet in tropical and subtropical systems and are crucial factors in the propagation of biodiversity across most natural ecosystems. Within the IAEA BIOMASS program during the late 1990s there was a specific investigation into the behaviour of radionuclides in fruits and the development of models to estimate that behaviour (Carini et al., 2005), irrespective of climatic zone. As part of that effort, an overview of the available information on radionuclide contamination of fruit (Carini, 1999) listed a number of factors in its conclusions that are believed to be generally pertinent to the behaviour of radioactivity in fruit trees, irrespective of climate. They are as follows:

- Fruit contamination arises from foliar uptake, root uptake, and remobilization from storage organs within the plants.
- Morphological differences between fruit trees and other crops can substantially influence radionuclide uptake. Factors such as shape and surface texture in fruit tree organs have the capacity to enhance interception rates of atmospherically transported radioactivity.
- The developmental stage of the tree and fruit will affect foliar adsorption and translocation processes.
- Remobilization can be both into and out of fruits, depending upon the radionuclide, the growth phase, and species, as well as upon agricultural practices.
- Biomass growth, and consequent mass dilution, can substantially influence radionuclide concentrations in fruit, depending upon the timing of contamination and fruit development. The lag between contamination and

harvest/consumption of fruit is also consequential for dose assessment due to the potential for radioactive decay.
- Fruit contamination is directly correlated to the ratio between the foliar surface area and the corresponding mass of fruit. Tropical forests typically have a large leaf area index (up to eight in the Amazon) to capture as much radiant energy as possible. However, very low values are evident in subtropical arid zones and savannahs.
- Soil type has a substantial influence upon root uptake and is more predictable than interspecies variability. This observation was generally consistent with the observations made across all plant and crop types by the soil-to-plant CRPs referred to previously (Table 5.6).
- Translocation within the plant is more important in the early years following contamination in agricultural fruit trees. Root uptake plays a more important role in semi-natural environments.
- Climatic conditions, as well as soil water and nutrition, affect both foliar and root accumulation.
- Agricultural practices can have a substantial influence on longer term radionuclide concentrations in fruit, such as by altering root and foliar uptake as a consequence of nutrient modification, or by removing leaf fall and thereby minimizing radionuclide cycling over time.
- Fruit preparation (e.g., peeling or canning) can substantially influence the dose derived from fruit, due to factors relating to both biodistribution and radioactive decay.

For tropical and subtropical fruits in particular, the list is also generally applicable. However, a couple of points are most relevant. As noted in Section 5.5.1, the influence of soil type will be markedly different in tropical systems. This is particularly evident in data provided in Table 5.11 on uptake by fruits from coral atoll soils by Robison and colleagues. The TF data for Cs are very high; whilst the Sr TF values are very low due to site-specific soil characteristics (see earlier discussion in Section 5.5). Leaf area indices also tend to be higher in tropical forest than in temperate systems (National Research Council, 2008). Hence, increased leaf area to fruit mass ratios are likely to occur in tropical fruit trees. This may lead to relatively increased concentrations in tropical fruit, particularly following atmospheric dispersion of radioactivity. Tropical fruit trees are also more likely to exist in semi-natural environments, particularly in areas of subsistence agriculture (although fruit tree plantations, such as banana and mango farms, are also very common) and, as such, may be more prone to enhanced radionuclide cycling.

Despite extended effort, the Fruits Working Group within the BIOMASS program was unable to accumulate more than a few data for tropical fruit TFs and these were for just for a few radionuclides. Much of the recorded data were obtained from researchers working within the IAEA/FAO/IUR CRP described earlier, but the complete set did extend beyond that. All the

TABLE 5.11 Best Estimates of TFs for Cs, Sr, Pu, and Am from Soil to Fruit under Tropical or Subtropical Conditions

Crop	Fruit Part	Element	TF	Soil Type[a]	Contamination Type[b]	Experiment Type[c]	% Water Content	Reference
Watermelon	Fruit	Cs	6.0E-4	Semi-arid	A	F	94	Othman and Yassine, 1995
Olive tree	Fruit	Cs	1.4E-3	CL	A	P	53	Skarlou et al., 1997
Olive tree	Fruit	Cs	1.0E-2	SL	A	P	59	Skarlou et al., 1997
Orange tree	Fruit	Cs	8.8E-4	CL	A	P	66	Skarlou et al., 1997
Orange tree	Fruit peel	Cs	7.8E-4	CL	A	P	68	Skarlou et al., 1997
Orange tree	Fruit	Cs	3.5E-2	SL	A	P	68	Skarlou et al., 1997
Orange tree	Fruit peel	Cs	2.5E-2	SL	A	P	68	Skarlou et al., 1997
Grape	Dregs	Cs	7.6E-4	CL	C	F (9yf)	87.4	CRP Topcouglu
Grape	Juice	Cs	2.1E-3	CL	C	F (9yf)	93.5	CRP Topcouglu
Mandarin	Fruit	Cs	2.4E-3	CL	C	F (9yf)	93.1	CRP Topcouglu
Mandarin	Fruit peel	Cs	2.2E-3	CL	C	F (9yf)	75.5	CRP Topcouglu
Lemon	Fruit	Cs	1.2E-3	C	F	F	86	CRP Avila
Lime	Fruit	Cs	<1.8E-2	LS	F	F	90.5	CRP Twining
Lemon (Meyer)	Fruit	Cs	<3.1E-2	L	F	F	90.5	CRP Twining
Banana	Fruit	Cs	1.1E-1	B	F	F	72	CRP Robison
Banana	Fruit	Cs	2.9E-2	LS	F	F	70	CRP Twining

continued

TABLE 5.11 Best Estimates of TFs for Cs, Sr, Pu, and Am from Soil to Fruit under Tropical or Subtropical Conditions—Cont'd

Crop	Fruit Part	Element	TF	Soil Type[a]	Contamination Type[b]	Experiment Type[c]	% Water Content	Reference
Banana	Fruit peeled	Cs	3.6E-3	C	F	F	75.7	CRP Avila
Banana	Fruit	Cs	1.1E-2	—	—	—	75.7	CRP Consultants
Plantano	Fruit peeled	Cs	7.3E-3	C	F	F	75.7	CRP Avila
Pineapple	Fruit	Cs	1.2E-2	SL	A	P	80	CRP Mollah
Mango	Fruit	Cs	2.5E-3	—	—	—	84	CRP Consultants
Mango	Fruit	Cs	<7.0E-2	LS	F	F	71	CRP Twining
Mango	Fruit	Cs	1.9E-3	C	F	F	84	CRP Avila
Avocado	Fruit	Cs	1.5E-2	LS	F	F	80	CRP Twining
Papaya	Fruit	Cs	1.6	—	F	F	80	Koranda and Robison, 1978
Papaya	Fruit	Cs	5.9E-1	B	F	F	90	CRP Robison
Papaya	Fruit peeled	Cs	4.7E-3	C	F	F	85	CRP Avila
Breadfruit	Fruit	Cs	1.4	—	F	F	80	Koranda and Robison, 1978
Breadfruit	Fruit	Cs	3.0E-1	B	F	F	76	CRP Robison

Pandanus	Fruit	Cs	1.6	—	F	F	80	Koranda and Robison, 1978
Pandanus	Fruit	Cs	3.2	B	F	F	81	CRP Robison
Copra nut	Fruit	Cs	3.8	B	F	F	44	CRP Robison
Coconut	Fruit meat	Cs	1.8	—	F	F	80	Koranda and Robison, 1978
Coconut	Fruit	Cs	2.0	B	F	F	74	CRP Robison
Guava	Fruit	Cs	1.8E-3	—	—	—	82	CRP Consultants
Kiwi fruit	Fruit	Cs	5.2E-3	LC	C	P	80	CRP Topcouglu
Pawpaw	Fruit	Cs	9.8E-2	LS	F	F	92	CRP Twining
Rambutan	Fruit	Cs	8.5E-2	LS	F	F	88	CRP Twining
Ruby grapefruit	Fruit	Cs	6.9E-3	LC	F	F	90	CRP Twining
Squash	Fruit	Cs	3.0E-1	B	F	F	88.5	CRP Robison
Watermelon	Flesh	Sr	5.4E-3	LC	A	F	92.6	CRP Yassine
Orange tree	Flesh	Sr	2.4E-2	C	F	F	86	CRP Avila
Lemon	Flesh	Sr	2.7E-2	C	F	F	86	CRP Avila
Banana	Fruit peeled	Sr	8.7E-3	C	F	F	75.7	CRP Avila
Plantano	Fruit peeled	Sr	1.3E-2	C	F	F	75.7	CRP Avila
Mango	Fruit	Sr	1.8E-3	C	F	F	84	CRP Avila

continued

TABLE 5.11 Best Estimates of TFs for Cs, Sr, Pu, and Am from Soil to Fruit under Tropical or Subtropical Conditions—Cont'd

Crop	Fruit Part	Element	TF	Soil Type[a]	Contamination Type[b]	Experiment Type[c]	% Water Content	Reference
Papaya	Fruit	Sr	8.0E-2	—	F	F	80	Koranda and Robison, 1978
Papaya	Fruit	Sr	1.3E-2	B	F	F	92	CRP Robison
Papaya	Fruit peeled	Sr	4.1E-3	C	F	F	85	CRP Avila
Breadfruit	Fruit	Sr	1.6E-1	—	F	F	80	Koranda and Robison. 1978
Breadfruit	Fruit	Sr	2.0E-2	B	F	F	74	CRP Robison
Pandanus	Fruit	Sr	1.0E-1	—	F	F	80	Koranda and Robison. 1978
Pandanus	Fruit	Sr	8.0E-2	B	F	F	79	CRP Robison
Copra nut	Fruit	Sr	1.5E-3	B	F	F	39	CRP Robison
Coconut	Fruit meat	Sr	6.0E-3	—	F	F	80	Koranda and Robison. 1978
Coconut	Fruit	Sr	2.7E-3	B	F	F	70	CRP Robison
Guava	Whole fruit	Sr	5.0E-3	C	F	F	82	CRP Avila
Papaya	Fruit peeled	Pu	7.2E-7	B	F	F	92	CRP Robison

Breadfruit	Fruit	Pu	4.3E-6	B	F	F	79	CRP Robison
Pandanus	Fruit	Pu	1.0E-5	B	F	F	85	CRP Robison
Copra nut	Fruit	Pu	2.6E-6	B	F	F	41	CRP Robison
Coconut	Fruit meat	Pu	2.0E-4	—	—	F	80	Adriano et al. 1980
Coconut	Fruit	Pu	6.9E-6	B	F	F	71	CRP Robison
Papaya	Fruit	Am	6.8E-7	B	F	F	92	CRP Robison
Breadfruit	Fruit	Am	5.9E-6	B	F	F	79	CRP Robison
Copra nut	Fruit	Am	5.2E-6	B	F	F	41	CRP Robison
Coconut	Fruit	Am	3.7E-6	B	F	F	71	CRP Robison

[a]*C - clay; L - loam; S - sand; B - coral-based. When used in combination, the first letter indicates the dominant soil type in the mixture.*
[b]*A - artificial; C - Chernobyl; F - weapons test fallout.*
[c]*F - grown in the field; P - grown in pots.*
From Carini (2001).

available data for soil-to-fruit transfer were recorded in Carini (2001). From that table, data for tropical species are listed in Table 5.11.

5.7. TROPICAL ANIMALS

Despite the fact that so little is known of tropical plants when compared to temperate species, there is even less information available for the radioecology of animals, in general, and for tropical environments, in particular. Most of what is known relates to animals in aquatic environments and is covered in Chapter 6. However, the available information for terrestrial tropical and subtropical environments is summarized in the following material. One additional set of TF data relate air concentrations of ^{131}I over South Africa, arising as fallout from atmospheric testing, to the concentration of ^{131}I in cow's milk. The estimated air-to-milk TF values ranged from 700 to 1200 (Basson et al., 1973).

You will note that much of the soil-to-animal information listed below relates to Australian data (Johansen and Twining, 2010), although some data from South America were also identified. Much of the Australian information came from 'grey' literature sources not generally available to the broader scientific community. Those data were typically derived from baseline or routine environmental surveys conducted for mining companies and were recorded in internal or consultancy reports. Other sources included government reports. A high likelihood exists that similar information may be found within other countries. To that end, the authors encourage any readers who may be aware of such data to take steps to make that information more readily available, as noted next.

At the time of writing, the IAEA EMRAS II program has been initiated with international support and interest. Its purpose is to improve capabilities in the field of environmental radiation dose assessment by means of acquisition of improved data for model testing; model comparison; reaching consensus on modelling philosophies, approaches, and parameter values; development of improved methods; and exchange of information. Its Web site is http://www-ns.iaea.org/projects/emras/emras2/default.htm. One of the EMRAS II working groups, 'Wildlife Transfer Coefficient Handbook' (WG 5), is focussing on the provision of relevant data via an online database, www.wildlifetransferdatabase.org, to support the production of a transfer handbook for wildlife. That group is led by Dr. B. Howard from the Centre for Ecology and Hydrology, Lancaster University, UK. Given the shortage of TF data for animals, these groups are the preferred contact points for reporting new or uncovered data. In addition, publication of data in the scientific literature is always recommended.

The recently released TECDOC 1616 (IAEA, 2009) and TRS No. 472 (IAEA, 2010; a revision of TRS No. 364, IAEA, 1994), mainly includes animal data on TFs in Northern Hemisphere agricultural systems. Often in those cases, the type and quantity of fodder or pasture consumed by the animals is known and the TFs are defined differently than described earlier (Section

5.4.1). They are defined as the proportion of daily consumption that is converted into an activity concentration in the meat at time of slaughter. However, most of the available data for tropical systems does not have access to the concentration of radionuclides in consumed vegetation nor to the quantity and variety of vegetation consumed. In that case there is no real basis for comparison with the IAEA data. Often in this field the ratio of radionuclides in animals over that in soils is given as a concentration ratio (C_r; see Section 5.4). We are using that convention in the following tables. Under those constraints, the data in Table 5.12 have been incorporated into the draft of a new TECDOC (IAEA, in prep.) covering wildlife.

The data in Table 5.12 cover 12 different types of terrestrial animals, some of which are composite samples, and eight radionuclides. However, some patterns may be discerned, even though they are based on very limited data. Taking flesh samples for mammal species as one of the major subsets from Table 5.12, it can be seen that the C_r values for any species can be across a range as high as three orders of magnitude (e.g., 2×10^{-6} to 4×10^{-3} for Ra in the fox or 2×10^{-3} to 3×10^{-1} for Po in the kangaroo). The variability between species reduces significantly when average values are considered, at least for U, Ra, and Pb from the U-series data.

Despite that consistency, when the comparisons are expanded to include other classes of organism across all measured radionuclides (Figure 5.16), it is possible that reptiles and insects have sequentially lower C_r values than mammals. Birds seem higher for Cs but lower for Th, U, and Ra. Some recent work on angiosperms has shown that phylogenetic differences in TF can be significant (e.g., Bennett and Willey, 2003; Willey and Fawcett, 2006) and similar discrimination is possible for animals.

To place the tropical data in context with other information, the default C_rs from the ERICA software for the same elements and types of animals are also plotted in Figure 5.16. This indicates that the limited tropical data are following the same general trends as the generic values but, if anything, they may be low in comparison. Polonium is an exception, having about two orders of magnitude higher median activity concentrations in Australian samples. However, substantially more data are required to validate those observations.

5.8. SUMMARY AND CONCLUSIONS

Tropical and subtropical radioecology in a wide variety of terrestrial environments is different from the more intensively studied areas of the planet. That is not to say that it is affected by any unique set of processes that are known to influence the fate and transport of radionuclides in the environment, or that the mechanisms underpinning bioaccumulation, distribution, and excretion of those radionuclides by the plants and animals that exist in low-latitude areas are different from those that control biodynamics in higher latitudes. On the contrary, from the available information it would appear that, on

TABLE 5.12 Soil-to-Animal Concentration Ratios (Fresh Weight Tissue /Dry Weight Soil) for Terrestrial Animals (Tropical Studies Emphasizing Australia).

Species	Tissue	^{241}Am	^{137}Cs	^{210}Pb	^{210}Po	$^{239/240}Pu$	^{226}Ra	U*	Th*	Notes	Reference
Chicken	Eggs		7.4 (5–11)							Field experiment, Goiânia, Brazil Goiânia TFs were derived from reported data.	Amaral et al., 1995
	Muscle		37 (21–55)								
	Viscera		21 (11–31)								
Kangaroo	Muscle			3.6 (1–9)			0.1 (0.01–0.5)	0.2 (<0.1–0.4)		Sampled in the vicinity of a uranium deposit (Yeelirrie), Western Australia. Uranium is U-tot. TFs were derived from reported data and did not include calcreted or mineralized soil values.	Williams, 1978
	Kidney			160 (45–370)			1.1 (<0.2–3)	1.4 (0.5–3)			
	Liver			150 (31–430)			2.4 (<0.2–7)	<0.1			
	Bone			57 (14–130)			190 (31–640)	<2.7			
Sheep (*Ovis aries*)	Muscle			2.2 (0.7–4)			0.2 (0.02–0.4)	0.2 (<0.04–0.4)			
	Kidney			19 (12–30)			0.9 (<0.2–1.5)	0.2 (0.08–0.4)			
	Liver			12 (4–29)			0.2 (0.03–0.5)	0.4 (<0.1–0.9)			
	Bone			52 (16–120)			47 (21–100)	12 (<2–56)			

Kangaroo	Muscle			<1.1	530 (330–740)		20 (<15–24)	4.5 (4–5)	3.9	Sampled in the vicinity of a uranium deposit (Manyingee), Western Australia. Uranium appears to be U-tot/^{238}U. Th is ^{230}Th. CRs were derived from reported data.	Lowson and Williams, 1985
	Viscera			<7.1	12 (4–21)		30 (<15–46)	6.3 (4–9)	18 (7–29)		
Sheep (*Ovis aries*)	Muscle			3.0 (<0.4–6)	180 (85–270)		<15	2.0 (1.9–2.0)	11 (7–14)		
	Viscera			1.6 (<0.4–3)	330 (250–420)		27 (<15–40)	7.5 (5–10)	4.3 (4–5)		
Pig (*Sus scrofa*)	Muscle			0.21	49		0.43	0.19	0.19	Sampled in the vicinity of a uranium mine (Ranger), Northern Territory, Australia. Samples gathered from floodplain downstream of mine operations. U is average of ^{234}U and ^{238}U. Th is average of ^{230}Th and ^{232}Th.	Martin et al., 1998
Water buffalo (*Bubalus bubalis*)	Muscle			0.16	2.3		0.26	0.15	0.03		
	Heart			1.6	3.7		0.24	0.07	0.11		
	Tongue			0.34	6.6		0.38	0.38	0.29		
	Kidney			1.4	190		2.8	0.11	0.11		
	Liver			3.1	31		0.63	0.05	0.09		
	Stomach wall				3.5				0.11		
	Intestines				11				0.75		
Goanna (*Varanus gouldi*)	Whole	1	10.6			0.7				Sampled in the vicinity of former nuclear weapons test site (Maralinga), South	Giles et al., 1990

continued

TABLE 5.12 Soil-to-Animal Concentration Ratios (Fresh Weight Tissue /Dry Weight Soil) for Terrestrial Animals (Tropical Studies Emphasizing Australia).—Cont'd

Species	Tissue	^{241}Am	^{137}Cs	^{210}Pb	^{210}Po	$^{239/240}Pu$	^{226}Ra	U*	Th*	Notes	Reference
Kangaroo (*Megeleia rufa*)	Cooked muscle	1 (0.5–12)				1 (0.4–9)				Australia. Cooked samples may include soil contamination due to methods of butchering, cooking, consumption, and storage. Hot particles were known to contaminate some samples. TFs are based on 2.5 cm deep soil samples. Most radioactivity was surface contamination.	
	Cooked viscera	10 (2–110)				9 (1–130)					
Lizard (*Pogona minor, Ctenophorus cristatus, and Ctenophorus nuchalis*)	Whole	5.6 (3.0–9.6)	0.6			3.8 (1.7–7.7)					
Rabbit (*Oryctolagus cuniculus*)	Raw muscle	2 (0.08–83)	13			2 (0.05–16)					
	Cooked muscle	8 (1–11)				7 (0.5–30)					
Witchetty grub (*Endoxyla leaucomochla*)	Whole	0.7 (0.3–10)				0.6 (0.1–7)					
Sheep (*Ovis aries*)	Muscle			2.7 (2–3)			<121		3.04 (2–4)	Sampled in the vicinity of a uranium deposit (Honeymoon), South Australia. TFs were derived from reported data.	Williams, 1980
	Viscera			23.8 (7–42)			<81		3.3 (1–7)		

Fox (*Vulpes vulpes*)	Muscle		9.5 (8–11)	1.4 (0.002–4)	3.4 (2–5)		Sampled in the vicinity of a uranium deposit (Beverly), South Australia. Uranium appears to be U-tot. Th is ^{230}Th. TFs were derived from reported data and include a reported assumption of 30% dry weight/fresh weight. TFs were derived using soil concentrations from Paralana Creek samples.	SAUC, 1982
	Viscera		32	3.8	6.7			
Kangaroo (*Macropus giganteus*)	Heart		14 (11–18)	0.7 (0.3–1.2)	3.8 (2–7)			
	Kidney		360 (270–540)	0.8 (0.2–2)	5.1 (1–13)			
	Liver		520 (300–890)	0.7 (0.1–1)	1.6 (1–2)			
	Muscle		4 (2–6)	0.6 (0.3–0.9)	2.2 (1–3)			
	Stomach contents		6.2	20.9 (6–38)	62 (25–110)			
Rabbit (*Oryctolagus cuniculus*)	Muscle			0.9 (0.6–1)	2.6 (2–3)			
	Stomach contents			41.7	106			
	Viscera			0.8 (0.5–1)	2.6 (2–4)			
Sheep (*Ovis aries*)	Kidney	24 (19–29)	110 (100–120)	2.5 (1–4)	2.9 (<1–5)	6.0		
	Liver	28 (14–41)	34 (18–50)	13 (1–24)	5.3 (<1–10)	3.4		
	Muscle	0.06 (<0.3–1)	4.1 (<1–7)	8.3 (0.007–24)	1.9 (<1–3)	1.3 (1.1–1.4)		

continued

TABLE 5.12 Soil-to-Animal Concentration Ratios (Fresh Weight Tissue /Dry Weight Soil) for Terrestrial Animals (Tropical Studies Emphasizing Australia).—Cont'd

Species	Tissue	^{241}Am	^{137}Cs	^{210}Pb	^{210}Po	$^{239/240}$Pu	^{226}Ra	U*	Th*	Notes	Reference
Beef (steer)	Flesh						0.25 (0.2–0.3)	0.06 (0.04–0.1)	0.06 (0.03–0.1)	Field study of farm animals near Pocos de Caldas, Brazil (see Chapter 1). All values estimated to 1 significant figure from graphical data. U is ^{238}U. Th is ^{232}Th. Data for La are also available.	Linsalata et al., 1991
	Bone						200 (100–300)	0.7 (0.3–1.0)	0.06 (0.04–0.08)		
Pig	Flesh						0.08 (0.03–0.1)	0.02 (0.01–0.04)	0.03 (0.02–0.2)		
	Bone						3.0 (2.0–5.0)	0.8 (0.2–2.0)	0.1 (0.04–0.2)		
Chicken	Flesh						0.4 (0.2–0.7)	0.07 (0.06–0.08)	0.1 (0.05–0.2)		
	Bone						100 (5.0–200)	1.1 (1.0–2.0)	0.2 (0.1–0.4)		

Values are means or reported single values followed by (range), if available to 2 s.f. (All values $\times 10^{-3}$).

The authors would like to particularly acknowledge the principals of Paladin Energy, Ltd. and Uranium One Pty., Ltd. for allowing publication of commercial data acquired for the Manyingee and Honeymoon uranium mine baseline field studies, respectively.

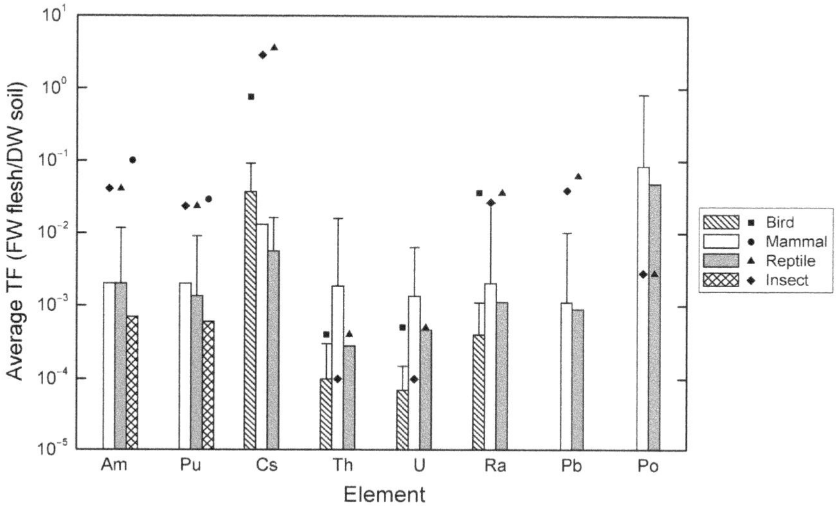

FIG. 5.16 Average soil DW-to-animal FW TFs (flesh for mammals and birds; whole animal for reptiles and insects) for a range of radionuclides in tropical environments. The error bars indicate the maximum range for data where more than 1 datum exists. The dots represent default TF values from the ERICA software for equivalent organisms.

average, tropical systems are similar to temperate zones and their responses to environmental change follow expected radiological trends. The differences only become apparent when one starts to plot individual tropical studies against the evenly spread results obtained in temperate regions; then the tropical examples can be seen to trend to more extreme outcomes, both high and low. The reasons for this cover a spectrum of factors.

- Tropical soils are more strongly weathered and have less nutrients and a range of conditions that adversely affect growth as a consequence of the higher temperatures and, in some areas, rainfall to which they are exposed. As for saturated soils, paddy conditions are unique to tropical areas. Different soil types give rise to different plant uptake characteristics and the soils in tropical regions are often substantially different to those in temperate regions.
- Soil parameters, such as reduced CEC, low pH, lower dissolved concentrations of stable nutrient analogues for the radionuclides, and other factors typical of tropical systems, can lead to enhanced bioaccumulation. On the other hand, high organic loads in tropical farms or forests or high levels of elements such as Ca in tropical coral soils, can dramatically reduce radionuclide accumulation. Hence, different management strategies can play a part.
- Increased rainfall and temperature leads to faster leaching of radionuclides from the soil profile and consequent lower bioaccumulation due to reduced bioaccessibility. Radionuclide turnover rates, due to a range of biotic and abiotic factors, are also generally higher in warmer climates, which can alter bioavailability. The same conditions can enhance the

ageing effect, which reduces bioavailability. Both conditions (lower bioaccessibility and reduced bioavailability) lead to reduced uptake into plants.
- Tropical organisms tend to grow more quickly under increased temperatures and rainfall regimes so as to better compete for limited resources, leading to faster metabolisms and uptake rates. But these conditions can also give rise to mass dilution and increased excretion, which will tend to reduce overall activity concentrations. Either process can push bioaccumulation toward extreme values.
- For plants, at least, warm, moist conditions can give rise to physiognomic features, such as deep or widely spread roots or large leaves, which can exacerbate bioaccumulation, in contrast to less 'ostentatious' vegetation. Those conditions can also favor the development of fungal communities that will increase the plant access to soil nutrients and radionuclides.

There is an inadequate quantity of data on terrestrial animals to determine whether distinct differences occur between temperate and tropical systems, although there is some indication that dissimilarities may exist. However, based on the available information for plants, tropical systems can be seen to have differences that need to be assessed on a case-by-case basis. The variables that may influence the overall radionuclide behaviour are so numerous, and the related biochemical responses so complex, that it is often very difficult to determine how radiological trends will occur prior to observation. Nonetheless, the relevant factors described within this chapter all have some influence on the overall outcome and are the natural consequence of occurring within tropical and subtropical environments.

ACKNOWLEDGMENTS

The principles and contractors of Paladin Energy, Ltd. and Uranium One Pty., Ltd. are gratefully acknowledged for allowing publication of commercial data on animals that contributes to a very limited international set. We would particularly like to acknowledge Drs. G. Voigt and H. Velasco for their help reviewing this chapter. We also thank Dr. M. Johansen, Australian Nuclear Science and Technology Organisation (ANSTO), for assistance in compiling the TF data on animals; J. Veliscek-Carolan for constructive comments on the draft versions of the chapter; Dr. Shigeo Uchida for help on the section relating to rice; and Dr. J. Wasserman for assistance on the section relating to soil organic matter. Drs. S. Hollins and C. Hughes, ANSTO, provided help with some of the figures.

REFERENCES

The references marked with an asterisk (*) provided TF data on rice in Tables 5.9a and 5.9b.

Abril, J.M., Fraga, E., 1996. Some physical and chemical features of the variability of kd distribution coefficients for radionuclides. J. Environ. Radioact. 30 (3), 253–270.

Absalom, J.P., Young, S.D., Crout, N.M.J., Nisbet, A.F., Woodman, R.F.M., Smolders, E., et al., 1999. Predicting soil to plant transfer of radiocesium using soil characteristics. Environ. Sci. Technol. 33 (8), 1218–1223.

Absalom, J.P., Young, S.D., Crout, N.M.J., Sanchez, A., Wright, S.M., Smolders, E., et al., 2001. Predicting the transfer of radiocaesium from organic soils to plants using soil characteristics. J. Environ. Radioact. 52, 31–43.

Adriano, D.C., Wallace, A., Romney, E.M., 1980. Uptake of transuranic nuclides from soil by plants grown under controlled environmental conditions. In: Hanson, W.C. (Ed.), Transuranic Elements in the Environment. US Dept. of Energy, pp. 336–360. DOE/TIC-22800.

Al-Oudat, M., Asfary, A.F., Mukhalallti, H., Al-Hamwi, A., Kanakri, S., 2006. Transfer factors of ^{137}Cs and ^{90}Sr from soil to trees in arid regions. J. Environ. Radioactiv. 90, 78–88.

Amaral, E.C.S., Paretzke, H.G., Campos, M.J., Pires do Rio, M.A., Franklin, M., 1995. Transfer of ^{137}Cs from soil to chicken meat and eggs. J. Environ. Radioact. 29 (3), 237–255.

Basson, J.K., van As, D., Vleggaar, C.M., 1973. Airborne and deposited radioactive fallout over southern Africa and the resultant biological uptake of ^{90}Sr, ^{137}Cs, and ^{131}I. Invited paper. In: Proceedings of the Regional Conference on Radiation Protection. Jerusalem, 5-8 March, 1973. Soreq Nuclear Research Centre, Israel.

Bennett, R., Willey, N., 2003. Soil availability, plant uptake, and soil to plant transfer of Tc-99 - A review. J. Environ. Radioact. 65 (2), 215–231.

*Binh, N.T., Quang, N.H., T, Y., Sieu, L.N., Ngo, N.T., Phuc, N.V., et al., 2003. The transfer of radionuclides in the terrestrial environments – Recent research results in monsoon tropical condition of Vietnam. In: Proceedings of the International Symposium on Transfer of Radionuclides in Biosphere – Prediction and Assessment. JAERI-Conf 2003-010 (2003).

Blanco Rodrıguez, P., Vera Tome, F., Lozano, J.C., 2002. About the assumption of linearity in soil-to-plant transfer factors for uranium and thorium isotopes and ^{226}Ra. Sci. Total Environ. 284, 167–175.

Blanco Rodrıguez, P., Vera Tome, F., Perez Fernandez, M., Lozano, J.C., 2006. Linearity assumption in soil-to-plant transfer factors of natural uranium and radium in *Helianthus annuus* L. Sci. Total Environ. 361, 1–7.

Boaretto, A.E., Roque, M.L., Villanueva, F.C.A., Nascimento Filho, V.F., Muraoka, T., Tagliaferro, F.S., 2006. Absorção de Césio-137 por feijoeiro cultivado em solos de região tropical. In: Gallardo Lancho, J.F. (Ed.), Medioambiente em Iberoamerica: Vision desde la Física Y la Química en los Albores del Siglo XXI.

Calmon, P., Fesenko, S., Voigt, G., Linsley, G., 2009. Quantification of radionuclide transfer in terrestrial and freshwater environments. J. Environ. Radioact. 100, 671–674.

Carini, F., 1999. Radionuclides in plants bearing fruit: An overview. J. Environ. Radioact. 46, 77–97.

Carini, F., 2001. Radionuclide transfer from soil to fruit. J. Environ. Radioact. 52, 237–279.

Carini, F., Atkinson, C.J., Collins, C., Coughtrey, P.J., Eged, K., Fulker, M., et al., 2005. Modelling and experimental studies on the transfer of radionuclides to fruit. J. Environ. Radioact. 84, 271–284.

Casadesus, J., Sauras-Yera, T., Vallejo, V.R., 2008. Predicting soil-to-plant transfer of radionuclides with a mechanistic model (BioRUR). J. Environ. Radioact. 99, 864–871.

Centofanti, T., Penfield, R., et al., 2005. Is the transfer factor a relevant tool to assess the soil-to-plant transfer of radionuclides under field conditions? J. Environ. Qual. 34 (6), 1972–1979.

*Choi, Y.H., et al., 2001. Parameters on the radionuclide transfer in crop plants for Korean food chain dose assessment. KAERI/TR-1993. Korea Atomic Energy Research Institute, p. 1–55 (2001) (in Korean).

*Choi, Y.H., Lim, K.M., Choi, H.J., Choi, G.S., Lee, H.S., Lee, C.W., 2005. Plant uptake and downward migration of ^{85}Sr and ^{137}Cs after their deposition on to flooded rice fields: lysimeter experiments with and without the addition of KCl and lime. J. Environ. Radioact. 78, 35–49.

Cronan, C.S., Aiken, G.R., 1985. Chemistry and transport of soluble humic substances in forested watersheds of the Adirondack Park, New York. Geochimica et Cosmochimica Acta 49 (8), 1697–1705.

Deckers, J., Driessen, P., Nachtergaele, F., Spaargaren, O., Berding, F., 2002. Towards a rationale in the world reference base for soil resources. 17th WCSS, 14–21 August 2002, Thailand. Paper 843. 14 Pp.

DeForest, D.K., Brix, K.V., Adam, W.J., 2007. Assessing metal bioaccumulation in aquatic environments: The inverse relationship between bioaccumulation factors, trophic transfer factors and exposure concentration. Aquat. Toxicol. 84, 236–246.

Delmas, J., Grauby, A., Paulin, R., Rinaldi, C., Ndoye, R., Seckgassama, S., 1987. Etude experimentale du transfert du ^{90}Sr et du ^{137}Cs d'un sol de la region du Cap Vert a l'arachide. Radioprotection 22, 333–340.

Dixon, J., Gulliver, A., Gibbon, D., 2001. Global challenges, potentials, and priorities. In: Hall, M. (Ed.), Farming Systems and Poverty: Improving Farmers' Livelihoods in a Changing World. FAO and World Bank, Rome and Washington D.C, pp. 307–357 (Chapter 8).

Duff, M.C., Ramsey, M.L., 2008. Accumulation of radiocesium by mushrooms in the environment: A literature review. J. Environ. Radioact. 99, 912–932.

*Duffa, C., Renaud, P., Goutelard, F., 2002. Activities and transfers of Pu and Am in rice samples from Camargue, France. J. Radioanal. Nucl. Chem. 252, 247–248.

Dupre´ de Boulois, H., Joner, E.J., Leyval, C., Jakobsen, I., Chen, B.D., Roos, P., et al., 2008. Role and influence of mycorrhizal fungi on radiocaesium accumulation by plants. J. Environ. Radioact. 99, 785–800.

Ehlken, S., Kirchner, G., 2001. Environmental processes affecting plant root uptake of radioactive trace elements and variability of transfer factor data: A review. J. Environ. Radioact. 58 (2–3), 97–112.

Eisenbud, M., 1987. Environmental radioactivity: From natural, industrial, and military sources. 3rd Ed. Academic Press (Orlando).

Food and Agriculture Organization, 2005. A System of Integrated Agricultural Concensuses and Surveys. World Programme for the Census of Agriculture 2010. Vol. 1. FAO, p. 161.

FAO/IAEA/IUR, 1998. Protocol for experimental studies on the uptake of radionuclides from soil by plants. International Atomic Energy Agency, Vienna.

Fontana, A., 2006. Caracterização Química e Espectroscópica da Matéria Orgânica em Solos do Brasil. Dissertação (Mestrado) – Universidade Federal Rural do Rio de Janeiro Instituto de Agronomia, p. 60.

Frissel, M.J., Deb, D.L., Fathony, M., Lin, Y.M., Mollah, A.S., Ngo, N.T., et al., 2002. Generic values for soil-to-plant transfer factors of radiocesium. J. Environ. Radioact. 58 (2–3), 113–128.

Garrett, S.D., 1981. Soil Fungi and Soil Fertility: An Introduction to Mycology, 2nd ed. Pergamon Press, Oxford.

Giles, M.S., Twining, J.R., Williams, A.R., Jeffree, R.A., Domel, R.U., 1990. Final Report to the Technical Assessment Group for the Maralinga Rehabilitation Project. Study No. 2. (Radioecology). Technical Assessment Group Study Program Reports, Vol. 1. Department of Primary Industries and Energy. Australian Government Printing Service, Canberra.

Gil-Garcıa, C., Rigol, A., Vidal, M., 2009a. New best estimates for radionuclide solid–liquid distribution coefficients in soils. Part 1: Radiostrontium and radiocaesium. J. Environ. Radioact. 100, 690–696.

Gil-Garcıa, C., Tagami, K., Uchida, S., Rigol, A., Vidal, M., 2009b. New best estimates for radionuclide solid–liquid distribution coefficients in soils. Part 3: Miscellany of radionuclides (Cd, Co, Ni, Zn, I, Se, Sb, Pu, Am, and others). J. Environ. Radioact. 100, 704–715.

Gri, N., 2000. ^{137}Cs and ^{90}Sr behaviour in contaminated soils of the Chernobyl area. (In French) CEA / IPSN Report CEA-R–5911.

Harries, J.R., KIrby, J.M., Payne, T.E., Smiles, D.E., 1998. Technical studies for site selection of a national low-level radioactive waste repository. 4. Vadose zone hydrology and radionuclide retardation. Report to the Department of Primary Industries and Energy, Bureau of Resource Science. Australian Nuclear Science and Technology Organisation report ANSTO/C536.

Hinsinger, P., Plassard, C., Tang, C., Jaillard, B., 2003. Origins of root-mediated pH changes in the rhizosphere and their responses to environmental constraints: A review. Plant and Soil 248, 43–59.

Hird, A.B., Rimmer, D.L., Livens, F.R., 1996. Factors affecting the sorption and fixation of cesium in acid organic soil. European Journal of Soil Science 47 (1), 91–104.

*IAEA, 1994. Handbook of Parameter Values for the Prediction of Radionuclide Transfer in Temperate Environments. International Atomic Energy Agency, Technical Report Series No. 364. IAEA, Vienna.

IAEA, 2009. Quantification of Radionuclide Transfer in Terrestrial and Freshwater Environments for Radiological Assessments. International Atomic Energy Agency Technical Report Series IAEA-TECDOC 1616. IAEA, Vienna.

IAEA, 2010. Handbook of Parameter Values for the Prediction of Radionuclide Transfer in Terrestrial and Freshwater Environments. International Atomic Energy Agency Technical Reports Series No. 472. IAEA, Vienna.

IAEA, 2012. Handbook of Parameter Values for the Prediction of Radionuclide Transfer to Wildlife. International Atomic Energy Agency Technical Report Series (in draft). IAEA, Vienna in press.

ICRU, 2001. International Commission on Radiation Units and Measurements Report 65. Quantities, Units and Terms in Radioecology. Journal of the ICRU 1 (2), 48.

*Ichikawa, R., Eto, M., Abe, M., 1962. Strontium-90 and cesium-137 absorbed by rice plants in Japan, 1960. Science 135, 1072.

IUR, 1989. VI Report of the Working Group Soil-to Plant Transfer Factors. IUR, Bilthoven, Netherlands, 1989.

IUSS Working Group WRB, 2007. World Reference Base for Soil Resources 2006, first update 2007. *World Soil Resources Reports No. 103* FAO, Rome.

Johansen, M.P., Twining, J.R., 2010. Radionuclide concentration ratios in Australian terrestrial wildlife and livestock: data compilation and analysis. Radiat. Environ. Biophys. 49 (4), 603–611. http://dx.doi.org/10.1007/s00411-010-0318-9.

*Komamura, M., Tsumura, A., 1994. The transfer factors of long-lived radionuclides from soil to polished rice measured by ICP-MS. Radioisotopes 43, 1–8 (Abstract in Japanese).

*Komamura, M., Tsumura, A., Yamaguchi, N., Kihou, N., Kodaira, K., 2006. Monitoring ^{90}Sr and ^{137}Cs in rice, wheat, and soil in Japan from 1959 to 2000. Miscellaneous publication of National Institute for Agro Environmental Sciences No. 28, pp. 1–56.

Koranda, J.J., Robison, W.L., 1978. Accumulation of radionuclides by plants as a monitor system. Environ. Health Perspect. 27, 165–179.

Kottek, M., Grieser, J., Beck, C., Rudolf, B., Rubel, F., 2006. World Map of the Köppen-Geiger climate classification updated. Meteorol. Z. 15, 259–263. http://dx.doi.org/10.1127/0941-2948/2006/0130.

*Krouglov, S.V., Filipas, A.S., Alexakhin, R.M., Arkhipov, N.P., 1997. Long-term study on the transfer of ^{137}Cs and ^{90}Sr from Chernobyl-contaminated soils to grain crops. J. Environ. Radioact. 34, 267–286.

*Lakshamanan, A.R., Venkateswarlu, 1988. Uptake of uranium by vegetables and rice. Water, Air, Soil Poll. 38, 151–155.

*Lauria, D.C., Sachett, I., Pereira, J.C.A., Zenaro, R., 1994. Determination of caesium-137 soil to plant concentration ratios for vegetables in Goiânia city. Brazil. J. Radioanal. Nucl. Ch. 182 (1), 91–96.

*Leung, J.K.C., Shang, Z.R., 2003. Uptake of ^{137}Cs and ^{90}Sr in rice plants. Health Phys. 84, 170–179.

*Li, J.G., Shang, Z.R., Yang, J.C., Tang, S.M., 2006. Radionuclide transfer parameter manual. Atomic Energy Press, Beijing (in Chinese, English version from Shang, Z. R.).

Lima, V.S.T.G. de., 1988. Comparative evaluation of endogenous and exogenous ^{226}Ra partitioning and uptake in farm soils of Pocos de Caldas, MG, Brazil: way of occurrence and plant availability. (M.Sc. Thesis, Universidade Federal, Rio de Janeiro, RJ (Brazil). Inst. de Biofisica). Report available from the Nuclear Information Center of Comissao Nacional de Energia Nuclear, RJ, Brazil. 218 p.

Linsalata, P., Morse, R., Ford, H., Eisenbud, M., 1991. Th, U, Ra and rare earth element distributions in farm animal tissues from an elevated natural radiation background environment. J. Environ. Radioact. 14, 233–257.

Lowson, R.T., Williams, A.R., 1985. A baseline radioecological survey, Manyingee uranium prospect. W.A. Report to Total Mining Australia Pty Ltd. AAEC/C47. Australian Atomic Energy Commission, Lucas Heights.

*Lu, J., Huang, Y., Li, F., Wang, L., Li, S., Hsia, Y., 2006. The investigation of ^{137}Cs and ^{90}Sr background radiation levels in soil and plant around Taiwan NPP. China. J. Environ. Radioact. 90, 89–99.

Luoma, S.N., Rainbow, P.S., 2005. Why is metal bioaccumulation so variable? Biodynamics as a unifying concept. Environ. Sci. Technol. 39 (7), 1921–1931. http://dx.doi.org/10.1021/es048947e, Publication Date (Web): 25 February 2005.

Martin, P., Hancock, G.J., Johnston, A., Murray, A.S., 1998. Natural-series radionuclides in traditional north Australian Aboriginal foods. J. Environ. Radioact. 40, 37–58.

McGee, E.J., Johanson, K.J., Keatinge, M.J., Synnott, H.J., Colgan, P.A., 1996. An evaluation of ratio systems in radioecological studies. Health Phys. 70 (2), 215–221.

*Mollah, A.S., Begum, A., Ullah, S.M., 1998. Determination of soil-to-plant transfer factors of ^{137}Cs and ^{90}Sr in the tropical environment of Bangladesh. Radiat. Environ. Biophys. 37, 125–128.

*Morishima, H., Koga, T., Kawai, H., Honda, Y., Katsurayama, K., 1977. Studies on the movement and distribution of uranium in the environment-distribution of uranium in agricultural products. J. Radiat. Res. 18, 139–150.

*Mousny, J.M., Myttenaere, C., 1981. Absorption of technetium by plants in relation to soil type, contamination level, and time. Plant Soil 61, 403–412.

*Muramatsu, Y., Uchida, S., Ohmomo, Y., 1993. Root-uptake of radioiodine by rice plants. J. Radiat. Res. 34, 214–220.

*Muramatsu, Y., Uchida, S., Sumiya, M., Ohmomo, Y., Obata, H., 1989. Tracer experiments on transfer of radio-iodine in the soil-rice plant system. Water Air Soil Poll 45, 157–171.

Muramatsu, Y., Uchida, S., Sumiya, M., Ohmomo, Y., 1990. Transfer of radioiodine from the environment to rice plants. In: Transfer of Radionuclides in Natural and Semi-Natural Environments. Elsevier, London.

*Myttenaere, C., Bourdeau, P., Masset, M., 1969. Relative importance of soil and water in the indirect contamination of flooded rice with radiocaesium. Health Phys. 16, 701–707.

*Narayana, Y., Radhakrishna, A.P., Somashekarappa, H.M., Karunakara, N., Balakrishna, K.M., Shiddappa, K., 1995. Distribution of some natural and artificial radionuclides in the environment of coastal Karnataka of south India. J. Environ. Radioact. 28 (2), 113–139.

National Research Council, 2008. Earth Observations from Space: The First 50 Years of Scientific Achievements. Committee on Scientific Accomplishments of Earth Observations from

Space, National Research Council ISBN: 978-0-309-11095-2, p. 142. http://www.nap.edu/catalog/11991.html.
*Noordijk, H., van Bergeijk, K.E., Lembrechts, J., Frissel, M.J., 1992. Impact of aging and weather conditions on soil-to-plant transfer of radiocesium and radiostrontium. J. Environ. Radioact. 15 (3), 277–286.
Norman, M.J.T., Pearson, C.J., Searle, P.G.E., 1995. The Ecology of Tropical Food Crops, 2nd ed. Cambridge University Press, Cambridge, U.K.
Othman, I., Yassine, T., 1995. Transfer of radiocaesium from soil to some crops in semi-arid areas of Syria. Proceedings of Environmental impact of radioactive releases. IAEA, location, pp. 630–632.
*Parekh, N.R., Poskitt, J.M., Dodd, B.A., Potter, E.D., Sanchez, A., 2008. Soil microorganisms determine the sorption of radionuclides within organic soil systems. J. Environ. Radioact. 99, 841–852.
Peel, M.C., Finlayson, B.L., McMahon, T.A., 2007. Updated world map of the Köppen-Geiger climate classification. Hydrol. Earth Syst. Sci. 11, 1633–1644.
Pereira, T.R., 2009. Influence of organic fertilization on the sorption mechanisms of ^{241}Am in tropical soils (In Portuguese). Report of Instituto de Radioprotecao e Dosimetria (IRD/CNEN-RJ), Rio de Janeiro, RJ (Brazil) 117 p, INIS-BR–8760.
*Quang, N.H., Nhan, D.D., Toan, P.V., Nga, V.T., 2003. Use of tracer technique to investigate the dependence of soil-to-plant transfer factor of Sr and Cs on the type of soils. In: United Papers of International Conference on Isotopic and Nuclear Analytical Techniques for Health and the Environment. IAEA-CN-103-032, 22/CD (2003).
Rabesiranana, N., Rasolonirina, M., Terina, F., Solonjara, A.F., Andriambololona, R., 2008. Top soil radioactivity assessment in a high natural radiation background area: The case of Vinaninkarena, Antsirabe—Madagascar. Appl. Radiat. Isot. 66 (11), 1619–1622.
*Radhakrishna, A.P., Somashekarappa, H.M., Narayana, Y., Shiddappa, K., 1996. Distribution of some natural and artificial radionuclides in Mangalore environment of South India. J. Environ. Radioact. 30, 31–54.
*Rahman, M.M., Chand, M.M., Koddus, A., Rahman, M.M., Zaman, M.A., Voigt, G., 2008. Transfer of radiocobalt from soil to selected plant species in tropical environments. J. Environ. Radioact. 98, 658–664.
*Rahman, M.M., Rahman, M.M., Koddus, A., Ahmad, G.U., Voigt, G., 2005. Soil-to-plant transfer of radiocaesium for selected tropical plant species in Bangladesh. J. Environ. Radioact. 82, 199–211.
Rainbow, P.S., 2002. Trace metal concentrations in aquatic invertebrates: why and so what? Environ. Pollut. 120, 497–507.
Riise, G., Bjornstad, H.E., Lien, H.N., Oughton, D.H., Salbu, B., 1990. J. Radioanal. Nucl. Chem. 142 (2), 531–538.
Robison, W.L., Conrado, C.L., Hamilton, T.F., Stoker, A.C., 2000. The effect of carbonate soil on transport and dose estimates for long-lived radionuclides at a U.S. Pacific Test Site. J. Radioanal. Nucl. Chem 243 (2), 459–465.
Robison, W.L., Brown, P.H., Stone, E.L., Hamilton, T.F., Conrado, C.L., Kehl, S., 2009. Distribution and ratios of ^{137}Cs and K in control and K-treated coconut trees at Bikini Island where nuclear test fallout occurred: Effects and implications. J. Environ. Radioact. 100, 76–83.
Rosholt, J.N., 1982. Mobilization and weathering. In: Ivanovich, M., Harmon, R.S. (Eds.), Uranium Series Disequilibrium: Applications to Environmental Problems. Clarendon Press, Oxford, pp. 167–180.
Russell, R.A., Holden, P.J., Payne, T.E., McOrist, G.D., 2004. The effect of sulfate-reducing bacteria on adsorption of ^{137}Cs by soils from arid and tropical regions. J. Environ. Radioact. 74, 151–158.

*Sachdev, P., Sachdev, M.S., Deb, D.L., 1998. The uptake and transfer of caesium-137, strontium-90, and zinc-65 from soil to food crops in tropical environment. J. Nuclear Agriculture and Biology 27, 1–9.

Sanchez, P.A., 2000. Linking climate change research with food security and poverty reduction in the tropics. Agriculture, Ecosystems, and Environment 82, 371–383.

Sánchez, P.A., Buol, S., 1975. Soils of the tropics and the world food crisis. Science 188, 598–603.

SAUC, 1982. Beverley Project: Draft Environmental Impact Statement. South Australian Uranium Corporation, Adelaide.

*Sasaki, T., Tashiro, Y., Fujinaga, H., Ishii, T., Gunji, Y., 2002. Determination of transfer factors of uranium, thorium, radium, and lead from soil to agricultural product in Japan for estimating internal radiation dose through ingestion. Japanese J. Health Phyis 37, 208–221. (Abstract in English).

Scardino, A.M.S., Helene, O.A.M., 1990. Mechanics of ^{137}Cs diffusion and transport in the environment. (Summary, in Portuguese) Report Socieda de Brasileira de Fisica, Sao Paulo, SP (Brazil), Proceedings of the 13th Workshop on Nuclear Physics in Brazil, 382 p, 1990, p. 327. INIS-BR–2702.

Selagen, P., 1971. Metallic Oxides and Hydroxides in Soils of the Warm and Humid Areas of the World: Formation, Identification, Evolution. In: Soils and Tropical Weathering. UNESCO, Paris (Chapter 2).

Sheppard, M.I., 1980. The Environmental Behaviour of Uranium and Thorium. AECL, Atomic Energy Canada Ltd, p. 6795.

Sheppard, M.I., Sheppard, S.C., 1985. The plant concentration concept as applied to natural uranium. Health Phys. 48, 494–500.

*Sheppard, S.C., Evenden, W.G., Pollock, R.J., 1989. Uptake of natural radionuclides by field and garden crops. Can. J. Soil Sci. 69, 751–767.

Simon, S.L., Graham, J.C., Terp, S.D., 2002. Uptake of ^{40}K and ^{137}Cs in native plants of the Marshall Islands. J. Environ. Radioact. 59, 223–243.

Simon, S.L., Ibrahim, S.A., 1987. The plant/soil concentration ratio for calcium, radium, lead and polonium: Evidence for non-linearity with reference to substrate concentration. J. Environ. Radioact. 5, 123–142.

Simon, S.L., Ibrahim, S.A., 1990. Biological uptake of radium by terrestrial plants. The Environmental Behaviour of Radium,Vol. 1. IAEA Technical Report Series No. 310, Vienna 1990 (Chapter 5–6).

Skarlou, V., Nobeli, C., Anoussis, J., Haidouti, C., 1997. Soil to plant transfer of ^{134}Cs for olive and orange trees. In: Gerzabek, M.H. (Ed.), Proceedings of the XXVII Annual Meeting of ESNA, WG 3, 183–188.

Sokolik, G.A., Ivanova, T.G., Leinova, S.L., Ovsiannikova, S.V., Kimlenko, I.M., 2001. Migration ability of radionuclides in soil-vegetation cover of Belarus after Chernobyl accident. Environ. Int. 26 (3), 183–187.

Staunton, S., 2002. Direct and indirect effects of organic matter on metal immobilisation in soil. Dev. Soil. Sci. 28, Part A, 79–97.

Staunton, S., Dumat, C., Zsolnay, A., 2002. Possible role of organic matter in radiocaesium adsorption in soils. J. Environ. Radioactiv. 58, 163–173.

Strakov, N.M., 1967. Principals of Lithogenesis, vol. 1. Oliver & Boyd, Edinburgh, U.K.

Supervising Scientist, 2009. Annual Report 2008–2009. Supervising Scientist, Darwin, Australia.

Syers, J.K., 1997. Managing soils for long-term productivity. Philos. Trans. Roy. Soc. Bio. Sci. 352 (1356), 1011–1021.

*Tagami, K., Uchida, S., 2009. Unpublished data.

Tagami, K., Uchida, S., 1996. Aging effect on technetium behaviour in soil under aerobic and anaerobic conditions. Toxicol. Environ. Chem. 56, 235–247.

Tagami, K., Uchida, S., 2004. Comparison of transfer and distribution of technetium and rhenium in radish plants from nutrient solution. Appl. Radiat. Isot. 61, 1203–1210.

Tagami, K., Uchida, S., 2005. A comparison of concentration ratios for technetium and nutrient uptake by three plant species. Chemosphere 60, 714–717.

*Takagi, H., Kimura, T., Kobayashi, H., Iwashima, K., Yamagata, N., 1985. Transfer of iodine from paddy soil to rice grain. Japanese J. Health Phyis. 20, 251–257. (Abstract in English).

Tamponnet, C., Martin-Garin, A., Gonze, M.A., Parekh, N., Vallejo, R., Sauras-Year, T., et al., 2008. An overview of *BORIS*: *B*ioavailability *o*f *R*adionuclides *i*n *S*oils. J. Environ. Radioact. 99, 820–830.

Tracy, B.L., Prantl, F.A., Quinn, J.M., 1983. Transfer of ^{226}Ra, ^{210}Pb, and uranium from soil to garden produce: assessment of risk. Health Phys. 44, 469–477.

*Tsukada, H., Hasegawa, H., Hisamatsu, S., Yamasaki, S., 2002a. Rice uptake and distribution of radioactive ^{137}Cs, stable ^{133}Cs, and K from soil. Environ. Pollut. 117, 403–409.

*Tsukada, H., Hasegawa, H., Hisamatsu, S., Yamasaki, S., 2002b. Transfer of ^{137}Cs and stable Cs from paddy soil to polished rice in Aomori, Japan. J. Environ. Radioact. 59, 351–363.

*Tsukada, H., Hasegawa, H., Takeda, A., Hisamatsu, S., 2007. Concentrations of major and trace elements in polished rice and paddy soils collected in Aomori, Japan. J. Radioanal. Nucl. Chem. 273, 199–203.

*Tsukada, H., Takeda, A., Takahashi, T., Hasegawa, H., Hisamatsu, S., Inaba, J., 2005. Uptake and distribution of ^{90}Sr and stable Sr in rice plants. J. Environ. Radioact. 81, 221–231.

Twining, J.R., Zaw, M., Russell, R., Wilde, K., 2004. Seasonal changes of redox potential and microbial activity in agricultural soils of tropical Australia: Implications for soil-to-plant transfer of radionuclides. J. Environ. Radioact. 76, 265–272.

Twining, J., Shotton, P., Tagami, K., Payne, T., Itakura, T., Russell, R., et al., 2003. Final report on a field study of soil-to-plant transfer of radioactive caesium, strontium and zinc in tropical northern Australia to the IAEA/FAO/IUR CRP on. Australian Nuclear Science and Technology Organisation technical report, ANSTO-E-734. Classification of Soils Systems on the Basis of Transfer Factors of Radionuclides from Soil to Reference Plants. Lucas Heights, Sydney, Australia, September, 2003, p. 27.

Uchida, S., 2007. Radionuclides in tropical and subtropical ecosystems. In: Shaw, G. (Ed.), Radioactivity in the Terrestrial Environment. Environmental Radioactivity Series, Elsevier (Chapter 9).

*Uchida, S., Tagami, K., 2007a. Soil-to-plant transfer factors of Cs-137 and stable Cs in various crops collected in Japan. J. Radioanal. Nucl. Chem. 273, 205–210.

*Uchida, S., Tagami, K., 2007b. Soil-to-crop transfer factors of radium in Japanese agricultural fields. J. Nucl. Radiochem. Sci. 8, 103–108.

*Uchida, S., Tagami, K., 2009. Tranfer of radium-226 from soil to rice: A comparison of sampling area differences. J. Nucl. Sci. Technol. 46, 49–54.

*Uchida, S., Tagami, K., Hirai, I., 2007. Soil-to-plant transfer factors of stable elements and naturally occurring radionuclides: (2) Rice collected in Japan. J. Nucl. Sci. Technol. 44, 779–790.

United Nations, 2005. Central Product Classification (CPC) Version 1.1. UN 776 pages, 2005.

Vandenhove, H., Gil-Garcıa, C., Rigol, A., Vidal, M., 2009. New best estimates for radionuclide solid–liquid distribution coefficients in soils. Part 2. Naturally occurring radionuclides. J. Environ. Radioact. 100, 697–703.

Vandenhove, H., Hees, M.V., Van Winckel, S., 2001. Feasibility of phytoextraction to clean up low-level uranium-contaminated soil. Int. J. Phytoremediation 3, 301–320.

Velasco, H., Juri Ayub, J., Sansone, U., 2008. Analysis of radionuclide transfer factors from soil to plant in tropical and subtropical environments. Appl. Radiat. Isot. 66, 1759–1763.

Walker, R., Gessel, S., Held, E., 1997. The ecosystem study on Rongelap atoll. Health Phys. 73, 223–233.

*Wang, J.J., Wang, C.J., Huang, C.C., Lin, Y.M., 1998. Transfer factors of ^{90}Sr and ^{137}Cs from paddy soil to the rice plant in Taiwan. J. Environ. Radioact. 39, 23–34.

Wasserman, M.A., Belém, L.M.J., 1996. Valores de transferência do ^{137}Cs de Latossolos para plantas comestíveis. In: Anais do VI Congresso Geral de Energia Nuclear. Rio de Janeiro, 1996.

Wasserman, M.A., 1998. The behaviour of caesium-137 in oxisols and in the Goiania soil. In: Goiania, Ten Years Later. IAEA, Vienna.

Wasserman, M.A., Perez, D.V., Bourg, A., 2002. Behaviour of ^{137}Cs in Some Brazilian Oxisols. Commun. Soil Sci. Plant Anal. New York 33 (7&8), 1335–1349.

Wasserman, M.A., Bartoly, F., Portilho, A.P., Rochedo, E.R., Viana, A.G., Perez, D.V., et al., 2008a. J. Environ. Radioact. 99 (3), 554–562.

Wasserman, M.A., Bartoly, F., Viana, A.G., Silva, M.M., Rochedo, E.R.R., Perez, D.V., et al., 2008b. J. Environ. Radioact. 99 (3), 546–553.

Whicker, F.W., Shultz, V., 1982. Radioecology: Nuclear energy and the environment. CRC Press, 1982.

Willey, N., Fawcett, K., 2006. A phylogenetic effect on strontium concentrations in angiosperms. Environmental and Experimental Botany 57 (3), 258–269.

Williams, A.R., 1978. The distribution of some naturally occurring elements in the environment of the Yeelirrie uranium deposit, Western Australia. Report 1. In: Brownscombe, A.J., Davy, D.R., Giles, M.S., Williams, A.R. (Eds.), Three Baseline Studies in the Environment of the Uranium Deposit at Yeelirrie, Western Australia. Australian Atomic Energy Commission, Lucas Heights, AAEC/E442.

Williams, A.R., 1980. A Report to Gutteridge, Haskins and Davey Pty Ltd on some background radiological data for the proposed Honeymoon uranium development, South Australia. Australian Atomic Energy Commission, Report AAEC/C3. Lucas Heights, August 1980.

*Yamagata, N., Kodaira, K., Matsuda, S., 1965. Cesium-137 in Japanese soil and rice. J. Radiat. Res. 6, 32–44.

*Yanagisawa, K., Muramatsu, Y., 1992. Transfer of technetium from soil to paddy and upland rice. J. Radiat. Res. 41, 397–402.

Yasuda, H., Uchida, S., 1995. Apparent non-linear relationship of radionuclide concentrations between crop and soil. Environmental impact of radioactive releases. Proceedings of a Symposium. Vienna, Austria, 8-12 May 1995: 259–267.

*Yunoki, E., Kataoka, T., Michiro, K., Sugiyama, H., Shimizu, M., Mori, T., 1993. Activity concentrations of ^{238}U and ^{226}Ra in agricultural samples. J. Radioanal. Nucl. Chem 174, 223–228.

Chapter 6

Radioecology of Tropical Freshwater Ecosystems: Mechanisms and Kinetics of Bioaccumulation and the Importance of Water Chemistry

Scott J. Markich[1] and John R. Twining[2]
[1]*Aquatic Solutions International, Australia*
[2]*Austral Radioecology, Australia*

6.1. INTRODUCTION

Radioecology describes the behaviour and effect of radionuclides in ecosystems and the biota within them. A radionuclide is an unstable form of an element that decays, emitting ionising radiation (as defined in Chapter 1). This chapter will extend discussion to include chemically analogous stable elements, encompassing metals (e.g., U and Pb) and metalloids (e.g., As and Po), particularly given that radionuclides are often used to better understand general aquatic processes. Freshwater is an important natural resource necessary for the survival of all organisms. It is defined here as water with less than 1000 parts per million (ppm) of dissolved salts (or $< 1‰$ salinity). From all water on the Earth, 2.75% is freshwater and only 0.01% occurs as surface water in rivers and lakes (Pidwirny, 2006).

The tropical zone occurs broadly between 23.5° N and 23.5° S of the equator, covering the tropical and subtropical regions of Australia and the south Pacific Islands, Papua New Guinea, South-east Asia, Africa, Mexico, and Central and South America, and is characterised not only by small daily and seasonal changes in temperature, but great seasonal changes in precipitation. About 40% of the world's human population lives within the tropical zone (based on 2010 statistics), which is predicted to increase to about 60% by 2060 through migration and high birth rates (GeoHive, 2010). Concomitant with economic and social development within the tropics and subtropics in the coming decades,

Radioactivity in the Environment, Vol. 18. http://dx.doi.org/10.1016/B978-0-08-045016-2.00006-0

is an expectation for substantially increased energy needs. From that, and recognising the current international drive to limit carbon emissions, it is likely that nuclear power will be a major component of that development. Indeed, new nuclear power reactors are currently being proposed, planned, and/or constructed in India, Bangladesh, Thailand, Indonesia, Malaysia, Vietnam, Brazil, and Mexico (WNA, 2012). The increased demand for U as a long-term alternative energy source to fossil fuels has focussed attention on the potential environmental impacts of radionuclide wastes associated with its use.

Bioaccumulation describes the net accumulation of a radionuclide into the tissues of an organism as a result of the balance of its uptake from all sources (e.g., food or water) and its loss across all routes (e.g., diffusion, faeces, or urine). It is a good integrative measure of an organism's radionuclide exposure in variously contaminated ecosystems, including freshwater systems. The content of a radionuclide in an organism is therefore the net difference between uptake and loss integrated over time. The activity concentration is expressed as content per unit weight (e.g., Bq/g, often as dry weight).

When discussing freshwater radioecology, we need to consider a number of contributing factors that influence the processes in relation to this specific environment. Bioaccumulation is influenced by the chemistry of the specific stable form of the radionuclide, exposure source, and species-specific physiology (Luoma and Rainbow, 2008). Radionuclide uptake from water by an organism is largely governed by water column chemistry. Physicochemical variables, such as pH, organic carbon concentration, water hardness, and alkalinity are the key drivers that influence the bioavailability of radionuclides in freshwater systems (Markich et al., 2001a; Luoma and Rainbow, 2008). The formation of inorganic and organic radionuclide complexes and sorption of radionuclides to particle surfaces has been shown to reduce radionuclide uptake. As a result, the relationship between radionuclide uptake and total (or dissolved) concentrations of radionuclides can be highly variable, depending on the ambient water chemistry. Mechanistically, the uptake rate of a radionuclide across the cell membrane/surface of an organism reflects the interaction between the bioavailable radionuclide concentration (in the water column or within the sediment interstitial waters) and the internal characteristics of the biological transport system. Species-specific cell membrane transport characteristics are also important in governing the uptake rates for each radionuclide.

Within the broad scope of radioecology in freshwater systems, this chapter focusses on the following:

- The influence of organisms on radionuclide content in freshwater ecosystems
- The influence of water chemistry on the biological uptake of radionuclides
- The mechanisms and kinetics of biological uptake and loss of radionuclides

Although these are generic processes across radioecology, examples will focus on a range of tropical freshwater organisms. This chapter aims to provide an overview of these important processes rather than an exhaustive or critical review.

6.2. ORGANISM EFFECTS AT AN ECOSYSTEM SCALE

At the outset of this chapter, it is important to appreciate that, in addition to the aquatic concentration of radionuclides having an effect on freshwater organisms, the organisms themselves can influence radionuclide concentrations in these environments. This is particularly so in fresh surface waters, as the volume of water is relatively limited (0.01% of the Earth's water) in comparison with marine systems (97%; Pidwirny, 2006). A case in point is wetland filtration, whereby freshwater plants and their associated microflora provide an effective mechanism to influence the distribution of radionuclides within their aquatic ecosystems. A tropical radioecological case example is Magela Creek, in a monsoon-affected area of northern Australia. As Magela Creek drains near surface uranium ore bodies, U-series radionuclides are normally present in the fresh surface waters and sediments at naturally elevated background levels (Sauerland et al., 2005; Frostick et al., 2011). The distribution of radionuclides in the system suggests some radioecological mechanisms may be in play. In a study of the transport of naturally occurring radionuclides in Magela Creek and its floodplain, Murray et al. (1993) concluded that no significant ($p > 0.05$) deposition of radionuclides takes place within a 12 km zone downstream of the major ore bodies. This zone is characterised by a sandy substrate with relatively few aquatic macrophytes. The authors further state that the great majority of the particulate material, with which many of the radionuclides are associated, is deposited as the flow rate of creek water reduces upon reaching and spreading out over the floodplain. However, deposition is clearly augmented by adsorption of dissolved radionuclides onto the abundant aquatic plant growth at the head of the floodplain. Indeed, Williams (1988) concluded that the major mechanism for ^{226}Ra accumulation from the dissolved phase at the head of the Magela floodplain is direct uptake onto macrophyte foliage, as well as adsorption to bed sediments. The plants persist over the dry season in remnant pools or quickly re-establish each year from seeds and other propagules (e.g., corms) as the floodplain becomes inundated (wet season), and grow at remarkable rates to ensure they reach and remain at the rising water surface in competition with the other vegetation. This leads to a proliferation of dense and diverse assemblages of aquatic vegetation over the wet season (Finlayson et al., 1989). The adsorbed activity of radionuclides is subsequently incorporated into the organically rich soils when the plants naturally decay over the dry season as part of the monsoonal climatic cycle.

Surface adsorption is an important factor in aquatic radioecology; it can substantially contribute to the overall concentration ratio between the organism and its supporting aquatic media and, from that, to the bioaccessible concentration to consumers (humans or other animals within the wider environment); it is related to surface area and tends to increase exponentially as organism size decreases, thereby becoming a more substantial proportion of total radionuclide concentration in smaller organisms (an allometric relationship). It can also comprise a radiation source (IAEA, 2010a), contributing to the external radiological dose of the organism in question (see Chapter 7). For all these reasons, surface binding of radionuclides to organisms is an important mechanism to consider.

As with most surface phenomena, adsorption can be a very rapidly equilibrating process in response to changing concentrations in water. It is also, for the most part, reversible. However, some surface adsorption can be slowly exchanged or even non-reversible. For example, in plants, epiphytic microflora can enhance uptake and many aquatic animals have an external mucous layer that acts as a barrier to infection or to assist in controlling ingress of toxins or loss of nutrients between the internal and external environments. Similarly, biofilms of exopolysaccharides can be formed on almost any surface in aquatic systems by bacteria and other microflora (e.g., Flemming et al., 2007). These layers can potentially bind or complex dissolved radionuclides, colloids, or small particulates, often to a very high degree (e.g., Schaller et al., 2010), thereby giving rise to slower exchange kinetics than normally anticipated for surface adsorption.

In aquatic macrophytes, there is also the apparent free space between the surface layer cells and the impermeable barrier within the tissues (Levitt, 1957) beyond which water is unable to permeate passively. This water-accessible zone within external plant tissues can provide large surface areas for exchange and complexation and may take longer to equilibrate than simple surface exchange, depending upon the tortuous diffusion path the external water needs to traverse in order to penetrate this zone of the plant (c.f. discussions on groundwater flow in Chapter 3). The apparent free space can increase dramatically in plant tissues as their surfaces start to deteriorate with age. This is one of the potential mechanisms invoked to explain the exponential increase in the adsorption of ^{226}Ra from the aquatic phase by waterlilies, *Nymphaea violacea*, as they aged under field conditions (Twining, 1989) and in the laboratory (Twining, 1988a). The semi-aquatic grass, *Pseudoraphis spinescens*, is a dominant species on the Magela Creek floodplain (Williams, 1979) and was assessed for ^{226}Ra biokinetics by Williams (1988). His study showed that ^{226}Ra was accumulated predominantly by surface adsorption and complexation onto the submerged foliage. Green tissues bound ^{226}Ra firmly (concentration ratio [C_R] of 2800; see Section 6.5.2) and released it back into radium-free water with a biological half time ($T_{b1/2}$) of seven days, longer than would be anticipated for simple surface adsorption processes. The older foliar tissue accumulated more ^{226}Ra ($C_R = 4900$), but released it 35 times more quickly than the green tissue ($T_{b1/2} = 0.2$ days), more in keeping with a physical, rather than a biological, process.

6.3. UPTAKE OF RADIONUCLIDES AT THE CELL SURFACE

To better understand how water chemistry influences the bioaccumulation of radionuclides or their stable elements, knowledge of uptake pathways at the cell surface is required. More than one uptake pathway is potentially available for the entry of radionuclides into multicellular organisms. A freshwater invertebrate, for example, can gain radionuclides from solution across the permeable body surface (e.g., gills) and from the diet via the digestive system (e.g., gut). Unicellular organisms generally take up radionuclides from solution only, but endocytosis of particulates is also possible. Radionuclides must cross the cell surface/membrane to be absorbed. Bioavailability describes a relative measure of that fraction of the total ambient radionuclide or stable element that an organism actually absorbs when encountering or processing environmental media, combined across all radionuclide sources (e.g., water and food). This chapter covers the importance of radionuclide uptake/absorption from water (dissolved phase) and food (particulate phase).

Protein carriers and major ion channels provide the major transport routes by which hydrophilic chemical species cross the hydrophobic cell membrane. Transport of most radionuclides or their stable elements from outside to inside a cell is not always energy dependent. An inward diffusion gradient is sustained by the rapid binding of the intracellular radionuclide to non-diffusible forms. Definable membrane traits govern the radionuclide and species-specific characteristics of radionuclide uptake. Calcium channels/transporters are widely considered to be the major active transport route for the uptake of divalent cations in aquatic organisms (Perry and Flik, 1988; Hogstrand et al., 1996; Tan and Wang, 2011). Divalent cations (with the exception of Cu; see below) are taken up as metabolic analogues of Ca (Markich and Jeffree, 1994; Hollis et al., 2000; Rogers and Wood, 2004; Komjarova and Blust, 2009). The uptake of Ca, and hence other divalent cations, by freshwater organisms occurs via active transport, where the rate of Ca/divalent cation uptake typically follows the Michaelis-Menten kinetic model (see Section 6.5) over a wide range of Ca/divalent cation concentrations. Radionuclides or their stable elements in solution bind to ligands, such as sulfydryl, carboxyl, amino, imidazole, or phosphoryl groups, at the cell surface of Ca channels (Simkiss and Taylor, 1995). Since the binding affinity of these ligands typically follows $Hg > UO_2 > Cu \sim Pb > Zn > Cd \sim Ni \sim Co \sim Mn > Ca$ (see Section 6.4.3), the ability of divalent cations to inhibit the uptake of Ca in the Ca channels, and thus impair Ca homeostasis, will progressively increase as the binding affinity of the cation increases. Copper (Cu^{2+}) is an unusual exception because it is widely believed to be reduced to Cu^{+} before being transported via sodium channels (Grosell and Wood, 2002; Handy et al., 2002). Environmentally relevant monovalent cations, such as silver (Ag^{+}) and cesium (Cs^{+}), are also known to be transported via Na/K channels, as their metabolic analogues (Bury and Wood, 1999; Heredia et al., 2002).

Only specific forms of an element are transported by the ion channels. Elements in fresh surface waters exist in a variety of physicochemical forms, including the free ion and complexes with organic and inorganic ligands in dissolved, colloidal, and/or particulate forms. A convincing body of experimental evidence supports the Biotic Ligand Model (BLM; see Section 6.4.5) — stemming from the Free Ion Activity Model (FIAM) and its extensions (Campbell, 1995; Brown and Markich, 2000) — which postulates that the bioavailability of a dissolved element (e.g., U) is best predicted by the activity of the free ion (e.g., UO_2^{2+}), rather than the total dissolved element concentration. To be taken up by an organism, a radionuclide or its stable element must first interact with a cellular ligand at the surface, forming an element-ligand complex. The free ion is a good predictor of the form of an element that binds to this cellular ligand. Radionuclide or stable element uptake is assumed to be proportional to the concentration of element in this cell-surface element-ligand complex. Provided the number of cellular ligands at the cell surface is approximately constant, the radionuclide or stable element uptake rate will vary as a function of the activity of the free ion. Some environmentally relevant elements, such as As, Cr, Mo, Po, Se, and V, do not follow the FIAM of uptake because their commonly dissolved forms are anionic. Their chemical behaviour offers potential for uptake at the cell surface via transport routes for sulphate (e.g., Se) or phosphate (e.g., As), as metabolic analogues (Riedel, 1985; Rahman et al., 2007; Fournier et al., 2010). Nonionic, nonpolar, dissolved chemical species (e.g., methylmercury), taken up via lipophilic (not hydrophilic) routes at the cell surface, will not be discussed here, but may form another important class.

6.4. INFLUENCE OF WATER CHEMISTRY ON RADIONUCLIDE BIOAVAILABILITY

6.4.1 Introduction

Knowledge of the distribution of a radionuclide or stable element amongst its various physicochemical forms (i.e., speciation) is essential in interpreting its interaction with the cell surface of organisms. This section assesses the effect of pH, dissolved organic carbon (DOC), alkalinity, and hardness—key physicochemical variables in freshwater systems—on the radionuclide uptake by, and chemical toxicity to, tropical freshwater organisms.

In general, there is a paucity of data for tropical organisms relative to their temperate counterparts. The majority of data are from northern Australia, with very limited data from Central America (Amazon Basin). Most studies have been conducted in very soft (low hardness, alkalinity, and conductivity) surface waters, which may enhance the bioavailability, and hence ecological risk, of radionuclides or their stable elements. There is no evidence from the literature to suggest that the ionic composition of fresh surface waters in the tropics is different to that of temperate regions. Moreover, there is a wide range of water chemistries within each climatic/geographical zone, driven largely by underlying bedrock composition. However, in tropical zones that are characterised by a distinct wet and dry season (latitudes broadly between 13° N and 13° S of the equator), seasonal increases in stream discharge during the wet season generally result in lower concentrations of major ions because of dilution, but this is often associated with increased concentrations of DOC (Allen and Castillo, 2007).

6.4.2 pH

6.4.2.1 General

In fresh surface waters, changes in pH may strongly influence the speciation of dissolved radionuclides or their stable elements (Franklin et al., 2000; Markich et al., 2000; Olguín et al., 2002; Wilde et al., 2006). Decreasing pH indicates increasing hydrogen ion concentrations (H^+) in solution, which may compete with other cations at cell-surface binding sites and thus reduce their uptake and accumulation. This may be offset by an increase in the free cation activity/concentration at lower pH, due to protonation of ligands in solution (replacing trace radionuclide-ligand complexes). The balance of these two contradictory effects typically determines cation uptake rates, and hence toxicity, at the cell surface. The following examples from the literature demonstrate the governing influence of pH on selected environmentally relevant cations for tropical freshwater organisms.

6.4.2.2 Biological Response Driven by Chemical Speciation

Markich et al. (2000) measured the uptake and toxicity of Mn or U with a bivalve (*Velesunio angasi*) exposed to a very soft water at pH 5.0, 5.5, and 6.0 at constant levels of alkalinity (4 mg $CaCO_3$/L), hardness (4 mg $CaCO_3$/L), and DOC (0.1 mg/L). For Mn, uptake and toxicity were independent of pH (at least for the range tested, which was relevant to the local river system used in the study). Both measured and calculated speciation of Mn over the pH range tested showed that Mn^{2+} was the dominant (99%) species and also independent of pH. This result can be interpreted as the bioavailable Mn^{2+} driving biological response, with no evidence of competition between H^+ (factor of 10 increase) and Mn^{2+} at the cell surface. In contrast, U uptake and toxicity were highly dependent on pH, and its speciation was more complex. At pH 5.0, UO_2^{2+} and UO_2OH^+ (94% combined) were the dominant uranyl species; whereas at pH 6.0, these species were less important (23% combined). Biological response was found to be proportional to a weighted function of the activities of $1.86 \times UO_2^{2+}$ and UO_2OH^+, with no evidence of competition between H^+ and these uranyl species at the cell surface.

Olguín et al. (2002) determined the effect of pH on the uptake of Cd by a floating plant (*Salvinia marina*). Cadmium uptake was independent of pH from 5.0 to 8.0. The calculated speciation of Cd over the pH range tested showed that Cd^{2+} was the dominant (90%) species and also independent of pH. This result can be interpreted as the bioavailable Cd^{2+} driving biological response, with no evidence of competition between H^+ (factor of 1000 increase) and Cd^{2+} at the cell surface.

6.4.2.3 Biological Response Driven by Proton Competition

Franklin et al. (2000) determined the effect of pH on the uptake and toxicity of Cu or U in a unicellular green alga (*Chlorella* sp.). Biological response was found to be highly pH dependent for Cu and U, with uptake and toxicity decreasing as pH decreased from 6.5 to 5.7. Calculation of Cu and U speciation showed that differences in the concentrations of the free ions (Cu^{2+} and $UO_2{}^{2+}$) were only minimal (<10%) between pH 5.7 and 6.5. The decreased uptake and toxicity at pH 5.7 was due to lower concentrations of cell-bound and intracellular Cu and U compared to those at pH 6.5. These results may best be interpreted as competition between H^+ and Cu^{2+} or $UO_2{}^{2+}$ at the cell surface.

Takasusuki et al. (2004) determined the effect of pH on the uptake and toxicity of Cu in a fish (*Prochilodus scrofa*). Biological response was found to be highly pH dependent, with Cu uptake and toxicity decreasing as pH decreased from 8.0 to 4.5. Calculation of Cu speciation showed that the Cu^{2+} concentration actually increased as the pH decreased (i.e., H^+ increased) from 8.0 to 4.5. These results may best be interpreted as competition between H^+ and Cu^{2+} at the cell surface.

6.4.2.4 Biological Response Driven by both Chemical Speciation and Proton Competition

Wilde et al. (2006) determined the effect of pH on the uptake and toxicity of Zn or Cu in a unicellular green alga (*Chlorella* sp.). Biological response was found to be highly pH dependent for both divalent cations, with uptake and toxicity decreasing as pH decreased from 7.5 to 5.5. Changes in speciation (Zn^{2+} or Cu^{2+}) alone did not explain the biological response; however, this was much improved once proton competition was considered.

6.4.3 DOC

A high percentage of DOC in fresh surface waters (ranging from <0.5 mg/L to >30 mg/L in organic-rich waters with a world average of 4.2 mg/L; Meybeck et al., 1996) exists in the form of humic substances (HS), which are formed from the decomposition of plant, animal, and microbial material (Perdue and Ritchie, 2003). Fulvic acid (FA) is the major component

(~90%) of dissolved HS (Tipping, 2002; Perdue and Ritchie, 2003) and binds strongly to cations such as Cu, Hg, Pb, and U in freshwaters with circumneutral pH (Perdue, 1998; Tipping, 2002; Tipping et al., 2011). Cation binding strength to FA (and hence DOC) is primarily governed by the oxygen containing carboxylic (–COOH) and phenolic (–OH) functional groups (Tipping, 2002; Perdue and Ritchie, 2003) and broadly follows the unified theory of metal ion complexation (UTMIC) (e.g., Hg > UO_2 > Cu ~ Pb > Zn > Cd ~ Ni ~ Co ~ Mn > Ca) developed by Brown and Sylva (1987). The UTMIC uses a single value of a cation (termed electronicity) to quantify earlier concepts of hard and soft acids and bases. The following examples from the literature demonstrate the influence of DOC on selected environmentally relevant cations for tropical freshwater organisms.

Markich et al. (2000) determined the influence of varying DOC concentration (0, 3.7, and 9.0 mg/L) on the uptake and toxicity of Mn or U with a bivalve at three pH levels (5.0, 5.5, and 6.0). They found that the uptake and toxicity of Mn was independent of increasing DOC concentration (Figure 6.1), which corresponded with Mn^{2+} as the dominant (94%) form

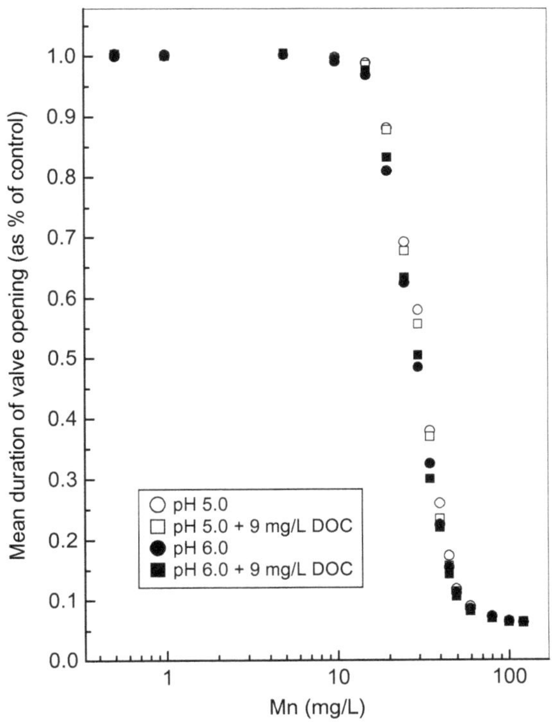

FIGURE 6.1 Concentration-response relationships of the duration of valve opening for the tropical freshwater bivalve, *V. angasi*, exposed to Mn at pH 5.0 and 6.0 without DOC, and pH 5.0 and 6.0 with 9 mg/L DOC. Each plotted point represents the mean response of six individuals. Error bars and curve fits are excluded for clarity. *From Markich et al. (2000).*

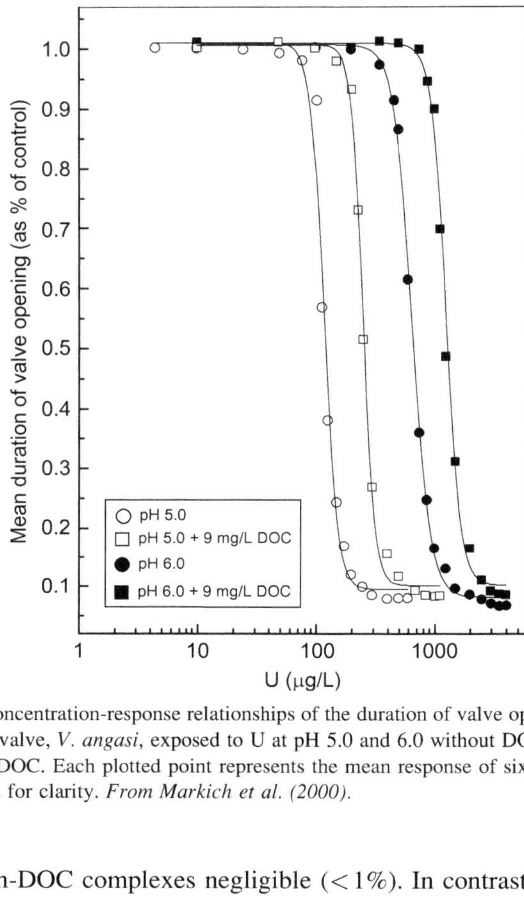

FIGURE 6.2 Concentration-response relationships of the duration of valve opening for the tropical freshwater bivalve, *V. angasi*, exposed to U at pH 5.0 and 6.0 without DOC and pH 5.0 and 6.0 with 9 mg/L DOC. Each plotted point represents the mean response of six individuals. Error bars are excluded for clarity. *From Markich et al. (2000).*

of Mn and Mn-DOC complexes negligible (<1%). In contrast, they reported that the uptake and toxicity of U decreased nonlinearly with increasing DOC concentration (Figure 6.2), which corresponded with a decrease in UO_2^{2+} and UO_2OH^+ and an increase in U-DOC complexation. The relative differences in the biological responses of the two cations are in accordance with the UTMIC.

Matsuo et al. (2005) determined the influence of varying DOC concentration (0, 20, 40, and 80 mg/L) on the uptake of Cd or Cu by the gills of an Amazonian fish (*Colossoma macropomum*) at pH 6.3. They found the uptake of Cd to decrease slightly (up to 25%) with increasing DOC (up to 80 mg/L), whereas the uptake of Cu reduced substantially (by ~70%) from 0 to 20 mg/L DOC, where the copper complexation capacity of the experimental system was reached and further increases in DOC did not have an influence on Cu uptake. Although no speciation data were available from the study, relative differences in the biological responses of the two cations are in accordance with the UTMIC, whereby Cu forms stronger complexes with DOC than Cd.

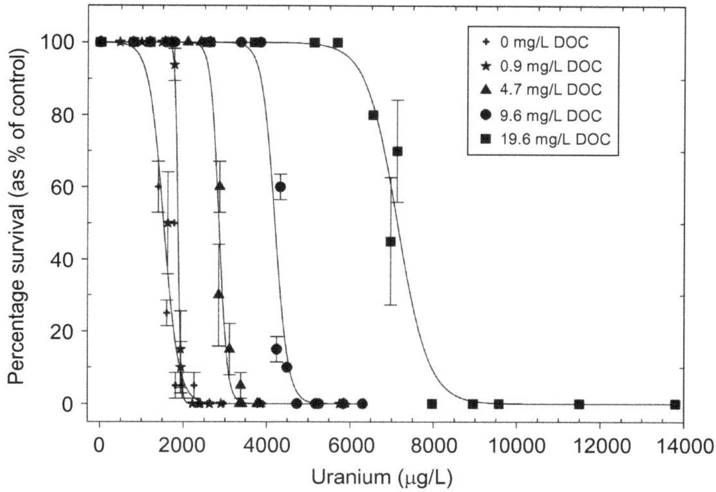

FIGURE 6.3 Concentration-response relationships for the percentage survival of the tropical freshwater fish, *M. mogurnda*, exposed to U at varying DOC concentrations. Each plotted point represents the mean response of four tests. Error bars indicate 95% confidence interval around the mean. *From Trenfield et al. (2011).*

Hogan et al. (2005) determined the influence of varying DOC concentration (0–8 mg/L) on the uptake and toxicity of U with a unicellular green alga (*Chlorella* sp.) at pH 6.5. They found that the uptake and toxicity of U decreased nonlinearly with increasing DOC concentration; the calculated concentration of UO_2^{2+} decreased with increasing DOC because of U-DOC complexation.

Trenfield et al. (2011) determined the influence of varying DOC concentration (0, 1, 5, 10, and 20 mg/L) on the toxicity of U to a green hydra (*Hydra viridissima*), a unicellular green alga (*Chlorella* sp.), and a fish (*Mogurnda mogurnda*; see Figure 6.3) at pH 6.0. They found that the toxicity of U decreased nonlinearly with increasing DOC concentration, where the calculated concentration of UO_2^{2+} decreased with increasing DOC because of U-DOC complexation (Figure 6.4); a result consistent with that of Markich et al. (2000) and Hogan et al. (2005).

Trenfield et al. (2012a) determined the influence of varying DOC concentration (0, 1, 2, 5, and 10 mg/L) on the toxicity of Al to a green hydra (*H. viridissima*), a unicellular green alga (*Chlorella* sp.), and a cladoceran (*Moinodaphnia macleayi*) at pH 5.0. They found that the toxicity of Al decreased nonlinearly with increasing DOC concentration, where the calculated concentrations of Al^{3+}, $AlOH^{2+}$, and $Al(OH)_2^{+}$ decreased with increasing DOC because of Al–DOC complexation.

Trenfield et al. (2012b) determined the influence of varying DOC concentration (0 and 20 mg/L) on the toxicity of U to a unicellular eukaryote

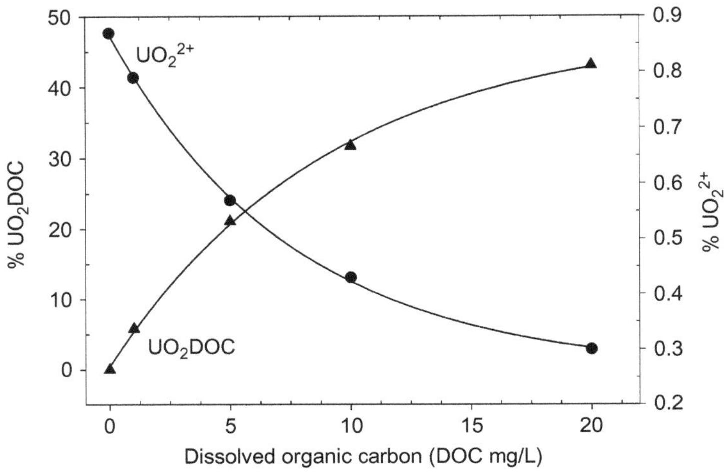

FIGURE 6.4 Calculated speciation of U in reconstituted Magela Creek water, showing a nonlinear increase in the % UO_2DOC and a decrease in the % UO_2^{2+}, with increasing DOC concentration. *From Trenfield et al. (2011).*

(*Euglena gracilis*) at pH 6.0. They found that the toxicity of U decreased with increasing DOC concentration, where the calculated concentration of UO_2^{2+} decreased with increasing DOC because of U–DOC complexation.

Duarte (personal communication) determined the influence of varying DOC concentration (1.5–8.3 mg/L) on the toxicity of Cu to nine species of Amazonian fish (*Carnegiella strigata*, *Hemigrammus rhodostomus*, *Hyphessobrycon socolofi*, *Paracheirodon axelrodi*, *Dicrossus maculatus*, *Corydoras schwartzi*, *Otocinclus hasemani*, *Apistogramma agassizi*, and *Apistogramma* spp.) at pH 6.0–7.0. The toxicity of Cu decreased linearly with increasing DOC concentration, where the calculated concentration of Cu^{2+} decreased with increasing DOC because of Cu–DOC complexation.

DOC is the best predictor of U toxicity to freshwater organisms, relative to other key physicochemical variables such as pH, hardness, and alkalinity (van Dam et al., 2012). Based on available U toxicity data from five tropical freshwater organisms (*Chlorella* sp., *H. viridissima*, *E. gracilis*, *M. mogurnda*, and *V. angasi*; see previously mentioned studies), a linear algorithm or relationship has been derived (Eq. 6.1) that may be used to adjust U water quality guidelines for the protection of freshwater ecosystems, depending on the surface water DOC concentration (over the range 0–20 mg/L DOC), as follows:

$$\text{DOC adjusted U guideline value} = GV + (GV * DOC * slope) \qquad (6.1)$$

where *GV* is the national (or site-specific) guideline value (typically derived from laboratory tests where DOC < 1 mg/L) and *DOC* is the DOC concentration (mg/L) of interest and *slope* is the relevant slope factor for acute (0.064) or chronic (0.090) toxicity, derived from cumulative probability distributions (van Dam et al., 2012). Simply, this relationship takes into account a 6.4% (acute) or 9.0% (chronic) decrease in U toxicity for every 1 mg/L increase in DOC. Work is currently underway to revise the existing site-specific U guideline for Magela Creek (Hogan et al., 2005), which drains the Ranger Uranium mine in northern Australia, and the default (low reliability) Australian and New Zealand U guideline for fresh surface water (ANZECC and ARMCANZ, 2000), to enable adjustment based on the fresh surface water DOC concentration (see Section 6.4.5 for further discussion on water quality guidelines).

6.4.4 Hardness and Alkalinity

6.4.4.1 General

Many studies that have investigated the influence of water hardness on the uptake and/or toxicity of radionuclides or their stable elements with freshwater organisms, have confounded the effects of true water hardness (Ca and/or Mg concentration) with alkalinity (carbonate concentration) and pH (proton concentration), since an increase in water hardness is frequently associated with an increase in alkalinity (where Ca and/or Mg are added as carbonate) and pH (Stumm and Morgan, 1996). It is important to separate the effects of hardness and alkalinity, since each variable has a different mechanism of toxicity. Calcium and/or Mg competitively inhibit the uptake, and hence, toxicity of many cations at the cell-membrane surface (Markich and Jeffree, 1994), whereas complexation of cations with carbonate in the aquatic medium reduces the concentration of the free ion (i.e., a change in speciation).

True water hardness and alkalinity may range from <20 mg/L as $CaCO_3$ in the very soft waters of the Orinoco, Tocantins, and lower Amazon rivers in South America (Edmond et al., 1995; Meybeck and Ragu, 1997; Mora et al., 2009); the Niger and Zaire rivers in west Africa (Ballio et al., 1996; Picouet et al., 2002); and Magela Creek and several other sandy, braided coastal streams in northern Australia (Klessa, 2000) to >300 mg/L as $CaCO_3$ in harder, alkaline, fresh waters such as the Panuco River in Mexico (Meybeck and Ragu, 1997). The following examples from the literature demonstrate the influence of true water hardness and alkalinity on selected environmentally relevant stable elements for tropical freshwater organisms.

6.4.4.2 Hardness

Twining (1988b) determined the influence of water hardness (4.0 and 20 mg/L as $CaCO_3$) on the uptake of ^{226}Ra on a waterlily (*N. violacea*) at a constant pH (6.3) and alkalinity (4 mg/L as $CaCO_3$). The uptake of ^{226}Ra decreased with increasing water hardness, where the predicted concentration of Ra^{2+} remained relatively constant over the range of water hardness. This indicates that Ca^{2+} and Mg^{2+} ions compete with Ra^{2+} for binding sites at the cell surface to decrease ^{226}Ra uptake with increasing water hardness. This is consistent with the metabolic analogue hypothesis proposed by Jeffree and Simpson (1986), where Ra is taken up in mistake for Ca via the Ca uptake pathways at cell surfaces.

Riethmuller et al. (2001) determined the influence of water hardness (6.6, 135, and 330 mg/L as $CaCO_3$) on the toxicity of U or Cu on a green hydra (*H. viridissima*) at a constant pH (6.0) and alkalinity (4 mg/L as $CaCO_3$). They found that the toxicity of U decreased with increasing water hardness, whereas the concentration of UO_2^{2+} remained relatively constant over the range of water hardness. This indicates that Ca^{2+} and Mg^{2+} ions compete with UO_2^{2+} for binding sites at the cell surface to decrease U toxicity with increasing water hardness. In contrast, the toxicity of Cu did not change with increasing water hardness, and the concentrations of Cu^{2+} remained constant. This indicates that Ca^{2+} and Mg^{2+} ions do not compete with Cu^{2+} for binding sites at the cell surface, and suggests that the mechanism of Cu uptake/binding at the cell surface is different from other divalent cations (such as U, Ra, Cd, Zn, Ni, and Pb), which are taken up at the cell surface as analogues of Ca via Ca transporters (Markich and Jeffree, 1994; Alsop and Wood, 1999; Niyogi and Wood, 2004; Rogers and Wood, 2004; Komjarova and Blust, 2009), such as Ca^{2+}-ATPase. Conversely, Cu is taken up at the cell surface via Cu-specific transporters, such as Cu-ATPase (Grossel and Wood, 2002; Hall and Williams, 2003) or via Na transporters, such as Na/K-ATPase, as an analogue of Na (Grossel et al., 2002; Handy et al., 2002). One popular hypothesis is that Cu^{2+} is reduced to Cu^{+} by reductases on the cell surface to facilitate uptake via Na^{+} transporters (Grossel et al., 2002; Handy et al., 2002).

Charles et al. (2002) determined the influence of water hardness (8, 40, 100, and 400 mg/L as $CaCO_3$) on the uptake and toxicity of U or Cu with a unicellular green alga (*Chlorella* sp.) at a constant pH (7.0) and alkalinity (8 mg/L as $CaCO_3$). They found that the uptake and toxicity of U decreased with increasing water hardness (Figure 6.5), where the concentration of UO_2^{2+} remained relatively constant over the range of water hardness. This indicates that Ca^{2+} and Mg^{2+} ions compete with UO_2^{2+} for binding sites at the algal cell surface to decrease U uptake and toxicity with increasing water hardness, a result consistent with that of Riethmuller et al. (2001). In contrast, the uptake and toxicity of Cu did not change with increasing water hardness (Figure 6.5), and the concentrations of Cu^{2+} remained relatively constant, indicating that Ca^{2+} and Mg^{2+} ions do not compete with Cu^{2+} for binding sites at the cell surface.

Matsuo et al. (2005) determined the influence of water hardness (10, 20, and 40 mg/L as $CaCO_3$; added as Ca only) on the uptake of Cd or Cu by the gills of an Amazonian fish (*Colossoma macropomum*) at a constant pH (6.3) and alkalinity (8 mg/L as $CaCO_3$). They found that the uptake of Cd decreased, but the uptake of Cu did not significantly ($p > 0.05$) change, with increasing water hardness. Speciation calculations of the experimental waters

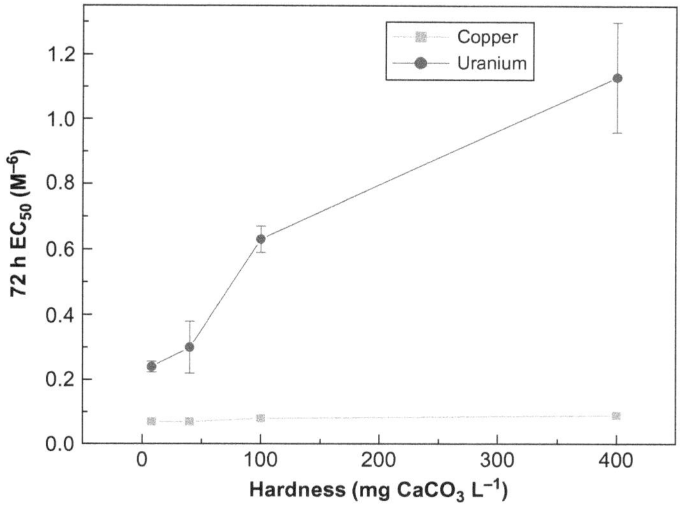

FIGURE 6.5 Comparative toxicity (measured as 72 h EC_{50} for cell growth rate) of U and Cu to a tropical freshwater unicellular alga (*Chlorella* sp.) plotted as a function of water hardness (mg as $CaCO_3$/L) at a constant pH (7.0) and alkalinity (8 mg as $CaCO_3$/L). Each plotted point represents the mean response of four independent tests. Error bars indicate 95% confidence interval around the mean. *Adapted from Charles et al. (2002).*

conducted by the present authors indicated that there was no change in the concentrations of Cd^{2+} or Cu^{2+}, respectively, with increasing water hardness. The results suggest that Ca^{2+} and Mg^{2+} ions do compete with Cd^{2+}, but not Cu^{2+}, for binding sites at the cell surface.

Markich et al. (2005) determined the influence of water hardness (44, 125, and 375 mg/L as $CaCO_3$) on the uptake and toxicity of Cu with a unicellular green alga (*Chlorella* sp.) at a constant pH (6.5) and alkalinity (22 mg/L as $CaCO_3$). The uptake and toxicity of Cu did not change with increasing water hardness, and the concentrations of Cu^{2+} remained relatively constant, indicating that Ca^{2+} and Mg^{2+} ions do not compete with Cu^{2+} for binding sites at the cell surface.

Markich et al. (2006) determined the influence of water hardness (35, 90, and 335 mg/L as $CaCO_3$) on the uptake and toxicity of Cu with a floating macrophyte (*Ceratophyllum demersum*) at a constant pH (7.0) and alkalinity (16 mg/L as $CaCO_3$). The uptake and sensitivity of Cu did not change with increasing water hardness, and the concentrations of Cu^{2+} remained relatively constant, indicating that Ca^{2+} and Mg^{2+} ions do not compete with Cu^{2+} for binding sites at the cell surface.

Several studies that have compared the protective effects of Ca and Mg have found Ca to be much more effective than Mg in reducing the uptake and/or toxicity of divalent cations by freshwater organisms (Carroll et al., 1979; Jeffree and Simpson, 1986; Jackson et al., 2000; Clifford and

McGeer, 2009). Hille et al. (1975) showed that the binding affinity of Ca to multidentate ligands (receptors) at the mouth of Ca channels exceeded that of Mg. The underlying reasons why Ca has a superior binding affinity over Mg for these multidentate ligands at the cell surface, and hence a greater protective effect, are related to differences in the basic chemical properties of the two elements, arising from differences in ionic radius (see Frausto da Silva and Williams (2001) for a more detailed discussion).

An obvious implication of a greater protective effect of Ca, relative to Mg, on the uptake and toxicity of divalent cations to freshwater organisms can be extended to national water quality guidelines for the protection of freshwater life, where hardness modified algorithms are used to provide guideline values for Cd, Cu, Ni, Pb, and Zn (e.g., ANZECC and ARMCANZ, 2000; USEPA, 2001; EC, 2006; CCME, 2011). We propose that Ca concentration in water may be a better choice of variable than water hardness (Ca + Mg). For example, it is possible for two freshwater bodies to have a comparable alkalinity and water hardness, yet differ markedly in their individual Ca and Mg concentrations, so that in one body water hardness may be dominated by Mg, whereas the other may be dominated by Ca. In light of the empirical evidence established from previous studies on the differences between these two hardness ions in reducing divalent cation uptake and toxicity, it would appear that a freshwater body dominated by Ca would offer a greater protective effect for freshwater organisms than a freshwater body dominated by Mg. This concept may be incorporated into the BLM (see Section 6.4.5).

6.4.4.3 Alkalinity

Markich et al. (1996) determined the influence of alkalinity (4 and 20 mg/L as $CaCO_3$) on the toxicity of U to a bivalve at a constant pH (5.0) and water hardness (4 mg/L as $CaCO_3$). They found that the toxicity of U decreased by 23% at the highest alkalinity level, which corresponded to a concomitant decrease in the concentrations of both UO_2^{2+} and UO_2OH^+ and an increase in the concentration of UO_2CO_3. This result supports the notion that changes in true alkalinity (carbonate concentration only) may influence U speciation, and hence, biological effect.

Riethmuller et al. (2001) determined the influence of alkalinity (4 and 102 mg/L as $CaCO_3$) on the toxicity of U or Cu to a green hydra (*H. viridissima*) at a constant pH (6.0) and water hardness (165 mg/L as $CaCO_3$). Interestingly, they found that the toxicity of U and Cu were both independent of alkalinity under the experimental conditions tested. However, the result can be explained because both concentrations of Cu^{2+} or UO_2^{2+}, and hence bioavailability, were not significantly ($p > 0.05$) different between the two alkalinity levels tested. Further work using a higher alkalinity level (> 250 mg/L as $CaCO_3$) would better define potential differences in U or Cu speciation, and hence, biological effect.

6.4.5 The BLM and Use in National Guidelines for Protecting Freshwater Ecosystems

BLMs allow chemical and biological interactions to be taken into account and relate, through water chemistry, radionuclide or stable element uptake and/or toxicity to a dissolved concentration (Paquin et al., 2002; Niyogi and Wood, 2004; Slaveykova and Wilkinson, 2005). Using an equilibrium geochemical modelling framework, the BLM incorporates the competition of the free ion with other naturally occurring cations (e.g., Ca^{2+}, Na^{+}, Mg^{2+}, H^{+}), together with complexation by abiotic ligands (e.g., DOC and carbonates) for binding with the biotic ligand, the site of uptake/toxic action on the organism (see schematic BLM for U in Figure 6.6). The use of BLMs in a compliance-based regulatory framework is an area that has received substantial attention (Niyogi and Wood, 2004). The United States Environmental Protection Agency (USEPA, 2007) adopted the BLM to revise its Cu criterion for freshwater, because the model can account for Cu speciation reactions and interactions with organisms under a wide range of water quality conditions. Criterion for Ni and Zn are currently being developed for future use. The BLM is also being considered as part of the current revision of the Australian and New Zealand water quality guidelines for protecting freshwater organisms.

6.4.6 Integrating the BLM with Bioaccumulation Kinetics

Moving one step forward, Veltman et al. (2010) reported that the BLM and bioaccumulation kinetics can be merged into a common mechanistic

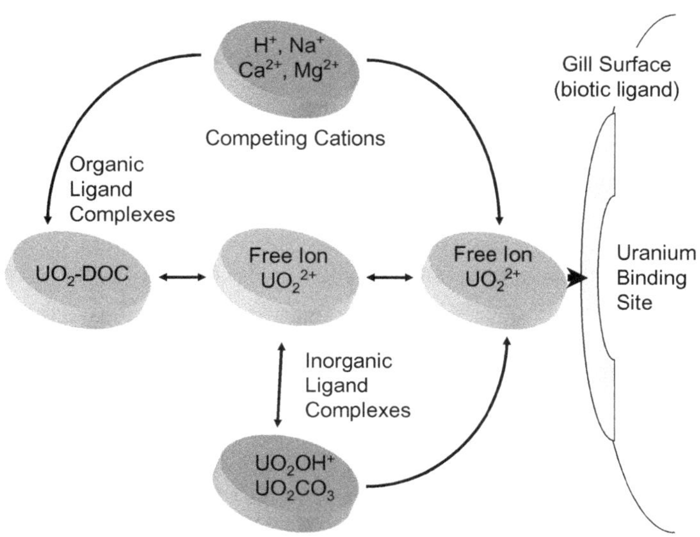

FIGURE 6.6 Schematic diagram of the BLM, using U as an example.

framework for radionuclide or stable element uptake that integrates the combined effect of chemodynamics (as speciation) and biodynamics (ligand affinity, competition, and species characteristics, such as size). Development of such a general principle for radionuclide influx has two key advantages. First, it is useful to quantitatively explain the complex environmental behaviour of radionuclide uptake, as the model defines both radionuclide- and species-specific parameters of radionuclide uptake rates (e.g., ligand affinity at cell membrane uptake sites and competition amongst radionuclides as a driver of uptake rate constants). Second, by integrating BLM and bioaccumulation kinetics, it is reported that several BLM variables (e.g., maximum uptake rate, internalisation rate, and transport protein capacity) are related to species size to the power -0.25. The current BLM does not explicitly recognise the strong potential influence of physiological characteristics, such as size. For example, the majority of interspecies variation in acute Ag and Cu toxicity can be attributed to size amongst tested species, whereby smaller organisms are more sensitive to Ag and Cu because of their higher sodium (metabolic analogue) uptake rate (Grossel et al., 2002).

6.4.7 Comparisons of Tropical and Temperate Freshwater Ecosystems

It is generally agreed that tropical freshwater ecosystems are characterised by higher temperatures, light intensity, and organic matter turnover, with faster oxidation-reduction activities, relative to temperate freshwater ecosystems (da Silva and Soares, 2010; Daam and Van der Brink, 2010). However, inter- and intra-specific relationships between species and ecological mechanisms do not show major differences along latitudinal gradients (da Silva and Soares, 2010). Although it is generally believed that species diversity increases from the poles to the tropics, this does not appear to hold true for all environmental compartments or species taxonomic groups. Although freshwater fish species richness, for example, does increase toward the equator (Lévêque et al., 2008; Oberdorf et al., 2011), freshwater plankton communities do not show a marked latitudinal trend in species diversity (Lewis, 1987; Fernando, 2002). Indeed, the species diversity of rotifers, cladocerans, and chironomids appear to be lower in the tropics compared to the higher latitudes (Daam and Van der Brink, 2010).

Growing concern about the risks of radionuclide or stable element contamination in tropical freshwaters has made temperate-to-tropical ecosystem extrapolation a focus of research over the last decade. Validation studies of the protective value of temperate chemical toxicity threshold values for tropical freshwaters have focussed mainly on a species (or species assemblage) level approach by comparing the chemical toxicity of species to various stable elements. Markich (unpublished) compared the toxicity of U to tropical and temperate freshwater fish and cladocerans. Only these taxonomic groups had sufficient sample sizes to make reasonable statistical comparisons. Comparisons were corrected for differences in water chemistry (e.g., hardness) where possible. The analyses showed that there were no significant ($p > 0.05$) differences in the toxicity of U for both taxonomic groups. Kwok et al. (2007) compared temperate and tropical freshwater species assemblages for Ag, As, Cd, Cr, Cu, Hg, Ni, Pb, and Zn using a species sensitivity distribution (SSD) approach. Tropical freshwater species were found to be less sensitive than their temperate counterparts to Cd, Cr, Cu, Hg, Ni, and Pb; however, the converse was true for Zn, with no apparent differences for Ag and As. The authors note that differences in the availability of toxicity data for different taxa could introduce bias into the SSD approach, and they state that sample sizes, although valid, were generally lower for tropical data. Based on these available studies, national water quality guidelines, derived largely using temperate species (Section 6.4.4.2), would generally be protective of tropical freshwater ecosystems, although further research is warranted. See Section 6.5.5 for further discussion of the effects of climatic zone, particularly in relation to radionuclide or stable element bioaccumulation.

6.5. MODELLING, KINETICS, AND MECHANISMS OF RADIONUCLIDE BIOACCUMULATION

6.5.1 Background

Organisms need to acquire sufficient macronutrients and micronutrients from their environment to satisfy their requirements for maintenance of health and to promote growth, development, and reproduction. Because of that, all organisms have evolved active and passive mechanisms to acquire their essential nutrients to achieve sufficient internal concentrations irrespective of the external media concentration. Ignoring photosynthesis in autotrophs, mineral nutrition strategies for freshwater organisms include feeding and digestion by animals, root uptake from sediments by macrophytes, and adsorption directly from the water via the gills, gut, or epidermis. However, irrespective of the means of acquisition, once nutrients are internalised, an excess can be as big a problem as deficiency (see Figure 6.7). Hence, radioecology can be a very useful tool to assist in interpreting toxicity, and vice versa. For example, many studies use radionuclides (e.g., ^{65}Zn, ^{54}Mn, ^{45}Ca ^{109}Cd, ^{75}Se, ^{67}Cu, ^{60}Co) as precise, sublethal tracers of stable elements to better understand mechanisms of uptake and toxicity, rather than because the radionuclides in themselves are of any particular radiological hazard.

Once inside an organism, a radionuclide or its stable element is distributed among different tissues and compartments. Some of these tissues accumulate radionuclides to a large extent (e.g., gills, bone, or liver in animals), while others accumulate radionuclides to a negligible extent (e.g., vacuoles in plants). More detailed discussion on biodistribution is provided in

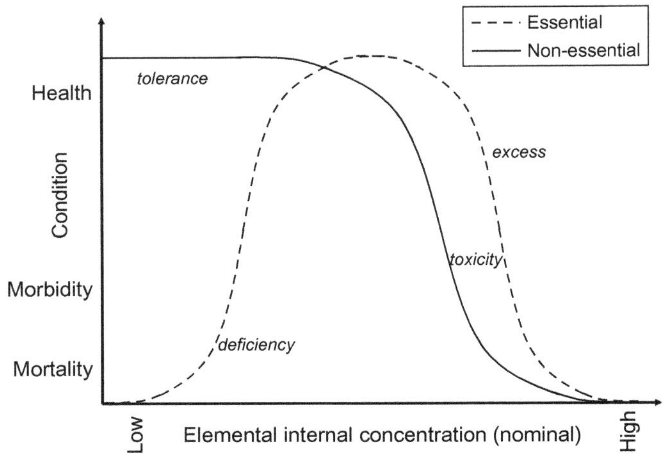

FIGURE 6.7 Relationship between internal biological concentration and organism condition/health for essential and nonessential elements. All elements can cause potential toxicity if their concentration falls outside the optimal range.

Section 6.5.8. Consistencies in organism uptake of radionuclides, based on their chemical properties (e.g., ionic charge and radius), permit the possibility of modelling bioaccumulation. Such models enable prediction of radionuclide concentrations in organisms consequent to their exposure in freshwater ecosystems. These bioaccumulation models have been described by Whicker and Schultz (1982) and Paquin et al. (2012); only a summary of the basic models is provided here (Section 6.5.2). It is important to note that the most widely used models only provide information concerning the kinetics of the uptake and loss process and allow the prediction of steady-state radionuclide body or tissue concentrations for a given exposure situation. They do not provide information concerning the toxic effects that may or may not be related to the accumulation of radionuclides in the tissues (see the following section). Nevertheless, several variations to kinetic-based, whole-body bioaccumulation models have been proposed to better explain field observations and incorporate physiological aspects, such as biologically active and inactive radionuclide pools and differences between slow and fast turnover radionuclide pools over time (e.g., Redecker and Blust, 2004; Croteau and Luoma, 2005). Physiologically based pharmacokinetic models incorporate substantially more detail than whole-body models. The explicit description of individual organs allows for the simultaneous representation of uptake and elimination kinetics for each. This increased specificity allows for more direct evaluation of target organ dose and provides a framework that is amenable to modelling different body sizes, species, and target organ toxicity (Thomann et al., 1997). More complex toxicodynamic models are required to couple radionuclide accumulation to toxicity/biological effect. These are comprehensively reviewed by Adams et al. (2011) and Paquin et al. (2012) and will not be discussed further here.

6.5.2 Basic Modelling

The first radioecological concept to consider is the C_R as defined by the International Commission on Radiation Units and Measurements (ICRU, 2001). This is 'the ratio of the activity density of a radionuclide in the receptor compartment to that in the donor compartment,' and they gave the term 'transfer factor' as an acceptable alternative. It is easily calculated by the following:

$$C_R = \frac{Radioactivity\ in\ biota\ (Bq/kg\ FW)}{Radioactivity\ in\ water\ (Bq/L)} \tag{6.2}$$

If the uptake pathway is via the sediment, the units of the divisor (denominator) will be in Bq/kg, and typically on a dry weight (DW) basis because sediment concentrations based on wet weight (WW) (c.f. in biota, fresh weight; FW) can be highly variable, depending on sample collection methodology. Note that it is always important to specify how the weight (and ratio) is measured. Failure to do so can lead to discrepancies by orders of magnitude between results, given that water will almost always comprise a major and variable component of any sample from a freshwater system. It should be noted that the C_R for biota is mathematically equivalent to the K_d for sediments (see Chapters 3 and 5) and is essentially a dimensionless proportionality parameter, as the mass and activity units cancel out.

Assuming a stable concentration of radionuclide in a freshwater medium, the simplest model reflects linear uptake with no saturation in the biota. This represents animals and plants that bioaccumulate at a constant rate irrespective of, or within, the period of observation. It is given by the following:

$$C_R = k * t \tag{6.3}$$

where k is the uptake rate coefficient and t is time (days). As t increases, so does the concentration in the organism and hence the C_R. This is the model best used to describe radionuclide accumulation (e.g., ^{226}Ra) by the tropical freshwater bivalve, *V. angasi* (see Section 6.5.8).

Next we look at an organism that is exposed to radionuclides from the external media, but also excretes that material over time. A basic bioaccumulation model is shown in Figure 6.8. The change over time of the radionuclide concentration in the organism (Q) is given by the following:

$$\frac{dQ}{dt} = \mu c - \lambda Q \tag{6.4}$$

where μ is the uptake rate coefficient, which is proportional to the concentration of radionuclide in the supporting medium (c), and λ is the loss rate coefficient, which is proportional to the radionuclide concentration in the

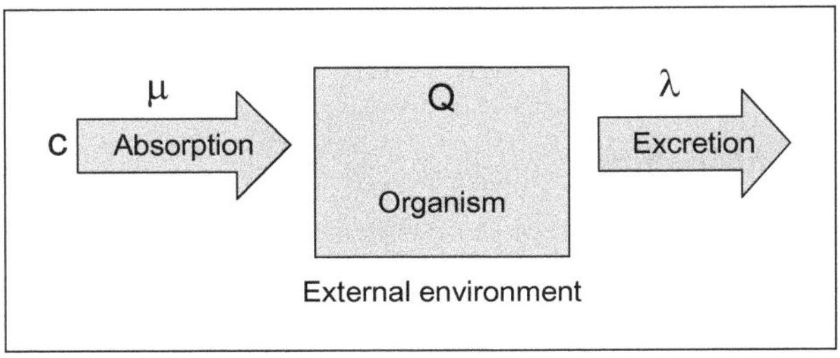

FIGURE 6.8 Basic model describing radionuclide bioaccumulation. Once a radionuclide enters the ecosystem, it will be taken up by an organism at a rate (μ), which is proportional to its concentration in the external environment (*c*). Having entered the organism, the radionuclide will then be excreted at a rate (λ), which is proportional to its concentration in the organism (*Q*). If the external radionuclide concentration remains constant, the uptake rate of the organism will remain constant. However, over time this will increase the radionuclide concentration in the organism and, hence the excretion or loss rate, until such time that the loss and uptake rates are equal, when the radionuclide concentration in the organism attains a steady state.

organism (Q). Solving this equation to estimate the concentration in the organism at any time (t) gives the following:

$$Q = \frac{\mu}{\lambda} c * \left(1 - e^{-\lambda t}\right) \tag{6.5}$$

As t increases, the latter term approaches zero and, from that, at equilibrium (when the uptake and loss rates become equal) the equation simplifies to the following:

$$Q = \frac{\mu}{\lambda} c \tag{6.6}$$

and the equilibrium, or steady state, concentration ratio, C_{Rss}, is equivalent to the ratio of the uptake and loss rate coefficients.

If we assume that the organism has migrated to an uncontaminated area, there will be no uptake and the animal or plant will begin to lose radionuclides, according to the model shown in Figure 6.8, with $c = 0$. Equation 6.4 then becomes the following:

$$\frac{dQ}{dt} = -\lambda Q \tag{6.7}$$

and Equation 6.5 becomes the following:

$$Q = Q_0 e^{-\lambda t} \tag{6.8}$$

This model is also known as a Michaelis-Menton, or first-order, single compartment exponential uptake and loss model. However, most organisms will take up radionuclides into more than one compartment, for example, at least surface adsorption plus a metabolised proportion. We can model that by using the same basic equations and simply adding different compartments together as follows, using our C_R notation and assuming, for example, that the organism has both linear (Eq. 6.3) and saturating (Eq. 6.5) uptake mechanisms at work.

$$C_{Rtot} = C_{Rl} + C_{Re} = (k * t) + C_{Rss} * \left(1 - e^{-\lambda t}\right) \tag{6.9}$$

Similar simple adaptations can be made to account for loss from more than one compartment or by more than one mechanism. For example, organisms with a rapid loss (*f*) from one compartment (e.g., excreted food or desorbed surface radionuclides) and slower loss (*s*) of metabolised radionuclide, as well as some radionuclides that are sequestered and not lost (*b*), can be modelled using the following:

$$Q_{tot} = Q_{0f}e^{-\lambda_f t} + Q_{0s}e^{-\lambda_s t} + b \tag{6.10}$$

The same simple models can be combined to estimate uptake and loss via multiple pathways. Here is an example for uptake and loss by a bivalve that can control its exposure to the aquatic phase by altering its filtration rate, which eats a variety of different food types and which gains mass as a consequence:

$$\frac{dQ}{dt} = a_w * FR * c_w + \sum_{f=1}^{n}(AE_f * IR_f * c_f) - (\sum_{p=1}^{n}(k_{ep}c_p) + gQ) \tag{6.11}$$

where Q = concentration in the mussel (Bq/g), t = time (days), a_w = absorption efficiency from the dissolved phase, FR = filtration rate (L/g/d), c_w = dissolved concentration (Bq/L), AE_f = absorption efficiency from particulate f (types 1 – n), IR_f = ingestion rate of particulate f (mg/g/d), c_f = concentration in particulate f (Bq/mg), k_{ep} = excretion rate constant from tissue p (days), c_p = concentration in tissue p (Bq/g), and g = growth rate constant.

There are additional refinements that can be made to any of these basic models to make them more relevant to specific forms of biota and to account for multiple and various exposure and elimination pathways. It is also possible to expand the model to account for trophic transfer and exchange between organisms. A good example of this type of modelling, using very similar terminology, is given in Smith et al. (2011) looking at bioaccumulation of $^{32/33}P$ in a river environment.

6.5.3 Databases and Their Underlying Assumptions

The uptake and loss rate parameters and the C_R values derived from these models will be specific to the organism studied. However, similar organisms tend to accumulate individual radionuclides to a similar degree and the International Atomic Energy Agency (IAEA, 2010a; Howard et al., 2013) has tabulated recommended equilibrium C_R values in order to be able to predict what may happen in any aquatic environment following the release of radionuclides into it. These generic values may be incorporated into models of radionuclide behaviour in lakes (Monte et al., 2003) and catchment scale models such as AQUASCOPE (Smith et al., 2002a, 2005) and MOIRA-PLUS (Monte, 2011). They may also be used for undertaking environmental dose assessment (see Chapter 7) using models such as RESRAD-BIOTA (USDOE, 2002) and ERICA (Howard and Larsson, 2008). The ERICA tool has its own set of C_Rs for aquatic ecosystems (Hosseini et al., 2008). It should be noted that there are still wide ranges in the C_R values of some radionuclides. For example, the calculated C_R values for ^{241}Am in edible freshwater plants (IAEA, 2010a, Table 55) range over four orders of magnitude from 7.5×10^0 to 3.9×10^4. This underlines the inherent variability in C_R estimation and points to the need to ensure that critical C_R values for individual site-specific assessments are well known and that the underlying assumptions are checked.

One of the key assumptions of the models, as applied to field sampling, is that equilibrium has been achieved. This is particularly risky when using C_Rs defined against a water concentration that may be highly variable. Apparently high values are derived when radionuclides have been diluted or flushed from a system or low values can be measured soon after the introduction of radionuclides into the water column. The assumption of equilibrium was tested by Pyle and Clulow (1998) for a range of U-series radionuclides in relation to uptake into fish tissues. They were able to demonstrate that equilibrium had been established for most of the radionuclides in a natural situation where the animals had been exposed to mine wastewaters for some considerable time. In contrast, ^{232}Th was probably not at equilibrium because of the growth dilution effect that resulted in lower tissue concentrations over time.

Many situations will not be in equilibrium, as a consequence of the exposure scenario (e.g., an accidental release, or a regular release that is pulsed) or because environmental conditions can vary dramatically (e.g., tropical monsoon areas experience very wet and very dry periods over annual cycles). Under these circumstances, the assumption of equilibrium must be taken more cautiously. The example in Section 6.5.7 of a single release of ^{133}Cs into a small pond (Pinder et al., 2011) clearly did not achieve equilibrium over the period of observation, given that the top predators were still at concentrations lower than their prey, despite having greater C_Rs. Smith et al. (2011) also noted that if the physical half-life of a radionuclide (e.g., ^{131}I with $T_{1/2}=8$ days and $^{32/33}$P with 14.3 or 25.3 days, respectively) was short in comparison with its biological half-life (within an organism) or ecological half-life (within a food web), then equilibrium may not be achieved for that radionuclide due to its decay rate following uptake. Some food chain exposure pathways can take months or longer for equilibrium to occur (e.g., Elliott et al., 1992; Smith et al., 2005; Pinder et al., 2011). Smith et al. (2011) evaluated this effect using $^{32/33}$P in a freshwater ecosystem. They concluded that the currently recommended C_R values for stable P are likely to be significant overestimates of radioactive P in many freshwater systems, particularly lowland rivers, as a consequence of radioactive decay and competitive inhibition of radioactive P uptake by stable P contaminants introduced by human activity.

Another assumption is that the C_R value listed for any organism is relevant for any use. However, most of the C_Rs derived have been for assessment of radiological dose to humans via the food ingestion pathway. They have focussed on tissues that are consumed by humans, such as the flesh of fish and invertebrates or edible parts of plants. Given the increasing need to undertake environmental dose assessments (see Chapter 7), there has been a concomitant increase in the need for whole body values to enable internal dose estimates to be calculated for organisms. We have already mentioned how different tissues of organisms can accumulate different radionuclides at very different rates and to variable extents (further discussed in Section 6.5.8). In order to overcome these discrepancies to some extent, Yankovich et al. (2010) compiled available data to convert tissue values to whole-body data for a range of different organism groups. Beresford (2010) discussed transfer of radionuclides to and between wildlife more thoroughly. An additional assumption to confirm is whether physical media within any freshwater ecosystem are the only sources of radionuclides to organisms living within it. This will not be true for migratory species (e.g., fish that migrate into a contaminated stretch of a river from upstream or the sea). Allochthonous input also needs to be considered. Monsoonal rains or other heavy irregular rainfall can readily wash surface contamination and nutrients into tropical lakes and streams. Food items originating outside the aquatic ecosystem can land on the water surface (e.g., insects) or fall from overhanging vegetation or even be harvested by species such as Archer fish (*Toxotidae* sp.). These inputs will influence apparent trophic transfer if the aquatic medium is the only considered source.

6.5.4 Biological Factors Influencing Bioaccumulation

6.5.4.1 Within Species Variability

Key factors that may influence radionuclide bioaccumulation within species include size, age, gender, reproductive status, and condition (health). However, their influence is not consistent amongst radionuclides or phyla. The following examples from the literature demonstrate the conflicting influence of organism size, age, or gender on the bioaccumulation of radionuclides or their stable elements by tropical freshwater organisms.

Liao et al. (2003) found a negative relationship between As uptake and body weight in the fish, *Oreochromis mossambicus*, whereas Ansari et al. (2006) found that Zn, Fe, Cu, Mn, and Ca tissue concentrations were all positively correlated with body weight or length for the fish, *Puntius chola*. Jeffree et al. (2005) reported negative relationships between length or age and Ba, Ca, Cd, Co, Cu, Fe, Mg, Mn, Ni, Pb, Sr, U, and Zn concentrations in the osteoderms of the crocodile, *Crocodylus johnstoni*. Jeffree et al. (2001) found negative relationships between length or age and Mg and Ti concentrations in the flesh of the crocodile *C. porosus*, but found positive relationships for Ba, Se, and Zn. Bollhöfer et al. (2011) reported a positive relationship between shell length or dry weight and ^{226}Ra, ^{210}Pb, and ^{228}Th concentrations in whole soft tissues of the bivalve, *V. angasi*. Dantas and Attayde (2007) identified body size as an important factor in P accumulation in fish species from both temperate and tropical climates. However, both positive (e.g., for *Plagioscion squamosissimus* – tropical) and negative (e.g., *Perca fluviatilis*—temperate) relationships between P and total length were reported, making it difficult to predict individual responses. Rowan et al. (1998) also found an effect of growth rate on Cs uptake in fish, with the larger adults having three different patterns of biomagnification. They were able to show that these patterns were a reflection of the ratio of Cs consumption rate to growth and elimination. Greater uptake occurred when the rate was high, so larger animals that still consumed strongly as they grew bigger would tend to have higher radionuclide uptake than larger animals whose intake declined relatively. This effect may in some way explain the disparate results in the literature (e.g., Dantas and Attayde, 2007).

The overall strength of size relationships with bioaccumulation has led to the development of generic models based on allometric relationships to estimate organism concentrations (e.g., Higley, 2010), particularly given the need to fill data gaps in dose assessment models as discussed further in Chapter 7. This is also relevant to efforts within the Environmental Modelling for Radiation Safety (EMRAS) programme (e.g., Beresford, 2010; IAEA, 2010b).

Costa and Hartz (2009) found that Cd accumulation in the liver and Zn accumulation in the flesh of the fish *Leporinus obtusidens* was greater in males compared to females, but there were no significant ($p > 0.05$) differences in Cr and Cu accumulation in both tissues amongst males and females. Jeffree et al. (2005) found no significant ($p > 0.05$) differences in the accumulation of Ba, Ca, Cd, Co, Cu, Fe, Mg, Mn, Ni, Pb, Sr, U, and Zn in the osteoderms of the crocodile, *C. johnstoni*, amongst males and females. Similarly, Schneider et al. (2010) reported no significant ($p > 0.05$) effect of gender on the accumulation of Hg by six species of turtles. Gender effects may also be influenced by the reproductive status/cycle of an individual. For example, Banks et al. (1999) reported that ovarian Zn levels in the fish *Ictalurus punctatus* were elevated during oocyte development in females (as part of the yolk), whereas hepatic Zn levels were elevated immediately after spawning.

The condition or health of an individual may also contribute to differences in the bioaccumulation of radionuclides or their stable elements. For example, Jeffree et al. (2001) found that Na and Fe concentrations in the flesh of the crocodile *C. porosus* increased as the condition of individuals deteriorated.

6.5.4.2 Between Species Variability

A proportion of the variability in C_Rs may be explained by accounting for differences in chemical speciation and consequent effects on bioavailability (as described in Section 6.4), but some is due to differences between species. As an example, Martin and Ryan (2004) classified species of fish from Kakadu National Park in tropical northern Australia into two separate groups or clusters based upon their bioaccumulation of naturally occurring radionuclides (Table 6.1). The reasons for the statistically significant differentiation between the groups are unclear as there are no obvious differences between the fish groups in terms of their feeding behaviour or habitat to explain the observations. Jeffree, Twining, and Markich (unpublished) also found clustering of fish species on the basis of radionuclide bioaccumulation (including ^{210}Po and ^{238}U) in the nearby Finniss River catchment. Because differences were observed in two separate catchments with contrasting surface water and sediment chemistry, this is a biological factor rather than a physicochemical one.

Jeffree (1991) reported that the accumulation of ^{226}Ra in the snapping turtle, *Elseya dentata*, was about two orders of magnitude lower than the bivalve *V. angasi* under similar experimental conditions. This is in accordance with field observations (Table 6.1; Martin and Ryan, 2004). Jeffree and co-workers (2001, 2005) compared the bioaccumulation of radionuclides in the osteoderms of two morphologically different crocodile species (*C. johnstoni* and *C. porosus*) from northern Australia. Interestingly, the bioaccumulation of Co, Cu, and Pb was significantly ($p \leq 0.05$) higher in *C. porosus* than

TABLE 6.1 Average Concentration Ratios (C_Rs; L/kg FW) of Radionuclides in the Flesh of Freshwater Organisms Collected in Kakadu National Park, Northern Australia

Organism	^{226}Ra	^{210}Pb	^{210}Po	U	Th	Species common or scientific names
Fish group 1	1200	160	1400	250	40	Bony bream, sleepy cod
Fish group 2	190	35	180	15	22	Fork-tailed catfish, Archer fish, Barramundi, Eel-tailed catfish, Freshwater mullet, Long tom, Saratoga, Tarpon
Bivalve	19,000	5100	10,000	100	500	*Velesunio angasi*
Turtle	250	120	1000	28	40	*Elseya dentata*
Shrimp	270	39	1200	150	250	*Macrobrachium rosenbergii*

From Martin and Ryan (2004).

C. johnstoni, whereas the converse was true for Ba, Fe, Mn, and Sr. Although there are well-documented differences in their osmoregulation, there were no significant ($p > 0.05$) differences in their Ca accumulation, which largely governs divalent cation accumulation (see Section 6.4.4.1).

6.5.5 Climatic Zone Differences

Basic radioecological processes are consistent irrespective of climatic zone, and bioaccumulation can generally be expected to follow normal patterns irrespective of location. However, factors such as temperature and growth rate have been known to affect uptake within any environment (e.g., Smith et al., 2002a, b). Given that tropical and subtropical waters are generally warmer than those in temperate zones and growth rates may also be faster, it begs the question of whether there may be systematic differences in bioaccumulation between tropical and temperate regions.

Dantas and Attayde (2007) found no significant ($p > 0.05$) differences in the accumulation of N and P between two temperate and six tropical fish species. Rowan and Rasmussen (1994) comprehensively reviewed field-collected bioaccumulation data for Cs in fish from both freshwater and marine ecosystems across latitudes of approximately 32–69° N (i.e., northern temperate data only). They showed that C_Rs for Cs were positively related to mean annual air temperature and to the thermal zone (i.e., epilimnetic > hypolimnetic) inhabited by the fish, and hypothesised that C_Rs for fish would increase in response to increasing habitat temperature.

The IAEA and the United Nations Food and Agriculture Organisation (FAO) established a Cooperative Research Project (CRP) to assess the 'transfer of radionuclides from air, soil, and fresh water to the food chain of man in tropical and subtropical environments.' The results of those efforts in relation to soil-to-plant transfer are summarised in Chapter 5. In relation to freshwater studies within that CRP, Twining et al. (1998) compiled data for a range of investigations from Australia (Twining et al., 1997), Bangladesh (Mollah et al., 1994, 1995, 1997), Thailand (Sinakhom et al., 1997), and Vietnam (Ngo and Binh, 1997) on $^{134/137}Cs$ and $^{85/90}Sr$ uptake by tropical freshwater fish or data from studies conducted in temperate laboratories under tropical conditions (Srivastava et al., 1990, 1994). The compiled data were then compared with the recommended C_R values based on temperate data as published by the IAEA (1994). The results are given in Table 6.2. These indicate that C_Rs for Sr and Cs in tropical fish were markedly lower than those expected for temperate fish. The results were consistent across a range of species measured across eight studies. Taking water quality into account, by incorporating specific water chemistry variables (i.e., the concentrations of K and suspended solids for Cs accumulation and the Ca concentration for Sr accumulation) into predictive accumulation models (IAEA, 1994), the apparent differences between the measured C_Rs in the flesh of tropical fish and the

TABLE 6.2 Results of Experiments to Determine ^{137}Cs and $^{85,90}Sr$ Transfer Factors (L/kg FW) and Biological Half-lives ($T_{1/2}$) in Tropical Freshwater Fish, and Parameters Measured to Assist Interpretation.

Species common name (*Scientific name*)	Size (SL or mass)	^{137}Cs			$^{85,90}Sr$			Temp (°C)	K (mg/L)	Ca (mg/L)	Suspended solids (mg/L)	Notes	Reference
		TF modeled (L/kg FW)	*TF measured (L/kg FW)*	$T_{1/2}$ *(d)*	*TF modeled (L/kg FW)*	*TF measured L/kg (FW)*	$T_{1/2}$ *(d)*						
Zebrafish (*Brachydanio rerio*)	0.24 g	1015	16	51	[b]			26	2.7		Nil	Whole fish	Srivastava et al., 1990
Goldfish (*Carassius auratus*)	2–6 g	1015	4	19				28	2.7		Nil	Whole fish	Srivastava et al., 1994
Tilapia (*Tilapia* sp.)	30–40 g[a]		26	12								Whole fish	Ngo and Binh, 1997
Carp (*Cyprinus carpio*)	10–30 g		40	31								Whole fish	
Catfish (*Clarias* sp.)	160–270 g	59	4	29	1	5	39	27	29	67	100	Flesh and skin	Sinakhom et al., 1997
		106	2	36	2	0.1	4	26.5	13	42	100	Whole fish	
		72	3	41				29.6	22	64	100	Flesh	
Singhi (*Heteropneustes fossiils*)	22–24 cm	175	6	93	37	10	62	28	6.9	3.7	110	Flesh	Mollah et al., 1994
Magur (*Clarias batrachus*)	20–25 cm		7	94		11	54	28	6.9	3.7	110	Flesh	
Climbing perch (*Anabas testudineus*)	16–19 cm		6	81		14	80	28	6.9	3.7	110	Flesh	

continued

TABLE 6.2 Results of Experiments to Determine ^{137}Cs and $^{85,90}Sr$ Transfer Factors (L/kg FW) and Biological Half-lives ($T_{1/2}$) in Tropical Freshwater Fish, and Parameters Measured to Assist Interpretation.—Cont'd

Species common name (*Scientific name*)	Size (SL or mass)	^{137}Cs			$^{85,90}Sr$			Temp (°C)	K (mg/L)	Ca (mg/L)	Suspended solids (mg/L)	Notes	Reference
		TF modeled (L/kg FW)	*TF measured (L/kg FW)*	$T_{1/2}$ *(d)*	*TF modeled (L/kg FW)*	*TF measured L/kg (FW)*	$T_{1/2}$ *(d)*						
Singhi	20–23 cm	174	6	110	26	15	73	27	7.2	4.8	118	Whole fish	Mollah
Magur	18–21 cm		9	104		19	65	27	7.2	4.8	118	Whole fish	et al.,
Climbing perch	14–17 cm		9	88		13	78	27	7.2	4.8	118	Whole fish	1995
Singhi	≥20 cm	174	16	160	26	3	7	27	7.2	4.8	118	Flesh	Mollah
Magur	≥20 cm		19	182		4	8	27	7.2	4.8	118	Flesh	et al.,
Climbing perch	≥20 cm		16	119		4	7	27	7.2	4.8	118	Flesh	1997
Silver perch (*Bidyanus bidyanus*)	18–25 cm	440	13	19	5	0.7	5	25	2	20	92	Flesh	Twining et al., 1997
Expected TF (range) from temperate data			2000 (30–3000)			60 (1–1000)							IAEA, 1994

FW, fresh weight; TF, transfer factor; SL, standard length (distance from the snout to the hypural plate). Also Included are the expected values and the results of applying predictive models from temperate data.

[a] *20 g fish (juveniles) were excluded from this analysis on the basis of failing tests of normality and heteroscedasticity.*

[b] *Blanks indicate that the data were not reported.*

From Twining et al. (1998) and IAEA (1994).

recommended C_Rs for temperate fish were greatly reduced. However, the tropical values were still lower than expected. This was not what was anticipated from the results of Rowan and Rasmussen (1994), which described an increase in C_R with habitat temperature, and which reinforces the need to check assumptions when undertaking predictive modelling for any ecosystem.

6.5.6 Ingestion, Egestion, and Biomagnification

For animals, ingested food is generally particulate in nature. At the gills, Fe (and, to a lesser extent, Pb and Cd) is known to be adsorbed in a particulate form by pinocytosis, which is an active process requiring ATP (Marigomez et al., 2002). Small external particles are engulfed into the cell by enclosure within a membrane vesicle that is subsequently pinched off within the cytoplasm. Within the digestive tract, only the dissolved, bioavailable fraction is absorbed across the gut membrane. The remainder is eliminated as faeces. Hence, non-bioavailable radionuclides in food have little influence, apart from any radiological dose that may be imparted to the organism (see Chapter 7), during the period of gut passage. The adsorption efficiency (AE) for radionuclides from particular food types is a parameter used in the radionuclide biokinetic models described in Section 6.5.2. If radionuclides associated with food are generally more bioavailable, then this will lead to efficient transfer to subsequent trophic levels. If that same radionuclide is a nutrient (or an analogue thereof), the organism will generally try to minimise excretion/elimination of that radionuclide. For some radionuclides, such as Cs, which is absorbed as an analogue of the nutrient K, this occurs at each trophic-level transfer leading to a phenomenon known as biomagnification. As a consequence of biomagnification, top-level predators in freshwater environments, such as piscivorous fish, can have much higher concentrations of Cs than are observed in organisms lower in the food chain. This effect is recognised in some of the predictive models (e.g., IAEA, 1994; Smith et al., 2005) used to estimate organism concentrations from media concentrations. However, there has been some variability in biomagnification observed both within and between species. Rowan et al. (1998) conclusively demonstrated the occurrence of Cs biomagnification in a range of habitats, but found that the degree of effect was influenced by the relative consumption rates of the organisms and how those rates varied with age. Given that Cs tends to be accumulated via ingestion, the proportion of food intake versus growth rate becomes a key component of Cs cycling in any organism.

Pinder et al. (2011) assessed Cs trophic transfer in Pond 4 at the Savannah River site and found good evidence in support of Cs biomagnification in at least two food webs in that ecosystem: one based on periphyton, the other on phytoplankton. Whilst the concentrations of Cs were different within these two food chains, the C_R values between trophic levels within the food chains were similar. This study also showed the effect of temporal removal of the added Cs on trophic transfer dynamics. The top predators in this system (the fishes *Lepomis macrochirus* and *Micropterus salmoides*) did not achieve concentrations as high as the mid trophic-level animals (*Chaoborus punctipennis*, a larval insect in the planktonic food web and *Helisoma trivolvis*, a snail in the periphyton food web). High C_R values were estimated for the fish using measured changes in Cs concentration over time. However, the biological half times for accumulation through the food web by the top predators were longer than the physical half-life of the Cs added to the pond, as it was adsorbed by the biotic and abiotic components of the ecosystem. This is another example of organisms driving the radioecology within an aquatic ecosystem, as discussed previously in Section 6.2.

6.5.7 Biphasic Uptake of Cs and Ra in Macrophytes

For freshwater macrophytes there are two clear exposure pathways for accumulation of radionuclides: from the water column via foliar uptake and from the sediments via root uptake. Both pathways may be important given that foliar uptake will be relatively easy for dissolved radionuclides, and sediments can be a major sink for radionuclides, thereby providing a potentially rich source for remobilisation into roots.

Waterlilies (*Nymphaea*) are a common tropical and subtropical freshwater macrophyte. These plants, together with other freshwater macrophytes, have been studied for $^{133/137}$Cs accumulation in the subtropical south-eastern United States (Kelly and Pinder, 1996; Pinder et al., 2006, 2011). Kelly and Pinder (1996) evaluated the relative importance of foliar, as distinct from root, uptake of ^{137}Cs for three floating leaved species: *Brasenia schreberi*, *N.odorata*, and *Nymphoides cordata*. These plants were transplanted into contaminated or uncontaminated sediments and then grown in a reservoir (Par Pond at the U.S. Department of Energy Savannah River Site) containing ^{137}Cs-contaminated surface waters. Their results showed rapid foliar bioaccumulation of ^{137}Cs for all plants. After 35 days, the ^{137}Cs concentration in the leaves of plants growing in uncontaminated sediment was not significantly ($p > 0.05$) different to those growing in the contaminated sediment, and was also similar to the ^{137}Cs concentration in naturally occurring plants in the pond. Hence, the results imply that ^{137}Cs in floating foliage of aquatic plants is most likely due to accumulation from the water phase rather than by uptake from sediments via roots. Similar results for *B. schreberi* and *N. odorata*, plus two submerged species, *Myriophyllum spicatum* and *Utricularia inflata*, were observed by Pinder et al. (2006) when adding stable ^{133}Cs to an adjacent pond at the same study site. Results for the emergent macrophytes, *Typha latifolia*, *Alternanthera philoxeroides*, and *Sagittaria latifolia* were more equivocal, with essentially no significant change in foliar ^{133}Cs observed. The authors also observed that ^{133}Cs concentrations at distances > 100 m into the floating leaved macrophyte beds were 50% of those in the open water, where the new

^{133}Cs was added, for periods of up to 100 days following the addition. This supports the notion that the foliage of fully aquatic plants acts as an efficient biofilter in a similar manner to that postulated in the Magela Creek wetlands (Section 6.2).

Root uptake of ^{226}Ra by the water lily *N. violacea* was measured in the field (Twining, 1989) and studied in the laboratory under controlled conditions (Twining, 1993a). This work was also of interest from a radiological perspective as several parts of the plant, including the rhizomes in the sediment, are consumed by traditional Aboriginals in northern Australia and serve as a potential exposure pathway for U-series isotopes from uranium mines (see Chapter 7). High concentrations of ^{226}Ra were found in roots and rhizomes, but the radioactivity was predominately restricted to the surfaces of the plant and very little penetrated into the tissues. Autoradiography showed that the alpha-emitting radioactivity was associated with a Fe-Mn oxide plaque on the surface of the tissues (Figure 6.9). The plaque was formed by precipitation as the redox environment shifted from the anoxic sediments to an oxidised environment at the root surface due to the presence of photosynthetically derived O_2 being circulated to the roots from the floating leaves via aerenchyma. As a consequence of the presence of this highly adsorptive physical barrier, there could be no direct pathway for ^{226}Ra uptake from sediment via roots to the foliage of *Nymphaea*. Rather, foliage was exposed indirectly by remobilisation of sediment-bound ^{226}Ra into a dissolved form in the water column and subsequent surface adsorption onto foliage. Pettersson et al. (1993) observed a relationship between concentrations of ^{226}Ra in sediment and waterlily foliage, whilst finding no correlation with ^{226}Ra concentrations in water. Twining (1993b) was able to explain this apparent discrepancy as being a consequence of sampling design and biokinetics. The ^{226}Ra concentrations in water were liable to relatively rapidly change given the monsoonal nature of the environment and were therefore highly variable over the sampling period, whilst ^{226}Ra concentrations in the sediment and waterlily remained more consistent.

6.5.8 Detoxification, Sequestration, and Biodistribution

Radionuclides absorbed by organisms may be processed in a variety of ways. If it is an essential element (e.g., Ca) or chemically analogous (e.g., Pb), the radionuclide or its stable element will be used in the various biochemical processes that occur within the tissues and be incorporated into cell components. For radionuclides and their stable elements that are in surplus, and therefore need to be reduced within tissues, the organism has mechanisms for detoxification. One of these is obviously excretion. For this, the radionuclide may be transported to detoxification and/or excretory organs for elimination. For example, a radionuclide may be transported to the hepatopancreas, liver, or kidneys in animals for excretion, or returned to the gills for passive release

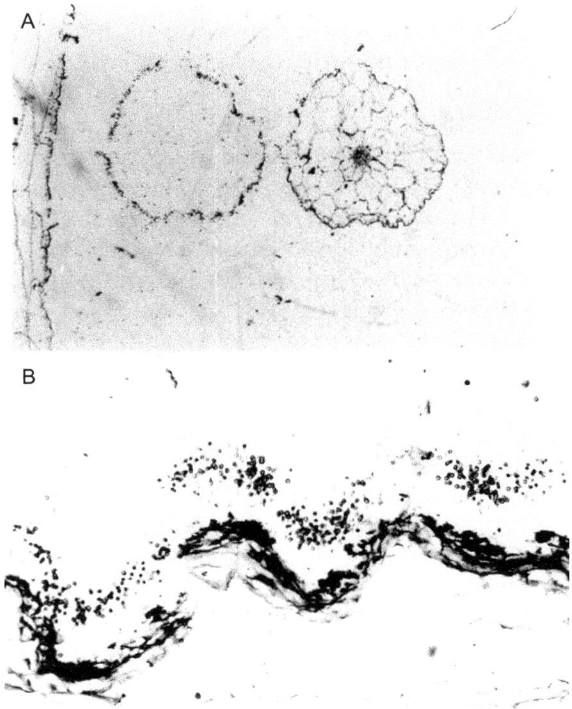

FIGURE 6.9 Autoradiographs of ^{226}Ra associated with waterlily roots. Alpha particles emitted from radium have induced defects in CR-39 sheets overlaying the root section over a period of exposure. The sheets were removed and defects enhanced using a hot caustic solution. The etched CR-39 was later relocated over the section and then displaced to show the pattern of alpha tracks with respect to the tissue structure. The patterns clearly show that ^{226}Ra was accumulated on the surface of the roots, but did not penetrate into the tissues. (A) Entire cross section. (B) Realigned closeup showing the alpha tracks associated with a MnO_2 plaque on the root surface. *From Twining (1988b).*

back into the water along thermodynamic gradients. Another option may be safe storage or sequestration. Within cells, this is achieved using a variety of processes. Sometimes this will involve complexation, for example, with metallothioneins, phytochelatins, or other ligands within the cell that have evolved to allow the organism to 'mop-up' excess toxins, such as U (e.g., Amiard et al., 2006; Pal and Rai, 2010). Alternatively, excess radionuclides or their stable elements may be actively transported across intracellular biological membranes and internalised (e.g., within the cell nucleus, lysosomes, or mitochondria in animal cells or vacuoles in plants). These latter options also hold the possibility of direct excretion from the cell vesicles, which can migrate to the cell membrane and then release their contents external to the cell. Some radionuclides are removed from the intracellular environment by incorporation into components of the organism that are periodically discarded. Examples include movement of radionuclides into the carapaces of crabs at ecdysis (Bergey and Weis, 2007), exoskeletons of isopods (van Hattum et al., 1996), and frustules of diatoms (Gehlen et al., 2002; Barua, 2007).

Many radionuclides and their stable elements are also stored within tissues for use when they are less abundant in their aquatic environment. For example, the freshwater bivalve, *V. angasi*, has a highly efficient uptake mechanism for Ca and evolved a mechanism for storage of any acquired Ca for later use. Any absorbed Ca that is not immediately used for general cell processes or by the mantle for laying down fresh layers of shell, is deposited in relatively insoluble extracellular calcium phosphate granules (Jeffree and Simpson, 1986). These granules thereby act as a sink for other essential and non-essential divalent cations, such as Ra, U, Pb, Cd, Mn, and Co (Jeffree and Simpson, 1986; Brown et al., 1996). Loss rates, which determine the natural rates of bioaccumulation, are governed by the relative solubility of a divalent cation within the granule (Jeffree et al., 1993; Brown et al., 1996; Markich et al., 2001b). The natural rates of accumulation of Ra, Mn, Co, Zn, Cu, and Ni in whole soft tissue are linearly and inversely related to their solubilities (log K_{sp}) as hydrogen phosphate salts (Figure 6.10). However, for U, Cd, and Pb, this linear inverse relationship does not hold (i.e., their rates of accumulation did not increase with decreasing solubility); these three divalent cations are so insoluble in the granules over the organism's lifetime (~50 years) that there is effectively no loss, and hence, no differential between their natural rates of accumulation (Figure 6.10). The very low loss rates of Ra and U in the edible flesh of *V. angasi*, a key component of the

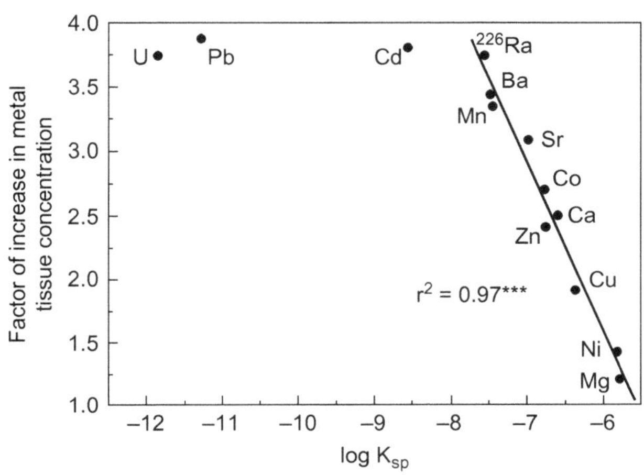

FIGURE 6.10 Factor of increase in tissue concentrations of a range of elements/radionuclides plotted against the logarithm of the solubility product (log K_{sp}) of each element/radionuclide with hydrogen phosphate (HPO_4) for a freshwater bivalve. *From Markich et al. (2001b).*

local Aboriginal traditional diet, makes these radionuclides, together with ^{210}Po, the most restrictive components in any radiological dose assessment. Markich et al. (2002a) also utilised the ability of *V. angasi* to sequester a range of stable elements within their shells as a natural archive of radionuclide exposure (see Section 6.5.9 for a more detailed discussion).

Whichever mechanism(s) is used to sequester or detoxify a radionuclide, the degree of complexation or solubility of the sequestered radionuclide will influence its bioavailability to consumers of the organism and thereby influence any potential biomagnification within the food web. As most bioaccumulated radionuclides tend to behave as chemical analogues of macronutrients and micronutrients, they tend to be distributed in a similar manner. For example, Cs, as a monovalent element, tends to concentrate in areas rich in K or Na, and divalent radionuclides, such as Ra, Ba, and Sr, tend to be co-located with Ca. This was clearly demonstrated by Twining et al. (1996) studying the accumulation of ^{137}Cs and ^{85}Sr in the freshwater fish *Bidyanus bidyanus*. The ^{85}Sr was preferentially distributed to the hard calcified tissues (scales and bone), whilst ^{137}Cs was mostly associated with metabolically active muscle (flesh) (Figure 6.11). Despite the consistencies between some chemically analogous elements, some key radionuclides are much less predictable in terms of their biological behaviour. Polonium isotopes, including ^{210}Po from the ^{238}U-series, are a good example. Polonium can exist in valence states of 2^-, 2^+, 4^+, or 6^+, which provide multiple covalent bonding possibilities. This natural chemical diversity leads to erratic bioaccumulation patterns for ^{210}Po when looking at between-species differences, as identified in Section 6.5.4.2.

6.5.9 Application of Bioaccumulation to Environmental Monitoring and Management

Markich et al. (2002a) sampled freshwater bivalves (*V. angasi*) in 1996 from the Finniss River system in northern Australia at 10 sites *a priori* exposed and non-exposed to mine pollution. Secondary ion mass spectrometry (SIMS) was used to measure Cu, Mn, Zn, U, Ni, Co, Pb, and Fe/Ca ratios across the annual shell laminations of the longest lived bivalves found at each site to evaluate the ability of the shells to archive measured annual radionuclide inputs and their temporal patterns. At sites not exposed to mine pollution, relatively constant and similar (baseline) signals were found for all radionuclides in the shell laminations dating as far back as 1965 (Figure 6.12). At sites impacted by mine pollution, relatively constant, but variably elevated, signals were evident for Cu, Mn, Zn, Ni, and Co in the shell, which extended back only to the end of remediation (1986). Since remediation, the temporal patterns of Cu, Zn, and Mn observed in the shells at the most polluted sites reflected those of the measured annual dissolved loads in the surface waters. The average concentrations of Cu, Mn, Zn, Ni, and Co in the shells decreased

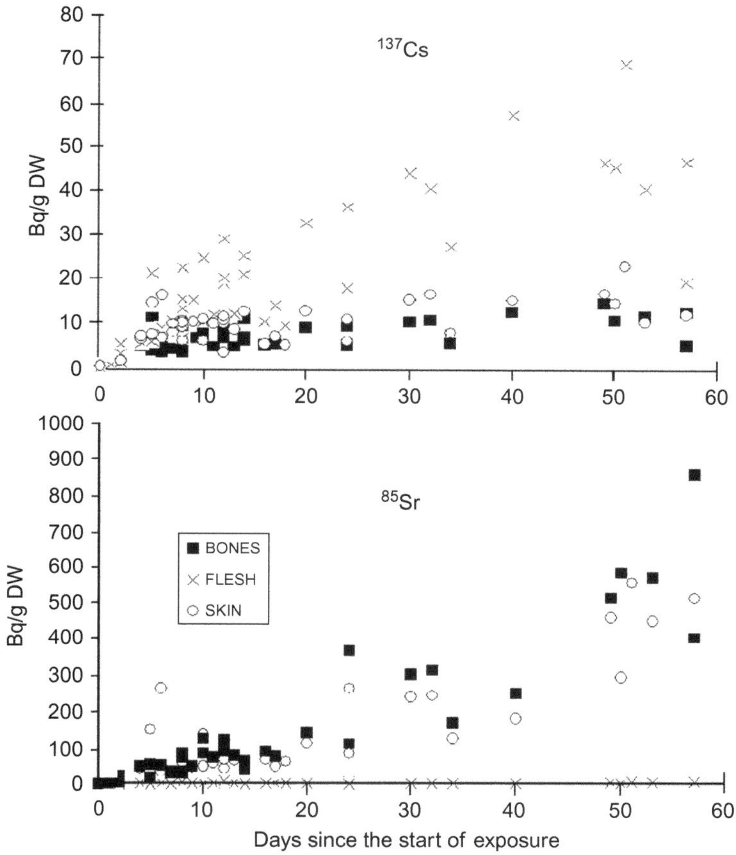

FIGURE 6.11 Preferential biodistribution of ^{137}Cs to flesh and ^{85}Sr to scales and bone in a tropical freshwater fish (*B. bidyanus*) following uptake from water. *From Twining et al. (1996).*

(3-fold to 13-fold) with increasing distance downstream of the mine site, until concentrations characteristic of the unimpacted sites were reached. This geographic pattern of decline in pollution signal in the shell with increasing distance downstream of the pollution input is consistent with the pattern established for water and sediment chemistry. Overall, the SIMS results supported the proposition that the shells of *V. angasi* can be used as archival indicators of radionuclide pollution in surface waters of the Finniss River over their lifetime.

Several studies with *V. angasi* (Jeffree, 1988; Brown et al., 1996; Bollhöfer et al., 2011) have confirmed the use of Ca tissue concentration to predict the tissue concentrations of other divalent cations, explaining up to 98% of the variability between individual bivalves. The use of Ca tissue concentration to effectively minimise the inherent variability between individuals

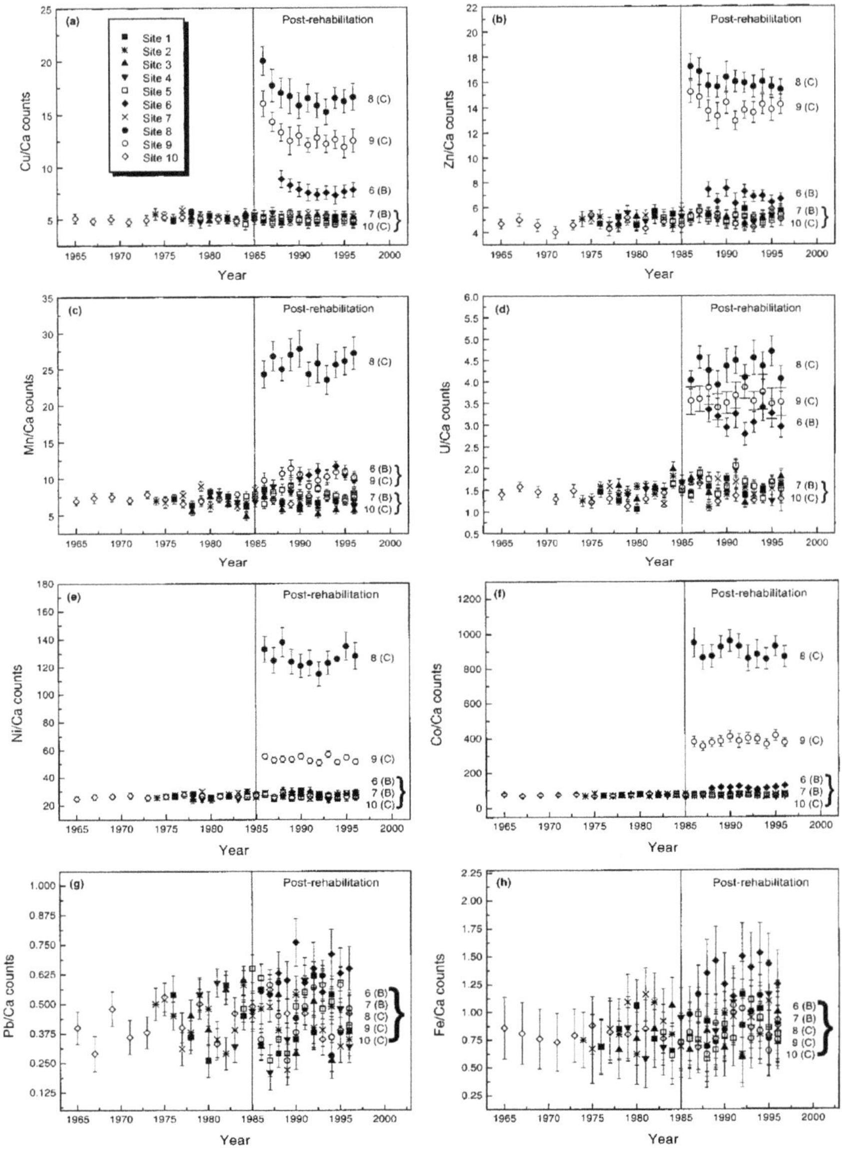

FIGURE 6.12 Element levels, normalised for calcium (Ca), in the shells of the oldest freshwater bivalves (*V. angasi*) from sites in the Finniss River surrounding the Rum Jungle uranium/copper mine (northern Australia). The data show a historical record of element levels in shell that are positively related to element levels in surface waters and sediments at each site. *From Markich et al. (2002a).*

improves the ability of an investigator to discern smaller spatial and/or temporal differences in the radionuclide or their stable element tissue concentrations of these bivalves, and thus to detect contamination. Markich (unpublished data) demonstrated the utility of *V. angasi* to effectively discern spatial (and temporal) differences in U, Cu, Pb, Zn, Co, and Ni contamination downstream of the former Rum Jungle uranium/copper mine (relative to upstream reference sites) using stable element/Ca concentrations in the whole soft tissue.

Markich et al. (2002b) determined the concentrations of Na, K, Ca, Mg, Ba, Sr, Fe, Al, Mn, Zn, Pb, Cu, Ni, Cr, Co, Se, U, and Ti in the osteoderms and flesh of crocodiles (*C. porosus*) from three adjacent catchments within the Alligator Rivers Region (ARR) of northern Australia. Despite considerable within-catchment variability, linear discriminant analysis showed that differences in stable element signatures in the osteoderms and flesh of *C. porosus* amongst the catchments were sufficient to classify individuals accurately to their catchment of occurrence. Using cross-validation, the accuracy of classifying a crocodile to its catchment of occurrence was 76% for osteoderms and 60% for flesh. These data suggest that osteoderms provide better predictive accuracy than flesh for discriminating crocodiles amongst catchments. Reasons for differences in the stable element signatures of crocodiles between catchments are generally not interpretable, due to limited data on surface water and sediment chemistry of the catchments or chemical composition of dietary items of *C. porosus*. From a wildlife management perspective, the provenance or source catchments of 'problem' crocodiles captured at settlements or recreational areas along the ARR coastline may be established using catchment-specific elemental signatures. If the incidence of problem crocodiles can be reduced in settled or recreational areas by effective management at their source, then public safety concerns about these predators may be moderated along with the cost of their capture and removal.

6.5.10 Physiological/Genetic Tolerance

Physiological and/or genetic tolerance of freshwater fish to stable elements has been reported in natural populations from the Finniss River system in northern Australia as a mechanism to explain their partial recovery/recolonisation since remediation of the former Rum Jungle uranium/copper mine, despite stable element concentrations in the sediments and surface water that exceed national guideline values. Gale et al. (2002) found that black-striped rainbowfish (*Melanotaenia nigrans*) from the contaminated East Branch of the Finniss River acquired tolerance to elevated Cu concentrations in the surface waters. The toxicity of Cu to *M. nigrans* living in the contaminated East Branch was 8-fold lower than populations from surrounding reference sites. Fish from the East Branch accumulated significantly ($p \leq 0.05$) less Cu (up to 50%) in all tissues compared to that of reference fish, when experimentally exposed to low and elevated (× 10) Cu concentrations (using $^{64/67}$Cu tracers); interestingly, the loss kinetics were similar for fish from contaminated and reference sites. Based on dissimilar allozyme frequencies and reduced heterozygosity in fish from the East Branch, the mechanism of Cu tolerance (based on reduced Cu uptake) may be genetically based. The selection of allozyme genotypes less sensitive to inhibition by Cu may allow fish from the East Branch to survive Cu concentrations that exceed the capacity of the exclusion mechanism.

Jeffree, Twining, and Markich (unpublished) found that bony bream (*Nematalosa erebi*) and black catfish (*Neosilurus ater*) exposed to the highest concentrations of Co, Cu, Mn, Ni, Pb, U, and Zn in surface water and sediment in the Finniss River, surrounding the former Rum Jungle uranium/copper mine, generally had the lowest tissue concentrations of these elements. The study indicated that communities of fish exposed to element concentrations high enough to induce regular fish kills, over five decades, may have evolved mechanisms that reduce their bioaccumulation. The possible influences of geochemical speciation of surface waters, sediments, and dietary exposure on stable element bioavailability, as well as fish migrations from other riverine regions, were either excluded or discounted as explanatory mechanisms. It may be concluded that populations of both fish species at the most contaminated sites have modified kinetics within their element bioaccumulation physiology. The working hypothesis of inhibited element uptake is firmly supported by: (1) concomitant inhibition of essential nutrient (e.g., Ca) uptake and (2) the results of Gale et al. (2002; see previous section). *A priori*, the response would seem to have clear adaptive value against element uptake and subsequent toxicity.

Results from both studies have implications for potential human exposures to contaminants via a fish diet under the likely prospect of growing populations and socio-economic development of tropical riverine systems.

6.6. CONCLUSIONS

It is generally agreed that tropical freshwater ecosystems are characterised by higher temperatures, light intensity, and organic matter turnover, with faster oxidation-reduction activities, relative to temperate freshwater ecosystems. However, the physicochemical range of tropical freshwaters is similar to that of freshwaters in other climatic regions, but the former can be strongly influenced by physical conditions, such as first flush events at the commencement of the wet season in monsoonal areas. Furthermore, there are no apparent differences in the toxicity of stable elements to freshwater organisms amongst climatic regions.

Radionuclides in (sub) tropical freshwater systems generally behave in a predictable manner, based on what is known from lakes and streams in the better studied temperate climes. Within the water column, fate and behaviour of radionuclides and their stable elements is typically governed by key physicochemical variables such as pH, redox potential, the concentrations of dissolved ions, and the presence and type of organic matter. The chemical form (or speciation) of a radionuclide or stable element is generally of greater biological importance (i.e., bioavailability) than the total concentration. This concept is currently being integrated into mechanistic frameworks (e.g., biotic ligand and bioaccumulation models) by national regulators for protecting freshwater ecosystems.

Given that the volume of fresh surface waters (i.e., rivers and lakes) is relatively small (0.01%) in comparison to seas and oceans, then the biota living within them can also substantially influence the chemistry of a radionuclide by effectively acting as "large particles" available for surface complexation. Sediments and muds behave in essentially the same manner to provide a sink for any radionuclides entering the system. There are several tropical examples, such as Magela Creek in northern Australia and the ponds downstream of the Savannah River site in South Carolina in the United States, where wetlands have acted as effective biofilters to remove radionuclides from the water column.

The physicochemical environment will determine the proportion of any radionuclide in the system that will be bioavailable. However, after radionuclides have been accumulated by organisms, their behaviours generally reflect their similarities to essential (macronutrients and micronutrients) and non-essential elements. Metabolic mechanisms tending toward homeostasis typically make internal organism chemistry less dynamic than that in the external water column. Again, this is to be expected, and radionuclide biokinetics generally follow the patterns observed for freshwater organisms in temperate climates. This constancy has enabled models to be developed to describe the rate and extent of radionuclide bioaccumulation. These models can be adapted to include various uptake pathways (from water, food, or sediment) via gills, skin, or gut and modes of excretion or dilution (e.g., diffusion, egestion, moulting, and growth), and different parameters can be estimated for each radionuclide and organism. Databases have been developed to incorporate the available knowledge in this regard and new data are always welcome, but there is a high degree of variability in the information. Key biotic factors known to influence radionuclide bioaccumulation are size, age, and gender. There are also differences within and between species that reflect the natural variability within any system.

Despite the similarities that exist between tropical and temperate freshwater systems, it should be noted that there is still a paucity of data for (sub) tropical freshwater organisms and systems, and hence there is the chance for exceptions to the consistency to exist. Data are presented in this chapter that show that the uptake of Sr and Cs by tropical freshwater fishes is much lower than would be expected based on the studies of freshwater temperate fishes. These observations point to the need to undertake additional site- and species-specific investigations on the radioecology of key radionuclides, whenever future nuclear developments in tropical systems are initiated.

ACKNOWLEDGMENTS

The authors are grateful to Dr Ross Jeffree and Prof. Murdoch Baxter for constructive comments on an earlier manuscript.

REFERENCES

Adams, W.J., Blust, R.J.P., Borgmann, U., Brix, K.V., DeForest, D.K., Green, A., et al., 2011. Utility of tissue residues for predicting effects of metals on aquatic organisms. Integr. Environ. Assess. Manag. 7, 75–98.

Allen, D.J., Castillo, M.M., 2007. Stream Ecology: Structure and Functioning of Running Waters. Springer, Dordrecht.

Alsop, D.H., Wood, C.M., 1999. Influence of waterborne cations on zinc uptake and toxicity in rainbow trout, *Oncorhynchus mykiss*. Can. J. Fish. Aquat. Sci. 56, 2112–2119.

Amiard, J.C., Amiard-Triquet, C., Barka, S., Pellerin, J., Rainbow, P.S., 2006. Metallothioneins in aquatic invertebrates: Their role in metal detoxification and use as biomarkers. Aquat. Toxicol. 76, 160–2002.

Ansari, T.M., Saeed, M.A., Raza, A., Naeem, M., Salam, A., 2006. Effect of body size on metal concentrations in wild *Puntius chola*. Pak. J. Anal. Environ. Chem. 7, 116–119.

ANZECC, ARMCANZ, 2000. Australian Water Quality Guidelines for Fresh and Marine Waters. Australian and New Zealand Environment and Conservation Council and Agriculture and Resource Management Council of Australia and New Zealand, Canberra.

Ballio, A., Bossa, F., Camoni, L., Di Giorgio, D., Flamond, M.C., Maraite, H., et al., 1996. Major and trace elements of river-borne material: The Congo Basin. Geochim. Cosmochim. Acta 60, 1301–1321.

Banks, S.D., Thomas, P., Baer, K.N., 1999. Seasonal variations in hepatic and ovarian zinc concentrations during the annual reproductive cycle in female channel catfish (*Ictalurus punctatus*). Comp. Biochem. Physiol. 124C, 65–72.

Barua, S., 2007. Microbial Diversity and Zinc Toxicity to Pseudomonas sp. from Coeur d'Alene River Sediment. M.Sc. Thesis. Montana State University, Bozeman, Montana, USA.

Beresford, N.A., 2010. The transfer of radionuclides to wildlife. Radiat. Environ. Biophys. 49, 505–508.

Bergey, L.L., Weis, J.S., 2007. Molting as a mechanism of depuration of metals in the fiddler crab, *Uca pugnax*. Mar. Environ. Res. 64, 556–562.

Bollhöfer, A., Brazier, J., Humphrey, C.L., Ryan, B., Esparon, A., 2011. A study of radium bioaccumulation in freshwater mussels, *Velesunio angasi*, in the Magela Creek catchment, Northern Territory, Australia. J. Environ. Radioact. 102, 964–974.

Brown, P.L., Markich, S.J., 2000. Evaluation of the free ion activity model of metal-organism interaction. Extension of the conceptual model. Aquat. Toxicol. 51, 177–194.

Brown, P.L., Sylva, R.N., 1987. Unified theory of metal ion complex formation constants. J. Chem. Res. 110–181.

Brown, P.L., Jeffree, R.A., Markich, S.J., 1996. Kinetics of ^{45}Ca, ^{60}Co, ^{210}Pb, ^{54}Mn, and ^{109}Cd in the tissue of the freshwater bivalve, *Velesunio angasi*: Further development of a predictive and mechanistic model of metal bioaccumulation. Sci. Total Environ. 188, 139–166.

Bury, N.R., Wood, C.M., 1999. Mechanism of branchial apical silver uptake by rainbow trout is via the proton-coupled Na^+ channel. Am. J. Physiol. 277, R1385–R1391.

Campbell, P.G.C., 1995. Interactions between trace metals and aquatic organisms: A critique of the free-ion activity model. In: Tessier, A., Turner, D.R. (Eds.), Metal Speciation and Bioavailability in Aquatic Systems. John Wiley and Sons, Chichester, pp. 45–102.

Carroll, J.J., Ellis, S.J., Oliver, W.O., 1979. Influences of hardness constituents on the acute toxicity of cadmium to brook trout *(Salvelinus fontinalis)*. Bull. Environ. Contain. Toxicol. 22, 575–581.

CCME, 2011. Canadian Water Quality Guidelines for the Protection of Aquatic Life. Summary Table. Canadian Council of Ministers for the Environment, Winnipeg. http://st-ts.ccme.ca.

Charles, A.L., Markich, S.J., Stauber, J.L., De Filippis, L.F., 2002. The effect of water hardness on the toxicity of uranium to a tropical freshwater alga (*Chlorella sp.*). Aquat. Toxicol. 60, 61–73.

Clifford, M., McGeer, J.C., 2009. Development of a biotic ligand model for the acute toxicity of zinc to *Daphnia pulex* in soft waters. Aquat. Toxicol. 91, 26–32.

Costa, S.C., Hartz, S.M., 2009. Evaluation of trace metals (cadmium, chromium, copper, and zinc) in tissues of a commercially important fish (*Leporinus obtusidens*) from Guaiba Lake, Southern Brazil. Braz. Arch. Biol. Techn. 52, 241–250.

Croteau, M.N., Luoma, S.N., 2005. Delineating copper accumulation pathways for the freshwater bivalve *Corbicula* using stable copper isotopes. Environ. Toxicol. Chem. 24, 2871–2878.

Daam, M.A., van der Brink, P.J., 2010. Implications of differences between temperate and tropical freshwater ecosystems for the ecological risk assessment of pesticides. Ecotoxicology 19, 24–37.

Dantas, M.C., Attayde, J.L., 2007. Nitrogen and phosphorus content of some temperate and tropical freshwater fishes. J. Fish Biol. 70, 100–108.

da Silva, E.M., Soares, A.M.V., 2010. Is there a distinct tropical ecotoxicology? Integr. Environ. Assess. Manage. 6, 313–314.

EC, 2006. Directive of the European Parliament and of the Council on Environmental Quality Standards in the Field of Water Policy and Amending Directive 2000/60/EC. Commission of the European Communities, Brussels.

Edmond, J.M., Palmer, M.R., Measures, C.I., Grant, B., Stallard, R.F., 1995. The fluvial geochemistry and denudation rate of the Guayana Shield in Venezuela, Colombia, and Brazil. Geochim. Cosmochim. Acta 59, 3301–3325.

Elliott, J.M., Hilton, J., Rigg, E., Tullett, P.A., Swift, D.J., Leonard, D.R.P., 1992. Sources of variation in post-Chernobyl radiocaesium in fish from two Cumbrian lakes (north-west England). J. Appl. Ecol. 29, 108–119.

Fernando, C.H., 2002. Zooplankton and tropical freshwater fisheries. In: Fernando, C.H. (Ed.), A Guide to Tropical Freshwater Zooplankton. Backhuys Publishers, Leiden, pp. 255–280.

Finlayson, C.M., Bailey, B., Cowie, F., 1989. Macrophyte Vegetation of the Magela Creek Floodplain, Alligator Rivers Region, Northern Territory. Research Report 5. Australian Government Publishing Service, Canberra.

Flemming, H.C., Neu, T.R., Wozniak, D.J., 2007. The EPS matrix: The house of biofilm cells. J. Bacteriol. 189, 7945–7947.

Fournier, E., Adam-Guillermin, C., Potin-Gautier, M., Pannier, F., 2010. Selenate bioaccumulation and toxicity in *Chlamydomonas reinhardtii*: Influence of ambient sulphate ion concentration. Aquat. Toxicol. 97, 51–57.

Franklin, N., Stauber, J.L., Markich, S.J., Lim, R., 2000. pH-dependent toxicity of Cu and U to a tropical freshwater green alga (*Chlorella* sp.). Aquat. Toxicol. 48, 275–289.

Frausto da Silva, J.J.R., Williams, R.J.P., 2001. The Biological Chemistry of the Elements: The Inorganic Chemistry of Life, 2nd ed. Clarendon, Oxford.

Frostick, A., Bollhöfer, A., Parry, D., 2011. A study of radionuclides, metals and stable lead isotope ratios in sediments and soils in the vicinity of the natural U-mineralisation areas in the Northern Territory. J. Environ. Radioact. 102, 911–918.

Gale, S.A., Smith, S.V., Lim, R.P., Jeffree, R.A., Petocz, P., 2002. Insights into the mechanisms of copper tolerance of a population of black-banded rainbowfish (*Melanotaenia nigrans*) (Richardson) exposed to mine leachate, using $^{64/67}$Cu. Aquat. Toxicol. 62, 135–153.

Gehlen, M., Beck, L., Calas, G., Flank, A.M., van Bennekom, A.J., van Beusekom, J.E.E., 2002. Unraveling the atomic structure of biogenic silica: Evidence of the structural association of Al and Si in diatom frustules. Geochim. Cosmochim. Acta 66, 1601–1609.

GeoHive, 2010. http://www.geohive.com.

Grossel, M., Wood, C.M., 2002. Copper uptake across rainbow trout gills: Mechanisms of apical entry. J. Exp. Biol. 205, 1179–1188.

Grossel, M., Nielson, C., Bianchini, A., 2002. Sodium turnover rate determines sensitivity to acute copper and silver exposure in freshwater animals. Comp. Biochem. Physiol. 133C, 287–303.

Hall, J.L., Williams, L.E., 2003. Transition metal transporters in plants. J. Exp. Bot. 54, 2601–2613.

Handy, R.D., Eddy, F.B., Baines, H., 2002. Sodium-dependent copper uptake across epithelia: A review of rationale with experimental evidence from gill and intestine. Biochim. Biophys. Acta 1566, 104–115.

Heredia, M.A., Zapico, R., García-Sánchez, M., Fernández, J.A., 2002. Effect of calcium, sodium, and pH on uptake and accumulation or radiocesium by *Riccia fluitans*. Aquat. Bot. 74, 245–256.

Higley, K.A., 2010. Estimating transfer parameters in the absence of data. Radiat. Environ. Biophys. 49, 645–656.

Hille, B., Woodhull, A.M., Shapiro, B.I., 1975. Negative surface charge near sodium channels of nerve: Divalent ions, monovalent ions, and pH. Philos. Trans. R. Soc. London B 270, 301–318.

Hogan, A., van Dam, R., Markich, S.J., Camilleri, C., 2005. Chronic toxicity of uranium to the tropical green alga (*Chlorella* sp.) in natural waters and the influence of organic carbon. Aquat. Toxicol. 75, 343–354.

Hogstrand, C., Verbost, P.M., Wendelaar Bonga, S.E., Wood, C.M., 1996. Mechanisms of zinc uptake in gills of freshwater rainbow trout: interplay with calcium transport. Am. J. Physiol. 270, R1141–R1147.

Hollis, L., McGeer, J.C., McDonald, D.G., Wood, C.M., 2000. Protective effects of calcium against chronic waterborne cadmium exposure to juvenile rainbow trout. Environ. Toxicol. Chem. 19, 2725–2734.

Hosseini, A., Thorring, H., Brown, J.E., Saxen, R., Ilus, E., 2008. Transfer of radionuclides in aquatic ecosystems – Default concentration ratios for aquatic biota in the ERICA tool. J. Environ. Radioact. 99, 1408–1429.

Howard, B.J., Larsson, C.M., 2008. The ERICA integrated approach and its contribution to protection of the environment from ionising radiation. J. Environ. Radioact. 99, 1361–1363.

Howard, B.J., Beresford, N.A., Copplestone, D., Telleria, D., Proehel, G., Fesenko, S., et al., 2013. The IAEA handbook on radionuclide transfer to wildlife. J. Environ. Radioact. in press.

IAEA, 1994. Handbook of Parameter Values for the Prediction of Radionuclide Transfer in Temperate Environments. Technical Report Series No. 364. International Atomic Energy Agency, Vienna.

IAEA, 2010a. Handbook of Parameter Values for the Prediction of Radionuclide Transfer in Terrestrial and Freshwater Environments. Technical Report Series No. 472. International Atomic Energy Agency, Vienna.

IAEA, 2010b. Modeling Radiation Exposure and Radionuclide Transfer for Non-human Species. Report of the Biota Working Group of EMRAS Theme 3. International Atomic Energy Agency, Vienna. http://www-ns.iaea.org/downloads/rw/projects/emras/final-reports/biota-final.pdf.

ICRU, 2001. Quantities, Units, and Terms in Radioecology. Report 65. International Commission on Radiation Units and Measurements, Bethesda, Maryland, USA.

Jackson, B.P., Lasier, P.J., Miller, W.P., Winger, P.W., 2000. Effects of calcium, magnesium and sodium on alleviating cadmium toxicity to *Hyalella azteca*. Bull. Environ. Contam. Toxicol. 64, 279–286.

Jeffree, R.A., 1988. Patterns of accumulation of alkaline-earth metals in the tissue of the freshwater mussel *Velesunio angasi* (Sowerby). Arch. Hydrobiol. 112, 67–90.

Jeffree, R.A., 1991. An experimental study of Ra-226 and Ca-45 accumulation from the aquatic medium by freshwater turtles (Fam. Chelidae) under varying Ca and Mg water concentrations. Hydrobiologia 218, 205–233.

Jeffree, R.A., Simpson, R.D., 1986. An experimental study of the uptake and loss of radium-226 by the tissue of the freshwater mussel *Velesunio angasi* under varying Ca and Mg water concentrations. Hydrobiologia 139, 59–80.

Jeffree, R.A., Markich, S.J., Brown, P.L., 1993. Comparative accumulation of alkaline-earth metals by two freshwater mussel species from the Nepean River, Australia: Consistencies and a resolved paradox. Aust. J. Mar. Fresh. Res. 44, 609–634.

Jeffree, R.A., Markich, S.J., Twining, J.R., 2001. Element concentrations in the flesh and osteoderms of estuarine crocodiles (*Crocodylus porosus*) from the Alligator Rivers Region, Northern Australia: Biotic and geographic effects. Arch. Environ. Contam. Toxicol. 40, 236–245.

Jeffree, R.A., Markich, S.J., Tucker, A.D., 2005. Patterns of metal accumulation in osteoderms of the Australian freshwater crocodile, *Crocodylus johnstoni*. Sci. Total Environ. 336, 71–80.

Kelly, M.S., Pinder, J.E., 1996. Foliar uptake of ^{137}Cs from the water column by aquatic macrophytes. J. Environ. Radioact. 30, 271–280.

Klessa, D., 2000. The Chemistry of Magela Creek. Supervising Scientist Report 151. Supervising Scientist, Darwin.

Komjarova, I., Blust, R., 2009. Effect of Na, Ca and Ph on simultaneous uptake of Cd, Cu, Ni, Pb, and Zn in the water flea *Daphnia magna* measured using stable isotopes. Aquat. Toxicol. 94, 81–86.

Kwok, K.W., Leung, K.M., Lui, G.S., Chu, V.K., Lam, P.K., Moritt, D., et al., 2007. Comparison of tropical and temperate freshwater animal species sensitivities to chemicals: Implications for deriving safe extrapolation factors. Integr. Environ. Assess. Manage. 3, 49–67.

Lévêque, C., Oberdorff, T., Paugy, D., Stiassny, M.L., Tedesco, P.A., 2008. Global diversity of fish (Pisces) in freshwater. Hydrobiologia 595, 545–567.

Levitt, J., 1957. The significance of "Apparent Free Space" (A.F.S.) in ion absorption. Physiol. Plant. 10, 882–888.

Lewis, W.M., 1987. Tropical limnology. Annu. Rev. Ecol. Evol. Syst. 18, 159–184.

Liao, C.M., Chen, B.C., Singh, S., Lin, M.C., Liu, C.W., Han, B.C., 2003. Acute toxicity and bioaccumulation of arsenic in tilapia (*Oreochromis mossambicus*) from a blackfoot disease area in Taiwan. Environ. Toxicol. 18, 252–259.

Luoma, S.N., Rainbow, P.S., 2008. Metal Contaminants in Aquatic Environments. Cambridge University Press, London.

Marigomez, I., Soto, M., Cajaraville, M.P., Angulo, E., Giamberini, L., 2002. Cellular and subcellular distribution of metals in molluscs. Micros. Res. Tech. 56, 358–392.

Markich, S.J., Jeffree, R.A., 1994. Absorption of divalent trace metals as analogues of calcium by Australian freshwater bivalves: An explanation of how water hardness reduces metal toxicity. Aquat. Toxicol. 29, 257–290.

Markich, S.J., Brown, P.L., Jeffree, R.A., 1996. The use of geochemical speciation modeling to predict the impact of uranium to freshwater biota. Radiochim. Acta 74, 321–326.

Markich, S.J., Brown, P.L., Jeffree, R.A., Lim, R.P., 2000. Valve movement responses of *Velesunio angasi* (Bivalvia: Hyriidae) to Mn and U: An exception to the free ion activity model. Aquat. Toxicol. 51, 155–175.

Markich, S.J., Brown, P.L., Batley, G.E., Apte, S.C., Stauber, J.L., 2001a. Incorporating metal speciation and bioavailability into water quality guidelines for protecting aquatic ecosystems. Australas. J. Ecotoxicol. 7, 109–122.

Markich, S.J., Brown, P.L., Jeffree, R.A., 2001b. Divalent metal accumulation in freshwater bivalves: An inverse relationship with metal phosphate solubility. Sci. Total Environ. 275, 27–41.

Markich, S.J., Jeffree, R.A., Burke, P.T., 2002a. Freshwater bivalve shells as archival indicators of metal pollution from a copper-uranium mine in tropical Australia. Environ. Sci. Technol. 32, 821–832.

Markich, S.J., Jeffree, R.A., Harch, B.D., 2002b. Catchment-specific element signatures in estuarine crocodiles (*Crododylus porosus*) from the Alligator Rivers Region, northern Australia. Sci. Total Environ. 287, 83–95.

Markich, S.J., Batley, G.E., Stauber, J.L., Rogers, S.C., Hyne, R.V., Bowles, K.C., et al., 2005. Hardness corrections for copper are inappropriate for protecting sensitive freshwater biota. Chemosphere 60, 1–8.

Markich, S.J., King, A.R., Wilson, S.P., 2006. Non-effect of water hardness on the accumulation and toxicity of copper in a freshwater macrophyte (*Ceratophyllum demersum*): How useful are hardness-modified copper guidelines for protecting freshwater biota? Chemosphere 65, 1791–1800.

Martin, P., Ryan, B., 2004. Natural-series radionuclides in traditional Aboriginal foods in tropical northern Australia: A review. SciWorld J. 4, 77–95.

Matsuo, A.Y.O., Wood, C.M., Val, A.L., 2005. Effects of copper and cadmium on ion transport and gill metal binding in the Amazonian teleost tambaqui (*Colossoma macropomum*) in extremely soft water. Aquat. Toxicol. 74, 351–364.

Meybeck, M., Ragu, A., 1997. River Discharges to the Oceans: An Assessment of Suspended Solids, Major Ions, and Nutrients. United Nations Environment Programme, Nairobi.

Meybeck, M., Friedrich, G., Thomas, R., Chapman, D., 1996. Rivers. In: Chapman, D. (Ed.), Water Quality Assessments: A Guide to the Use of Biota, Sediments, and Water in Environmental Monitoring, 2nd ed. Chapman and Hall, London, pp. 243–318.

Mollah, A.S., Begum, A., Akther, S., Malek, M.A., Ullah, S.M., Khan, Z.H., 1994. Studies on Radionuclide Transfer From Soil and Freshwater to the Foodchain of Man in Tropical Environment of Bangladesh. Report to the IAEA/FAO CRP on Transfer of radionuclides from air, soil, and fresh water to the food chain of man in tropical and subtropical environments. 12-16 December 1994. Damascus.

Mollah, A.S., Begum, A., Ullah, S.M., Khan, Z.H., 1995. Studies on Radionuclide Transfer From Soil and Freshwater to the Foodchain of Man in Tropical Environment of Bangladesh. Report to the IAEA/FAO CRP on Transfer of radionuclides from air, soil and fresh water to the food chain of man in tropical and subtropical environments. 6-10 November 1995. Istanbul.

Mollah, A.S., Begum, A., Ullah, S.M., Khan, Z.H., 1997. Studies on Radionuclide Transfer From Soil and Freshwater to the Foodchain of Man in Tropical Environment of Bangladesh. Report to the IAEA/FAO CRP on Transfer of radionuclides from air, soil, and fresh water to the food chain of man in tropical and subtropical environments. 16-20 June 1997. Vienna.

Monte, L., 2011. Regional-scale application of the Decision Support System MOIRA-PLUS: An example of assessment of the radiological impact of the Chernobyl accident on the fresh water ecosystem in Italy. J. Environ. Radioact. 102, 73–83.

Monte, L., Brittain, J.E., Hakanson, L., Heling, R., Smith, J.T., Zheleznyak, M., 2003. Review and assessment of models used to predict the fate of radionuclides in lakes. J. Environ. Radioact. 69, 177–205.

Mora, A., Alfonso, J.A., Sánchez, L., Calzadilla, M., Silva, S., Labrecque, J., et al., 2009. Temporal variability of selected dissolved elements in the lower Orinoco River. Venezuela. Hydrol. Processes 23, 476–485.

Murray, A.S., Johnston, A., Martin, P., Hancock, G., Marten, R., Pfitzner, J., 1993. Transport of naturally occurring radionuclides by a seasonal tropical river, northern Australia. J. Hydrol. 150, 19–39.

Ngo, N.T., Binh, N.T., 1997. Factors of Radionuclide Transfer from Air, Soil and Freshwater to the Foodchain of Man in Monsoon Tropical Condition of Vietnam. Report to the IAEA/FAO CRP on Transfer of radionuclides from air, soil and fresh water to the food chain of man in tropical and subtropical environments, 16-20 June 1997. Vienna.

Niyogi, S., Wood, C.M., 2004. Biotic ligand model, a flexible tool for developing site-specific water quality guidelines for metals. Environ. Sci. Technol. 38, 6177–6192.

Oberdorf, T., Tedesco, P.A., Hugueny, B., Leprieur, F., Beauchard, O., Brosse, S., et al., 2011. Global and regional patterns in riverine fish species richness: A review. Int. J. Ecol. 2011, Article ID 967631, 12.

Olguín, E.J., Hernández, E., Ramos, I., 2002. The effect of both different light conditions and the pH value on the capacity of *Salvinia minima* BAKER for removing cadmium, lead, and chromium. Acta Biotechnol. 22, 121–131.

Pal, R., Rai, J.P., 2010. Phytochelatins: Peptides involved in heavy metal detoxification. Appl. Biochem. Biotechnol. 160, 945–963.

Paquin, P.R., Gorsuch, J.W., Apte, S.C., Bowles, K.C., Batley, G.E., Campbell, P.G.C., et al., 2002. The biotic ligand model: A historical overview. Comp. Biochem. Physiol. 133C, 3–36.

Paquin, P., Redman, A., Ryan, A., Santore, R., 2012. Modeling the phyiology and toxicology of metals. In: Wood, C.M., Farrell, A.P., Brauner, C.J. (Eds.), Homeostatis and Toxicology on Non-Essential Metals. Elsevier, Amsterdam, pp. 429–484.

Perdue, E.M., 1998. Metal binding by humic substances in surface waters - experimental and modeling constraints. In: Allen, H.E., Garrison, W., Luther, G.W. (Eds.), Metals in Surface Waters. Ann Arbor Publishers, Chelsea, pp. 169–190.

Perdue, E.M., Ritchie, 2003. Dissolved organic matter in fresh waters. In: Holland, H.D., Turekian, K.K. (Eds.), Treatise on Geochemistry, Elsevier-Pergamon, Oxford, pp. 273–318.

Perry, S.F., Flik, G., 1988. Characterization of branchial transepithelial calcium fluxes in freshwater trout *Salmo gairdneri*. Am. J. Physiol. 254, R491–R498.

Pettersson, H.B.L., Hancock, G., Johnston, A., Murray, A.S., 1993. Uptake of uranium and thorium series radionuclides by the waterlily, *Nymphaea violacea*. J. Environ. Radioact. 19, 85–108.

Picouet, C., Dupré, B., Orange, D., Valladon, M., 2002. Major and trace element geochemistry in the upper Niger river (Mali): Physical and chemical weathering rates and CO_2 consumption. Chem. Geol. 185, 93–124.

Pidwirny, M.J., 2006. Fundamentals of Physical Geography, 2nd ed. Electronic online resource. http://www.physicalgeography.net/fundamentals/8b.html.

Pinder, J.E., Hinton, T.G., Whicker, F.W., 2006. Foliar uptake of cesium from the water column by aquatic macrophytes. J. Environ. Radioact. 85, 23–47.

Pinder, J.E., Hinton, T.G., Taylor, B.E., Whicker, F.W., 2011. Cesium accumulation by aquatic organisms at different trophic levels following an experimental release into a small reservoir. J. Environ. Radioact. 102, 283–293.

Pyle, G.G., Clulow, F.V., 1998. Radionuclide equilibria between the aquatic environment and fish tissues. J. Environ. Radioact. 40, 59–74.

Rahman, M.A., Hasegawa, H., Ueda, K., Maki, T., Okumura, C., Rhaman, M.M., 2007. Arsenic accumulation in duckweed (*Spirodella polyrhiza* L.): A good option for phytoremediation. Chemosphere 69, 493–499.

Redecker, E.S., Blust, R., 2004. Accumulation and toxicity of cadmium in the aquatic polychaete *Tubifex tubifex*: A kinetic modeling approach. Environ. Sci. Technol. 38, 537–543.

Riedel, G.F., 1985. The relationship between chromium (VI) uptake, sulphate uptake, and chromium (VI) toxicity in the estuarine diatom *Thalassiosira pseudonana*. Aquat. Toxicol. 7, 191–204.

Riethmuller, N., Markich, S.J., van Dam, R.A., Parry, D., 2001. Effects of water hardness and alkalinity on the toxicity of uranium to a tropical freshwater hydra (*Hydra viridissima*). Biomarkers 6, 45–51.

Rogers, J.T., Wood, C.M., 2004. Characterization of branchial lead-calcium interaction in the freshwater rainbow. J. Exp. Biol. 207, 813–825.

Rowan, D.J., Rasmussen, J.B., 1994. Bioaccumulation of radiocesium by fish: The influence of physicochemical factors and trophic structure. Can. J. Fish. Aquat. Sci. 51, 2388–2410.

Rowan, D.J., Chant, L.A., Rasmussen, J.B., 1998. The fate of radiocesium in freshwater communities – Why is biomagnification variable both within and between species? J. Environ. Radioact. 40, 15–36.

Sauerland, C., Martin, P., Humphrey, C.L., 2005. Radium-226 in Magela Creek, northern Australia: Application of protection limits from radiation for humans and biota. Radioprotection 40, S451–S456.

Schaller, J., Weiske, A., Mkandawire, M., Dudel, E.G., 2010. Invertebrates control metals and arsenic sequestration as ecosystem engineers. Chemosphere 79, 169–173.

Schneider, L., Belger, L., Burger, J., Vogt, R.C., Ferrara, C.R., 2010. Mercury levels in muscle of six species of turtles eaten by people along the Rio Negro of the Amazon Basin. Arch. Environ. Contam. Toxicol. 58, 444–450.

Simkiss, K., Taylor, M.G., 1995. Transport of metals across membranes. In: Tessier, A., Turner, D.R. (Eds.), Metal Speciation and Bioavailability in Aquatic Systems. John Wiley and Sons, Chichester, pp. 1–44.

Sinakhom, F., Supaokit, P., Punnachiya, M., Chantaraprachoom, N., Prasertchewchan, N., Yamkate, P., 1997. Transfer Factors of ^{137}Cs and ^{85}Sr by Freshwater Fish in Tropical Environments. Report to the IAEA/FAO CRP on Transfer of radionuclides from air, soil and fresh water to the food chain of man in tropical and subtropical environments. 16-20 June 1997. Vienna.

Slaveykova, V.I., Wilkinson, K.J., 2005. Predicting the bioavailability of metals and metal complexes: Critical review of the biotic ligand model. Environ. Chem. 2, 9–24.

Smith, J.T., Konoplev, A.V., Bulgakov, A.A., Comans, R.N.J., Cross, M.A., Kaminski, S., et al., 2002a. AQUASCOPE Technical Deliverable. Simplified models for predicting ^{89}Sr, ^{90}Sr, ^{134}Cs, ^{137}Cs, ^{131}I in water and fish of rivers, lakes and reservoirs. Centre for Ecology and Hydrology, Natural Environment Research Council, Lancaster.

Smith, J.T., Kudelsky, A.V., Ryabov, I.N., Daire, S.E., Boyer, L., Blust, R.J., et al., 2002b. Uptake and elimination of radiocaesium in fish and the "size effect" J. Environ. Radioact. 62, 145–164.

Smith, J.T., Belova, N.V., Bulgakov, A.A., Comans, R.N.J., Konoplev, A.V., Kudelsky, A.V., et al., 2005. The "AQUASCOPE" simplified model for predicting Sr-89, Sr-90, I-131, and

Cs-134, Cs-137 in surface waters after a large-scale radioactive fallout. Health Phys. 89, 628–644.

Smith, J.T., Bowes, M.J., Cailes, C.R., 2011. A review and model assessment of ^{32}P and ^{33}P uptake to biota in freshwater systems. J. Environ. Radioact. 102, 317–325.

Srivastava, A., Denschalag, H.O., Kelber, O., Urich, K., 1990. Accumulation and discharge behavior of Cs-137 by zebra fish (*Brachydanio rerio*) in different aquatic environments. J. Radioanal. Nucl. Chem. 138, 165–170.

Srivastava, A., Reddy, S.J., Urich, K., Denschalag, K.A., 1994. Uptake and release kinetics of ^{134}Cs by goldfish (*Carassius auratus*) and ^{137}Cs by zebra fish (*Brachydanio rerio*) in controlled aquatic environment. J. Radioanal. Nucl. Chem. 182, 63–69.

Stumm, W., Morgan, J.J., 1996. Aquatic Chemistry: Chemical Equilibria and Rates in Natural Waters, 3rd ed. John Wiley and Sons, New York.

Takasusuki, J., Araujo, M.R., Fernandes, M.N., 2004. Effect of water pH on copper toxicity in the neotropical fish, *Prochilodusscrofa* (Prochilodondidae). Bull. Environ. Contam. Toxicol. 72, 1075–1082.

Tan, Q.G., Wang, W.X., 2011. Acute toxicity of cadmium in *Daphnia magna* under different calcium and pH conditions: Importance of influx rate. Environ. Sci. Technol. 45, 1970–1976.

Thomann, R.V., Shrkreli, F., Harrison, S., 1997. A pharmocokinetic model of cadmium in rainbow trout. Environ. Toxicol. Chem. 16, 2268–2274.

Tipping, E., 2002. Cation Binding by Humic Substances. Cambridge University Press, London.

Tipping, E., Lofts, S., Sonke, J.E., 2011. Humic ion-binding model VII: A revised paramaterisation of cation-binding by humic substances. Environ. Chem. 8, 225–235.

Trenfield, M.A., Ng, J.C., Noller, B.N., Markich, S.J., van Dam, R.A., 2011. Dissolved organic carbon reduces uranium bioavailability and toxicity. 2. Uranium[VI] speciation and toxicity to three tropical freshwater organisms. Environ. Sci. Technol. 45, 3082–3089.

Trenfield, M.A., Markich, S.J., Ng, J.C., Noller, B.N., van Dam, R.A., 2012a. Dissolved organic carbon reduces the toxicity of aluminium to three tropical freshwater organisms. Environ. Toxicol. Chem. 31, 427–436.

Trenfield, M.A., Ng, J.C., Noller, B.N., Markich, S.J., van Dam, R.A., 2012b. Dissolved organic carbon reduces uranium toxicity to the unicellular eukaryote *Euglena gracilis*. Ecotoxicology 21, 1013–1023.

Twining, J.R., 1988a. Radium accumulation from water by foliage of the water lily, *Nymphaea violacea*. Verh. Internat. Verein. Theor. Angew. Limnol. 23, 1954–1962.

Twining, J.R., 1988b. Radium-226 Accumulation by the Water Lily *Nymphaea violacea*. M.Sc. Thesis. Macquarie University, Sydney, Australia.

Twining, J.R., 1989. Principal coordinate analysis of the distribution of radium-226 between water, sediment, and the water lily, *Nymphaea violacea* (Lehm), in the vicinity of a uranium mine in the Northern Territory. Australia. J. Environ. Radioact. 10, 99–113.

Twining, J.R., 1993a. A study of radium uptake by the water lily *Nymphaea violacea* (Lehm) from contaminated sediment. J. Environ. Radioact. 20, 169–189.

Twining, J.R., 1993b. Letter to the Editor. Re:- Paper by Petterssen et al. J. Environ. Radioact. 19, 85–108.

Twining, J.R., Ferris, J.M., Markich, S.J., 1996. Bioaccumulation of ^{85}Sr and ^{137}Cs by an Australian sub-tropical freshwater teleost (*Bidyanus bidyanus*). Sci. Total Environ. 192, 245–257.

Twining, J.R., Ferris, J.M., Markich, S.J., 1997. Measurement of Transfer Factors for Radionuclides, including ^{137}Cs and ^{85}Sr, in Freshwater Fish and Plant Crops from Tropical and Sub-tropical Australia. Report to the IAEA/FAO CRP on Transfer of radionuclides from air, soil and fresh water to the food chain of man in tropical and subtropical environments. 16-20 June 1997. Vienna.

Twining, J.R., Markich, S.J., Ferris, J.M., 1998. Determinants of radiological dose from Sr and Cs accumulated by tropical freshwater fish. In: Proceedings of Pacific Basin Nuclear Conference. May 3-7. Banff, Canada, p. 8.

USDOE, 2002. A Graded Approach for Evaluating Radiation Doses to Aquatic and Terrestrial Biota. Technical Standard DOE-STD-1153-2002. United States Department of Energy, Washington DC. http://web.ead.anl.gov/resrad/home2/biota.cfm.

USEPA, 2001. 2001 Update of Ambient Water Quality Criteria for Cadmium. EPA 822-R-01-001. United States Environmental Protection Agency, Washington DC.

USEPA, 2007. Aquatic Life Ambient Freshwater Quality Criteria – Copper. EPA-822-R-07-001. United States Environmental Protection Agency, Washington DC.

van Dam, R.A., Harford, A.J., Trenfield, M.A., Markich, S.J., Humphrey, C.L., Hogan, A.C., et al., 2012. Re-analysis of uranium toxicity data for selected freshwater organisms and the influence of dissolved organic carbon. Environ. Toxicol. Chem., in press.

van Hattum, B., van Straalen, N.M., Govers, H.A.J., 1996. Trace metals in populations of freshwater isopods: Influence of biotic and abiotic variables. Arch. Environ. Contain. Toxicol. 31, 303–318.

Veltman, K., Huijbregts, M.A., Hendriks, A.J., 2010. Integration of biotic ligand models (BLM) and bioaccumulation kinetics into a mechanistic framework for metal uptake in aquatic organisms. Environ. Sci. Technol. 44, 5022–5028.

Whicker, F.W., Schultz, V., 1982. Radioecology: Nuclear Energy and the Environment, vol. 2. CRC Press, Boca Raton.

Wilde, K.L., Stauber, J.L., Markich, S.J., Franklin, N.M., Brown, P.L., 2006. The effect of pH on the uptake and toxicity of copper and zinc to a tropical freshwater alga (*Chlorella* sp.). Arch. Environ. Contam. Toxicol. 51, 174–185.

Williams, A.R., 1979. Vegetation and stream pattern as indicators of water movement on the Magela floodplain, Northern Territory. Aust. J. Ecol. 4, 239–247.

Williams, A.R., 1988. Radium Cycling in a Tropical Wetland. Macquarie University, M.Sc. Thesis. Sydney, Australia.

WNA, 2012. World Nuclear Power Reactors and Uranium Requirements. World Nuclear Association, London. http://www.world-nuclear.org/info/reactors.html.

Yankovich, T.L., Beresford, N.A., Wood, M.D., Aono, T., Andersson, P., Barnett, C.L., et al., 2010. Whole-body to tissue-specific concentration ratios for use in biota dose assessments for animals. Radiat. Environ. Biophys. 49, 549–565.

Chapter 7

Radiological Consequences Modelling

Gerhard Pröhl[1], John R. Twining[2], and Jagoda Crawford[3]
[1]*International Atomic Energy Agency, Vienna, Austria*
[2]*Austral Radioecology, Oyster Bay, New South Wales, Australia*
[3]*Australian Nuclear Science and Technology Organisation Institute for Environmental Research*

7.1. INTRODUCTION

The release of radionuclides into the environment is subject to evaluation according to the principles of radiation protection. For most of the period during which dose assessment methodologies were developed, emphasis was placed on conditions in North America, Europe, and Northern Asia with their predominantly temperate environments where, still today, most nuclear installations, including nuclear power plants, fuel fabrication, mining, and milling facilities, are located. A variety of models has been developed to assess radiation exposures arising from both routine and unplanned releases of radionuclides to the environment (US NRC, 1977; Whicker and Kirchner, 1987; Whicker et al., 1990b; Müller and Pröhl, 1993; Brown and Simmonds, 1995). As a consequence of the location of the majority of nuclear facilities, the development of such models focussed on the environmental and agricultural conditions that predominate in the northern temperate zone, such as intensively managed farmland, pronounced seasonality of plant growth, soil types, and living habits.

More and more nuclear installations are being constructed in tropical and subtropical countries. This raises the need to set up appropriate dose assessment models for tropical environments to evaluate whether radioactive releases comply with local radiation safety standards. However, despite the obvious differences among different sites and different climates, due to the lack of data specific to tropical environments, parameter values determined in temperate environments are often used for tropical environments as well. Furthermore, radioecological models should be available that: (1) estimate

Radioactivity in the Environment, Vol. 18. http://dx.doi.org/10.1016/B978-0-08-045016-2.00007-2

activity concentrations as a function of time in foods, animal feeds, and environmental media; (2) assess radiological consequences of releases; and (3) provide the possibility of estimating the mitigating effect of countermeasures to support emergency management. It has also been recognised with increased priority in recent times that radiological impacts on the broader ecosystem need to be considered in addition to the more specific assessments on humans.

Very little information can be found on modelling of radionuclide transfer in tropical rain forests, which are concentrated in a band around the equator between the two tropics and usually characterised by dense canopy, understorey, and forest ground. Rainfall rates in these areas are high, and can be over 2000 mm/a. There is a lack of suitable data for use in simulation of radionuclide dispersion and fate after an accidental deposition on a tropical rain forest; for example, no tropical equivalents to accidents such as the Chernobyl accident; the B-52 bomber collision near the village of Palomares, Spain, in 1966 (Espinosa et al., 1998, 2000); or the re-entry to the atmosphere and burn-up of the Cosmos 954 Satellite over Canada in 1978 (Eaton, 2002). Even the recent Fukushima incident, while occurring outside of Europe and North America, was still well within the temperate zone.

The assessment of radionuclide transport through the environment and of the resulting exposure to the human population and to flora and fauna is the result of a complex interaction of transfer processes and environmental conditions. These hold true for any climate; however, climate does influence the crops cultivated and their growing periods, the development of soils, the time spent on outdoor activities by the population, and their consumption habits, amongst other factors.

There are numerous papers on the processes that lead to the transfer of radionuclides in the environment, and they are discussed in earlier chapters of this book. As such, in this chapter, these processes are not described in detail, but how they are influenced by climatic conditions will be reviewed. Similarly, terrestrial systems have been the focus of many modelling approaches. Because of this, the discussion in this chapter focusses initially on terrestrial dose assessment and subsequently provides a separate section on aquatic pathways.

After the consideration of radiological consequences to humans, the more recent deliberations on environmental dose assessment and, from that, radioecological risk assessment will be discussed later in this chapter. As noted for human dose assessment, there are only limited data for radiological behaviours in tropical and subtropical ecosystems. Nonetheless, the basic principles and processes, and models that are developed from them, are equally pertinent to tropical systems, as long as the parameters used are based on the biota and exposure scenarios likely to exist within such environments.

7.2. RADIONUCLIDE TRANSFER FOLLOWING ATMOSPHERIC DISPERSION

The potential terrestrial exposure pathways for radionuclides in the environment to humans are summarised in Figure 7.1; aquatic exposures are considered in detail in Section 7.4. Radioactivity released to the atmosphere is dispersed by the wind (see Chapter 2). Downwind contamination of vegetation and soil occurs by dry and wet deposition. Activity intercepted by plants is, for the most part, subsequently lost by weathering processes. Deposition on farmland leads to contamination of the soil, which in turn represents a long-term source for contamination of plants via root uptake. Another long-term source of plant contamination arises from resuspension of contaminated soil particles and their re-deposition onto plants. Radioactivity may be bound or fixed to soil components (see Chapters 3 and 5), or it may be lost from soil via physical decay, erosion processes, and migration into deeper soil layers from which the radionuclides are not available for root uptake. Contaminated plants are fed to animals, leading to a contamination of meat, milk, and eggs. For grazing animals, the involuntary intake of activity due to soil ingestion may be relevant. Exposure of humans occurs due to the following:

- Internal exposure due to inhalation of contaminated air or of resuspended contaminated soil particles
- Transfer of radionuclides in food chains and subsequent internal exposure of humans due to ingestion of contaminated foodstuffs
- External exposure from radionuclides in the air (immersion)
- External exposure from radionuclides deposited on the ground

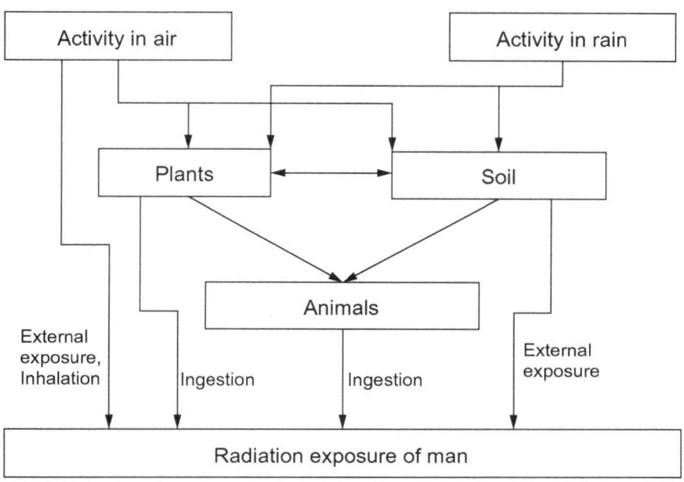

FIGURE 7.1 Simplified scheme of terrestrial radiological exposure pathways to humans subsequent to releases of radionuclides to the atmosphere.

Many factors must be considered in the assessment of all exposure pathways. Most complex is the assessment of the ingestion exposure, which is the result of many processes. Activity concentrations have to be estimated in many different foods, which are dependent on time since deposition. In detail, the following transfer processes have to be taken into account:

- Dry deposition of radionuclides onto soil and vegetation
- Interception of wet-deposited radionuclides by vegetation
- Growth dilution and weathering loss from vegetation
- Systemic transport of radionuclides in plants subsequent to deposition on the foliage
- Migration and fixation of radionuclides in soil
- Uptake of radionuclides by plants from the soil
- Intake of radionuclides by domestic animals
- Transfer of radionuclides to animal-derived food products such as meat, milk, and eggs
- Modification of activity concentrations in foods during processing and culinary preparation
- Proportional consumption of different foods

The large numbers of parameters needed to estimate radionuclide transfers to plant- and animal-derived foodstuffs and the resulting exposures to humans are summarised in Table 7.1. The degree of dependency of the parameters on site conditions, elemental chemistry, or the characteristics of the radionuclide are also indicated. In general, site-specific parameters characterise the exposure scenario. These values describe the environmental conditions, identify relevant crops and animal products, specify agricultural practices (such as the growing periods of plants and harvest times of crops), define feeding regimes, and quantify human habits. Such data are independent of nuclides and specific elements.

In general, factors controlling element-dependent processes are widely applicable, but the parameter values may be influenced by the environmental conditions at any specific site. For example, interception of wet deposits by vegetation is a result of the interaction of the element, rainfall, and the type and degree of development of the canopy. As described in more detail in the following sections, these relationships are likely to be universally valid. However, the interception for a specific situation is determined by the actual site conditions, which are defined by rainfall and the seasonally dependent crop development. So, the resulting radionuclide transfer value is the result of element-specific transfer processes and actual environmental conditions.

From this, it may be concluded that there is no specific *tropical radioecology*, since the transfer processes are determined by universally valid relationships. However, it is also obvious that radioecology has to take into consideration the specific climatic, environmental conditions, and agricultural practices to achieve reliable and traceable results. Hence, generic parameter

TABLE 7.1 Important Processes and Parameters to Consider in Radioecological Modelling

Process/Parameters	Specificity		
	Site	*Radionuclide*	*Element*
Plant food products			
Crop types	+++		
Growth periods	+++		
Harvest times	+++		
Leaf-area indices	+++		
Yields	+++		
Depth of ploughing layer	++		
Foliar contamination			
Deposition velocities	+		+++
Interception of wet deposit	+		+++
Weathering half-life	+		+
Translocation factors			+++
Uptake of radionuclides from soil			
Transfer factors soil-plant	++		+++
Fixation, migration in soil	++		+++
Contamination from resuspension	+++		++
Animal food products			
Animal feeding regime	+++		
Inhalation rates for animals	+		
TF feed-animal food products	+		+++
Biological half-lives	+		+++
Food processing and storage			
Processing of food and animal feeds	++		++
Storage times for foods and animal feeds	++		
Activity intake			
Age-dependent intake rates	+++		
Seasonality of food consumption	+++		
Age-dependent inhalation rates			

continued

TABLE 7.1 Important Processes and Parameters to Consider in Radioecological Modelling—Cont'd

Process/Parameters	Specificity		
	Site	*Radionuclide*	*Element*
Internal exposure			
Dose coefficients for ingestion		+++	
Dose coefficients for inhalation		+++	
External exposure			
Time dependence of gamma dose rate	+	+++	+++
Occupational habits	+++		
Shielding from buildings	++	++	
Dose coefficients for ground exposure		+++	
Dose coefficients for cloud exposure		+++	
Physical half-life		+++	

values defined in, and suitable for, temperate zones are not necessarily appropriate for tropical and subtropical systems.

7.2.1 Regional Differences in Site-Specific Parameters

7.2.1.1 Characteristics of Tropical Areas

Geographically, the tropics are defined as the region between the Tropic of Cancer and the Tropic of Capricorn; this is between 23°N and 23°S. In climatology, the tropics are defined as the region where the mean temperature of the coldest month is above 20°C. The subtropical area defines the transition between the tropical and the temperate zone; it is usually considered to be between 23°–25° and 23°–40° North and South, respectively. The subtropics are characterised by warm, tropical summers, but the winters are non-tropical.

In the tropics, the differences in temperature during the year are relatively small. There are no pronounced seasons as are typical for the temperate zone, where plant growth in winter is especially influenced and limited by low temperatures and short days. In the tropical zone, growing seasons are often defined by the fluctuations of rainfall during the year, which are due to the seasonality of the intertropical conversion zone that follows the zenith of the sun. So, the intertropical convergence zone moves back and forth across the equator in a semi-annual pattern. It also induces a diurnal cycle, with cumulus clouds developing around noon causing heavy rainfall in mid to late afternoon. The movement of the intertropical convergence zone affects rainfall patterns in the tropical zone, resulting in the wet and dry seasons of the tropics instead of the cold and warm seasons of higher latitudes. The resultant wet seasons are often referred to as monsoon periods within the tropics (see Chapter 2).

Two-thirds of global agricultural land is located in the tropical zone (see Chapter 5). About five billion people live in this area — the majority of the global population. In general, the variety of crops in tropical areas is much greater than that in the temperate zone. Important crops are rice, maize, sorghum, soybean, manioc, yam, sugar cane, and palm trees. Furthermore, there is an enormous variety of fruit and vegetables. For illustration, food intake characteristics of the continents in the tropical zone are compared in Table 7.2. It should be noted that Oceania includes Australia with food intake habits that are similar to Europe and North America. The list of food items is not complete, but the data underline the differences of the food spectra and the importance of specific foodstuffs for any region.

The staple foods in a region reflect the relationship between climate and the requirements of the crop regarding temperature and precipitation. For example, the consumption of rice is especially important for South Asia, where its cultivation is favoured by high temperatures and precipitation. Millet has some significance in the hot dry areas of Africa and South Asia. Maize is important in South America and Africa, and wheat is important for both temperate and subtropical areas.

Similar differences can be seen among the animal food products. Beef and milk are important in South America, Oceania, and Europe. The consumption of poultry is popular in South America and Oceania. The intake of seafood is closely related to the distance of an area from the sea.

7.2.1.2 Seasonality

Seasonal weather variations control the growth periods of many crops and the grazing regimes of animals. Although in tropical countries crops grow during the whole year, the regularities of dry and wet periods (e.g., monsoon) cause site-specific growth periods (defined by the sowing/planting and harvest dates).

In temperate environments, one growing period with one harvest or harvest period is generally adequate for all crops. However, in tropical areas, some crops have more than one growing and harvest period. As an example, Table 7.3 summarises the months of harvest of rice for a number of countries in the tropical, subtropical, and temperate zones. At any time, somewhere on the globe, rice is harvested. In temperate climates only one rice harvest is possible, but in tropical countries more harvests per year are possible and rice may even be more or less continuously harvested. The vegetation periods in temperate climates are determined by temperature and day length; in tropical areas the distribution of rainfall has the greatest impact.

TABLE 7.2 Comparison of the Intake of Some Food Items in Regions within the Tropical Zone and Europe

Item	Unit	South America	Caribbean	Africa	South Asia	Oceania	Europe
Energy	$kcal\ d^{-1}$	2884	2647	2436	2425	3000	3353
Protein	$g\ d^{-1}$	77	64	61	57	99	100
Fat	$g\ d^{-1}$	85	59	49	51	113	123
Wheat	$kg\ a^{-1}$	59	48	47	63	60	111
Potatoes	$kg\ a^{-1}$	29	14	13	17	43	93
Vegetables	$kg\ a^{-1}$	47	76	54	58	90	117
Fruits, other	$kg\ a^{-1}$	21	32	11	15	42	25
Tomatoes	$kg\ a^{-1}$	15	22	15	5	16	26
Sweet potatoes	$kg\ a^{-1}$	2	16	11	0	17	0
Rice	$kg\ a^{-1}$	45	77	29	112	26	7
Bananas	$kg\ a^{-1}$	22	24	5	10	32	7
Millet	$kg\ a^{-1}$	0	0	12	9	0	0
Maize	$kg\ a^{-1}$	26	17	41	6	3	7
Yams	$kg\ a^{-1}$	1	10	19	0	9	0
Beef	$kg\ a^{-1}$	28	6	5	2	33	16
Pork	$kg\ a^{-1}$	9	7	0	0	18	35
Poultry meat	$kg\ a^{-1}$	25	17	4	1	28	18
Milk	$kg\ a^{-1}$	89	47	27	43	72	91
Cheese	$kg\ a^{-1}$	1	1	1	0	7	12
Eggs	$kg\ a^{-1}$	6	4	2	1	5	12
Crustaceans	$kg\ a^{-1}$	0	0	0	0	2	1

TABLE 7.3 Harvest Periods of Rice in (Arbitrarily) Selected Countries

Country	Month of Harvest of Rice											
	Jan	*Feb*	*Mar*	*Apr*	*May*	*Jun*	*Jul*	*Aug*	*Sep*	*Oct*	*Nov*	*Dec*
India										x	x	x
Indonesia				x	x	x						
Japan										x		
Vietnam				x	x							
Australia	x	x										
Algeria										x	x	x
Zaire	x	x	x	x	x	x	x	x	x	x	x	x
Tanzania							x	x				
Brazil		x	x	x	x	x	x	x	x	x		
Venezuela	x	x							x	x	x	
Honduras									x	x	x	
Cuba							x	x	x	x	x	x
USA							x	x	x	x	x	
Italy									x	x		

From Franke (1975).

7.2.1.3 Crops

Tropical regions are characterised by an enormous variation of crops and fruit, which are cultivated year round, providing the possibility for several harvests per year. Many tropical countries produce and export a wide range of agricultural products such as cereals, soybeans, rice, sugar, cocoa, coffee, oils, and fruit to the rest of the world. However, The World Bank and FAO recognise that much of the agricultural production in developing economies across the tropics is used for local consumption (Dixon et al., 2001; see Chapter 5; http://faostat.fao.org/site/609/default.aspx#ancor).

The assessment of exposures to any local population needs to consider crop characteristics. As an example of the variability of possible scenarios, a summary of the key crops that are considered in some radioecological models applied in different parts of the world is given in Table 7.4. It compares food products taken into account in models as developed for Hong Kong (Poon et al., 1997), for Brazil (Rochedo et al., 2001, Rochedo, 2005), and for central Europe (Müller and Pröhl, 1993). Although Brazil and Hong Kong represent tropical areas, Hong Kong is characterised by pronounced seasonal weather variations due to monsoon activity. For comparison, the main crops for middle Europe, as an example for the temperate zone, are given. For the tropical zone, preferable periods for plant growth can occur more than once a year; sometimes these periods even overlap. Outside the tropical zone, the seasons are more pronounced, so the growth periods of the crops are associated with distinct periods of time to match the specific requirements of the crops for temperature and precipitation.

7.2.1.4 Feeding Regimes

Animal husbandry is closely linked to environmental and climatic conditions. Highly productive arable land is preferred for the cultivation of food crops. Grazing animals often use grassland, areas with low productivity, or areas remote from the market. Feeding regimes for animals also depend on the availability of food-processing by-products and the low quality fraction of food crops.

Activity intake by domestic animals with feed is very much affected by local conditions. The intake of radionuclides by animals is dependent on the animal species; their mass, age, productivity level and growth rate; and the digestibility of their feed. Furthermore, the use of feeds for animal nutrition depends on the season and the agricultural conditions and practices within any specific region. Hence, reliable predictions can only be expected if the situation for which the estimate made is sufficiently well known.

Typical feeding regimes, as assumed in radiological models for Hong Kong (Poon et al., 1997) and Middle Europe (Müller and Pröhl, 1993), are compared in Table 7.5 for the animal food products considered and the most important local feeds used. The selection of animals reflects consumption habits of these areas; the local crops are important in animal husbandry, as well as for human consumption.

Cow's milk is more typical for temperate climates; the typical feeds are grass or forage, which cannot be consumed by humans directly. Dairy cattle are also kept in tropical areas, which is the consequence of the culinary assimilation by tropical countries of Western diets. In tropical areas, there is a clear preference for poultry, which is easy to produce and much more effective in terms of amount of feed per unit meat compared to large animals. Typical stock feeds are soybean and sorghum in Southeast Asia and maize in Brazil.

7.2.2 Transfer Processes

7.2.2.1 Deposition and Interception

Dry deposition is the result of the interaction of meteorology, the chemical and physical characteristics of the contaminant, and the properties of the canopy. Radionuclides may be released from nuclear facilities to the terrestrial environment as particles or gases.

TABLE 7.4 Comparison of Crops Considered in Radioecological Models of Hong Kong, Brazil, and Middle Europe

Crops	Growth and Harvest Periods					
	Hong Kong		*Brazil (Rio de Janeiro)*		*Middle Europe*	
	Sowing	*Harvest*	*Sowing*	*Harvest*	*Sowing*	*Harvest*
Rice	Jan June	May–July Oct–Dec	Aug–Nov	Jan–Apr	–	–
Wheat		–			Oct	Aug
Barley		–			Oct	July
Potato	–	–	Jan–Mar Apr–Jun Aug–Sep Nov–Dec	May–Jun Sep–Oct Dec Feb–Mar	May	Aug–Sep
Maize	Feb Aug	July Dec			May	Oct
Leafy vegetables	Feb Sep	Mar–Sep Oct–Apr	Jan–Dec	Jan–Dec	Mar	May–Oct
Root vegetables	Mar Sep	Jun–Aug Nov–Feb			Apr	Aug–Oct
Fruit vegetables	Feb Aug	May–Oct Nov–Apr			Apr	Aug–Oct
Beans	Mar Sep	Jun–Sep Nov–Mar	Dec–Feb May–Jun Aug–Oct	Mar–May Sep–Oct Dec–Jan	–	–
Fruit	Jan Jan	Jul-Aug Dec–Feb	Jul–Aug	Oct–Dec	Apr	Jul–Oct
Grass/forage		whole year			Mar	Apr–Oct
Soya bean	Mar	Jun/Sep			–	–
Sorghum	Mar	Jun–Sep			–	–
Manioc	–	–	May–Sep	Jan–Dec	–	–

The data show that there is a specific seasonality for different regions in tropical areas, which has to be properly addressed when evaluating nuclide exposure pathways.
From Poon et al. (1997), Rochedo (2005), and Müller and Pröhl (1993).

TABLE 7.5 Comparison of Feeding Regimes Considered in Radioecological Models of Hong Kong and Middle Europe

Animal Food Product	Main Local Feed Component	
	Hong Kong/South China	*Middle Europe*
Cow's milk	Forage, soya bean	Maize, silage, grass
Sheep milk	–	Grass
Goats	–	Grass
Pork	Soya bean, forage	Wheat/barley
Beef	Grass, forage	Maize silage, grass
Duck	Sorghum/forage	–
Pigeon	Sorghum	–
Quail	Sorghum	–
Chicken	Soya bean, forage	Wheat
Eggs	Soya bean	Wheat
Veal	–	Processed milk

From Poon et al. (1997) and Müller and Pröhl (1993).

Usually, dry deposition of radionuclides from the atmosphere to soil and vegetation is quantified by the deposition velocity v_g (m s^{-1}), which is defined as the ratio of the activity deposited on a specific surface (Bq m^{-2}) $A_{d,r}$ and the time-integrated activity concentration in air $C_{a,r}$ of radionuclide r (Bq s m^{-3}):

$$v_g = \frac{A_{d,r}}{C_{a,r}} \tag{7.1}$$

The deposition velocity is an empirically determined quantity that depends on a variety of factors, such as the particle size, the characteristics of the surface/air interface, meteorological conditions, and the chemical form of the deposit. Dry deposition is controlled by the transport in the gas phase or by sorption at surfaces. The deposition per unit time is directly proportional to the radionuclide concentration in the near-surface air. For large particles, gravitation is the dominating process, whereas diffusion is the most important process for small particles. For particles, a minimum deposition velocity is observed for diameters between 0.1 and 1 μm. The deposition of smaller particles is more effective due to diffusion (Jonas, 1984). For particle sizes >1 μm, deposition increases with increasing particle size because the effects of gravitational settling (sedimentation) and impaction become more important.

The size spectrum of contaminated particles is influenced by the source of the radioactivity in air. Routine releases are usually filtered, and the cut-off point of the filter device controls the maximum particle size. Unplanned releases during accidents may be unfiltered, and the whole particle size spectrum may be released. However, with increasing distance from the release point, the size spectrum approaches that of natural aerosols, $\sim$0.1–1 μm. Particles < 0.1 μm coagulate or "attach" to larger particles; particles larger than about 1 μm are lost from the near-surface atmosphere due to gravitational settling.

Gaseous radionuclides can be classified by their relative physicochemical reactivity. Noble gases, such as radon, krypton, or xenon, do not interact with other substances or surfaces, therefore their deposition is negligible. Once released, noble gases are, more or less, globally distributed in the atmosphere, if their half-lives are long enough. Reactive gases (e.g., $^{131}I_2$, $^{14}CO_2$, and $^{35}SO_2$) are characterised by intensive interactions with plants. These substances may enter plants through the stomata and undergo effective chemical reactions with specific compounds of the plant.

Deposition velocities generally increase with increasing wind speed due to a more intensive contact of air with the tree canopy. Furthermore, deposition is generally more effective on wetted surfaces due to a more pronounced sticky effect. A study using ^{203}Pb-labelled aerosols showed that rough or hairy leaf surfaces were up to approximately eight times more efficient in collecting aerosols than smooth leaf surfaces (Smith and Jones, 2000).

For both particles and gases, the deposition velocity is influenced by the kind of plant canopy. In general, deposition is more effective for well-developed canopies, since the area of interface between vegetation and atmosphere is increased (Heinemann and Vogt, 1980; Jonas, 1984; Roed, 1987). So, dry deposition of radionuclides is subject to the seasonality of plant development. In models that describe routine releases, this problem can be circumvented by the choice of long-term mean values, which average appropriately over diurnal and seasonal variations of the canopy development. However, seasonality is important for model applications to single releases, due to the possibility of pronounced seasonality of the standing biomass. This may also occur in wet-dry tropics but, in general, the reduced seasonality of tropical and subtropical areas compared with more temperate regions would tend to even out this potential fluctuation.

Radioactivity in air is washed out by rain and snow. Some fraction of radionuclides deposited with precipitation is retained by the vegetation and the rest falls through the canopy to the ground. Although most of the activity retained is eventually weathered off to soil and retained only temporarily by vegetation, the fraction that is initially intercepted is a very important quantity in all radioecological models. Direct deposition may cause relatively high activity concentrations in animal feedstuffs and foods consumed by humans. A proportion of the radioactivity may also penetrate the plant cuticle and be translocated to the consumed parts of the plant.

In general, simple approaches are used to estimate interception of wet-deposited radionuclides in assessment models (IAEA, 2001). The interception fraction is often assumed to be constant for all elements, all plants, and all precipitation events. In some models, the interception of wet-deposited activity is modelled as a function of the biomass density according to the approach of Chamberlain (1970). However, investigations in recent years have demonstrated that interception of wet-deposited radionuclides is actually much more complex (Hoffman et al., 1992; Kinnersley et al., 1997; Pröhl, 2009). For single deposition events, interception of wet deposits is the result of the interaction of the development of the plant, the water storage capacity of the foliage, the amount of rainfall, and the element (Müller and Pröhl, 1993). These relationships are considered in radioecological models developed for accident consequence assessment, for example, in Hong Kong (Poon et al., 1997) and Brazil (Rochedo et al., 2001).

7.2.2.2 Weathering and Dilution

Once deposited on vegetation, radionuclides are lost from plants due to removal by wind and rain either through leaching, by cuticular abrasion, or by defoliation. The increase of biomass during growth phases does not cause a loss of activity; however, it does lead to a decrease in activity concentration due to effective dilution. Since growth is subject to seasonal variations, the postdeposition decrease of the activity concentration of plants also depends on the season.

All the processes that control activity loss from plants subsequent to deposition occur simultaneously. Therefore, the exact contribution of each process to the removal of activity from plants is difficult to quantify. The total loss is expressed by the weathering half-time T_w, which is the result of all the processes described earlier (including growth dilution). The weathering half-time is empirically determined by regression of the radionuclide concentration in vegetation with time.

Weathering is simulated in assessment models by a single exponential function. However, observations indicate that weathering loss is most rapid immediately after deposition (Ertel et al., 1989), and a multi-exponential loss function may be more appropriate, particularly if leaf fall is a substantial component of the weathering loss. Although the effects of wind and rain cause the postdeposition activity loss, long-term retention is not, in general, correlated with weather conditions (Miller and Hoffman, 1983).

The chemical form of the contaminant seems to be of less importance than physical processes for weathering loss. Observations after the Chernobyl accident (Kirchner, 1994) resulted in weathering half-times for iodine with a median of approximately eight days (90% confidence interval, 4.5–15 days). For cesium, a median of 10 days (90% confidence interval, 5.4–20 days) was observed. The shorter half-times were observed primarily in regions with fast growing vegetation; whereas the longer half-lives were found in Scandinavia, where the growth rate was lower due to the later spring in this area (Kirchner, 1994). In general, longer weathering half-times are observed for slowly growing or dormant vegetation (Miller and Hoffman, 1983). However, weathering is determined by several processes that have an effect over several weeks and that partly compensate each other. Therefore, the overall uncertainty of weathering is less than that for deposition and interception, which depend on specific particle and canopy conditions at the time of deposition. The data currently available suggest that weathering might be faster in warm and humid climates with high growth rates.

7.2.2.3 Translocation

Translocation describes the movement and distribution of radionuclides within the plant subsequent to foliar deposition and absorption by the leaves. For radioecological assessments, translocation is especially important for plants that are partly used as human food or animal feed, such as fruit, cereals, and potatoes. Translocation is often quantified by the translocation factor ($m^2\,kg^{-1}$), defined as the ratio of the activity concentration in the edible part ($Bq\,kg^{-1}$) to the total activity retained by the plant canopy per unit ground area ($Bq\,m^{-2}$).

The most important factors influencing translocation are the physiological behaviour of radionuclides in the plant and the time at which the deposition occurs during the growth period. Concerning the physiological behaviour, elements and, hence radionuclides, are differentiated according to their mobility in plants (see Chapter 5).

An example of pronounced element dependence of translocation is given in Figure 7.2. In this case, translocation is characterised by the mobility factor for a number of elements; the mobility factor is defined as the ratio of the activity in the edible part to the mean activity in the plant. The higher this ratio, the more effective is the systemic transport in the plant. The data were determined in an experiment with beans that were continuously exposed during the growth period (Shinonaga et al., 1999). Samples of premature and mature seeds were taken on days 37 and 75 of growth, respectively. The mobility factors differ by up to an order of magnitude between different sampling times for one element. Between the elements, the difference is more than a factor of 1000.

In view of the potential importance of foliar deposition for food contamination, the database for modelling foliar uptake and translocation is poor. To overcome the data gaps associated with this contamination process, more experiments are needed to investigate the distribution of contamination as a function of the element, the stage of development at which deposition occurs, the time after contamination, the plant species, and the chemical form of the contaminant.

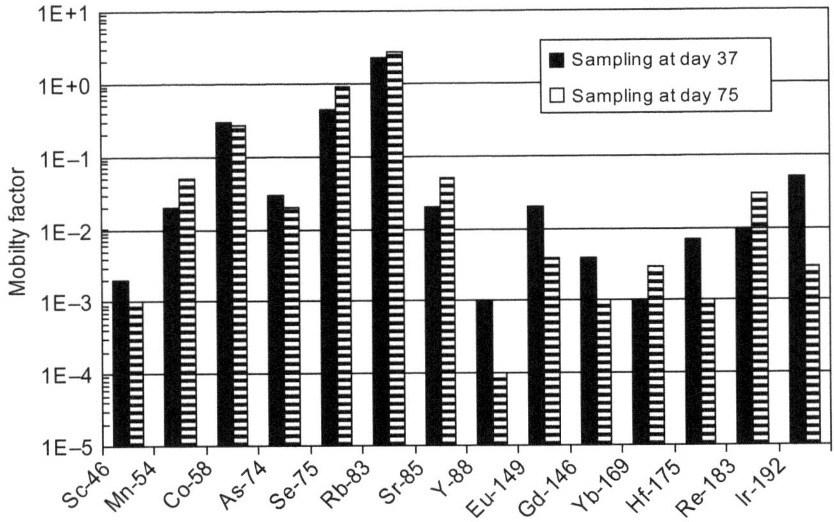

FIGURE 7.2 Mobility factor for soybean as a function of element and sampling day. *From Shinonaga et al. (1999), with permission.*

There are two transport systems in the plant, the xylem and the phloem; mobile elements can be transported in either system. The xylem transports water and minerals from the soil to upper plant parts. The driving force for transport in the xylem is transpiration; therefore, the transport direction is from the bottom to the top of the plant. Phloem transports the sugars generated by the process of photosynthesis from the leaves to the rest of the plant; hence, its transport is typically in the other direction. There is no discrimination between elements to be transported in the xylem once they have reached that system (e.g., by uptake from soil).

The translocation of radionuclides subsequent to a deposition on foliage is the result of a complex interaction of the element, the plant, and the stage of development at the time of contamination. All these factors are not primarily dependent on the climate. Of course, in tropical areas, there is a wide range of crops with specific growth characteristics that influence the rates of systemic transport in the plant. However, physiological behaviour of elements in the plants is universal, rather than site specific. Choi et al. (2002) performed translocation experiments in Korea under a subtropical climate in summer. In this study, ^{54}Mn, ^{57}Co, ^{85}Sr, and ^{103}Ru were applied to rice and radish plants at different times during the vegetation period. A pronounced dependence on element, crop, and time of application was observed, but overall the results agree well with those obtained for cereals reported for temperate climates (IAEA, 1994; Colle et al., 2009).

7.2.2.4 Uptake of Radionuclides from Soil

Soil is the main reservoir of long-lived radionuclides deposited on terrestrial ecosystems. The behaviour of radionuclides in soils controls their migration in the soil profile and possible transport to the groundwater, as well as long-term food contamination. The concentration of radionuclides in soil solution is the result of physicochemical interactions with the soil matrix, where ion exchange is the dominant mechanism. By its very nature, ion exchange is a competitive mechanism. The concentrations and compositions of major and other competitive elements in the soil are thus of primary importance in determining the distribution of any radionuclide between soil particles and soil solution (see Chapters 3 and 5).

The sorption and desorption of radionuclides and their possible uptake by plants are the result of the complex interactions between soil type, pH, redox potential, sorption capacity, clay content, organic matter, and management practice (IAEA, 2009; Chapter 5).

For quantification of the root uptake of radionuclides, an empirically derived soil-plant transfer factor (TF_{sp}) is generally applied, defined as the ratio of the activity concentration in plant C_p, and the activity concentration in soil C_s:

$$TF_{sp} = \frac{C_p}{C_s} \tag{7.2}$$

However, this parameter has several disadvantages, the most important of which are the following:

- The depth of the rooting zone is, more or less, arbitrarily defined to a depth of 20 cm.
- The activity concentration in the soil is used to estimate plant concentrations without consideration of the chemical speciation in soil.
- The experimentally determined transfer factors (TFs) vary considerably.

Despite the conceptual simplicity of the TF, its inherent limitations should be acknowledged and taken into account. It is generally accepted that the variation of radionuclides in plants is not entirely explained by varying concentrations in soil alone. The radionuclide concentration in plants is the result of the complex interaction of ecological parameters, such as the type of soil, content and type of clay minerals, and agricultural practices and climate.

Frissel et al. (2002) studied the transfer of cesium and analysed the literature. No indications were found that the climate directly influences the uptake of cesium. However, they noted that soil is the result of the interaction of the original rock, the climate, and anthropogenic factors. Soils contain a number of radionuclide-adsorbing components in the silt and clay fractions. For example, under tropical conditions with high precipitation, the weathering of rocks is accelerated, with implications on the soil texture and the type of clay minerals. Clay decomposes faster; 2:1 clay minerals decay and release silicon and transform into 1:1 clay minerals. During this degradation process, clay loses the property of cation sorption since the layers of 2:1 clay minerals are negatively charged; whereas 1:1 clay minerals are not negatively charged and therefore have no specific ability for cation sorption.

Furthermore, in warm and humid climates, the microbiological activity is very high and the degradation of organic matter is much faster (Scheffer and Schachtschabel, 2002). The higher turnover of organic matter may facilitate the mobilization of radionuclides incorporated in the humus and may enhance their root uptake and migration into deeper soil layers.

Typical soils in the tropical and subtropical area are latosols originating from silicate rocks. Due to high temperatures and high rainfall, clay degradation is much more advanced. Cations and silicic acid are mobilised during this process and leached into deeper soil layers. This causes an accumulation of iron and aluminium oxides in the upper soil layer. Intensive weathering also leads to a rapid degradation of organic matter. Latosols are frequently found in many regions of the tropical and subtropical zone with a pronounced seasonality of precipitation (Birkeland, 1999). The fertility of such soils is generally low (Scheffer and Schachtschabel, 2002). They are often characterised by a considerable acidification, a low content of organic matter, and a low cation exchange capacity. Such conditions tend to enhance availability of radionuclides (Frissel et al., 2002), especially for cesium, strontium, and the transuranic elements, which are potentially relevant for the long-term consequences of reactor accidents. Summarising, there is no direct effect of climate on the uptake of radionuclides from the soil, but persistently high temperatures and high precipitation cause long-term developments of the soil that favour the availability of radionuclides (see Chapter 5). A compilation of soil-plant TFs determined under temperate, tropical, and subtropical conditions is given in IAEA (2009). A comparison of the data is possible for a limited number of plant-element combinations (Table 7.6). Although this comparison is neither complete nor statistically sound, these few examples indicate that in tropical conditions higher root uptake factors may be expected.

In recent years a number of experiments were performed to determine TFs for tropical and subtropical environments in particular (Wang et al., 1997; Tsukada and Nakamura, 1998; Frissel et al., 2002; Rahman and Voigt, 2004; Twining, et al., 2004; Wasserman et al., 2008). The test plants cover species such as cereals, potatoes, and vegetables that grow in temperate and tropical climates, as well as those that are cultivated predominantly in tropical and subtropical areas, such as rice, sorghum, and a number of specific vegetables. The investigations were carried out on sites with a wide variation of soil conditions. The results do not indicate any systematic impact of the climatic conditions on the transfer of radionuclides from soil to plants. However, the tropical data tend toward more extreme values (Chapter 5). For example, Wasserman et al. (2005) investigated the root uptake of ^{137}Cs and ^{90}Sr for

TABLE 7.6 Comparison of Soil-Plant Transfer Factors Determined in Tropical, Subtropical, and Temperate Environments

Element	Plant	TF, Soil-plant (Geom. Mean) (Bq kg^{-1} dry plant per Bq kg^{-1} dry soil)		
		Tropical	*Subtropical*	*Temperate*
Cobalt	Leafy vegetables	0.92	0.11	0.17
	Tubers	0.37	–	0.11
	Root crops	0.12	0.0013	0.054
Cesium	Leafy vegetables	0.98	0.038	0.06
	Tubers	0.43	0.065	0.056
	Root crops	0.43	0.015	0.042
Strontium	Leafy vegetables	3.6	0.98	0.76
	Tubers	0.68	0.45	0.16
	Root crops	2.0	0.041	0.72
Uranium	Cereals	0.018	–	0.0062
	Leafy vegetables	0.048	–	0.002
	Tubers	0.02	–	0.005
	Root crops	0.047	–	0.0084

From IAEA (2009).

Brazilian soils. On intensively weathered soils, higher TFs were determined than those collected for temperate climates in the IUR database.

Transfer from soil to fruit for various radionuclides (mainly Cs, Sr, Pu, Am, I, Ru, Ce, Cm, U) were compiled for both temperate and tropical environments (Carini, 1999, 2001). Significant differences in the transfer between temperate and tropical environments were observed. Higher values for cesium and lower values for americium and plutonium were reported for tropical environments. However, it was concluded that the differences were due to different experimental soils rather than to the impact of climate.

Another interesting example of the importance of soil properties for root uptake of cesium is reported for coral sands frequently found on many Pacific islands (Robison et al., 1997). Some of these islands were contaminated during the 1950s and 1960s as a direct consequence of atmospheric nuclear weapons tests. Soils on these coral atolls are composed almost entirely of alkaline earth carbonates with varying amounts of organic carbon. Due to the absence of clay, the cation exchange capacity is almost entirely determined by the organic matter content (Walker et al., 1997). The fertility of coral sands depends largely on the management of the organic matter, which is subject to a rapid decomposition in the tropical climate. Furthermore, the natural potassium content is very low. These factors cause extraordinarily high and very variable TFs, especially for cesium. TFs for cesium, reported in Robison et al. (1997), vary over nearly three orders of magnitude from 0.2 to 144 Bq/kg plant (d.w.) per Bq/kg soil (d.w.). The average values for a variety of plant species range from 0.7 to 44 Bq/kg plant (d.m.) per Bq/kg soil (d.w.).

A special case in tropical and subtropical countries is the cultivation of rice. It is the most important food for several billion people. There are different cultivars of rice, which are classified into paddy, upland, and floating rice. Most of the rice is produced on flooded fields (or paddies) covered by a water layer of 5–15 cm. Flooded soils have anaerobic conditions where the redox potential is low and reduced radionuclide species dominate (e.g., iodide, selenide, and sulphide). Soil-plant TFs for rice are summarised in IAEA (2009). In general, the TFs for rice are in the same order of magnitude as for other cereals.

More detailed discussions on soil-to-plant transfer in tropical and subtropical systems are provided in Chapter 5. A special section relating to rice, in addition to tropical fruit, is included.

7.2.2.5 Migration in Soil

The vertical migration of radionuclides in the soil column is driven by various transport mechanisms such as convection, dispersion, diffusion, and bioturbation. Details are expanded upon in Chapter 3.

Long-term consequences of downward migration, however, differ considerably depending on whether convective or diffusive transport dominates. For convective-driven migration, the input moves down the soil as a marked peak, with a given velocity, and shows some broadening with time as a result of dispersive mixing. For diffusive transport, on the other hand, the concentration maximum always remains at the surface, with depth profiles showing a close to exponential decrease with depth. Consequently, for this type of transport, which is typical in soils of low hydraulic conductivity, the bulk of the activity deposited from the atmosphere remains within the rooting zone of plants.

Agricultural practices have a major impact on radionuclide behaviour. Depending on the intensity and type of soil cultivation, a mechanical redistribution of radionuclides in the soil may occur. This causes a rather uniform distribution of radionuclides in the tilled horizon of arable soils.

The classical approach to estimate the degree of migration in soil is based on the distribution coefficient K_d (which describes the tendency of a radionuclide to adsorb to soil from solution) according to the following:

$$\lambda_s = \frac{v_a}{L \cdot (1 + \frac{\rho}{\Theta} \cdot K_d)} \tag{7.3}$$

where λ_s is the migration rate (1/a), v_a is the water velocity in the soil (m/a), L is the thickness of the soil layer considered (m), ρ is the soil density (kg/m^3), Θ is the water content of the soil, and K_d is the distribution coefficient (m^3/kg). A key quantity in this equation is the water velocity in soil, which depends on annual rainfall and its distribution during the year.

For strongly bound radionuclides (e.g., Cs, Pu, Am), this approach may overestimate residence times in the upper soil layers, since the only downward transport process is due to sorption and desorption; this means that only transport of radionuclides dissolved in soil water is taken into account. However, recent investigations (Bunzl et al., 1989, 1992, 1994, 1995a, b, 1997; Shinonaga et al., 2005; Whicker and Ibrahim, 2006) indicate that this approach causes pronounced overestimations of the residence time in soil. From measurements of depth profiles of ^{137}Cs, ^{239}Pu, ^{241}Am, and ^{237}Np in undisturbed soils from both Chernobyl and weapons fallout, residence times are derived that are much shorter than those derived from the distribution coefficient K_d. It is interesting to note that the residence times of the radionuclides investigated are relatively similar, although their chemical properties are different. Obviously, element-independent transport mechanisms, such as the transport of radionuclides attached to clay particles or bound to soil colloids, play a more important role than previously assumed.

The migration of radionuclides to deeper soil layers is closely connected with root uptake. In general, only radionuclides dissolved in the soil solution or weakly sorbed on the soil are available for root uptake. However, this radionuclide fraction is, in general, also subject to greater leaching. Therefore, the possible tendency to higher TFs for intensively weathered soils in humid tropical areas also implies higher leaching rates. This hypothesis is consistent with the findings of Wasserman et al. (2005) who found an increased extractability of ^{137}Cs and ^{90}Sr on soils rich in kaolinite (a weathered 1:1 clay mineral) and Fe-Al-oxides.

The migration of strontium is somewhat faster than cesium since a larger fraction of strontium is generally present in a more mobile form. Observations indicate that strontium migrates approximately two times faster than cesium. The distribution coefficient determined for barium, tellurium, manganese, and zinc indicate that, similar to strontium, their migration as dissolved compounds is more important than for cesium.

7.2.2.6 Resuspension

Resuspension refers to the removal of deposited material from the ground to the atmosphere as a result of wind, traffic, soil cultivation, and other activities. Potentially, resuspension is a persistent source for radioactivity in air subsequent to deposition on the ground. Furthermore, resuspension may lead to redistribution of activity to previously uncontaminated areas and to recontamination of cleaned or decontaminated surfaces. Due to the complexity of the process, relatively simple, empirically derived approaches are applied to quantify resuspension in assessment models. In a widely used approach, resuspension is quantified by use of a resuspension factor K (m^{-1}), defined as the ratio of the activity in air ($Bq\ m^{-3}$) to the surface deposit ($Bq\ m^{-2}$). The resuspension factor can be easily derived from measurements; however, it must be decided at what height above ground the air concentrations are to be measured and to what soil depth the activity is assumed to be available for resuspension.

Resuspension is influenced by various factors such as time since deposition; the soil type, texture, and moisture; the vegetation cover; amount and type of human activity; and the wind speed. The long-term resuspension factor K can be described by a time-dependent function (Garland et al., 1992):

$$K = A \cdot e^{-B \cdot t} + C \tag{7.4}$$

where $A + C$ gives the initial value of the resuspension factor (m^{-1}), B is the rate of decrease of the resuspension factor per unit time, t (d^{-1}), and C is the final resuspension factor (m^{-1}).

The impact of climate on resuspension is considerable. Under European conditions, values for the initial resuspension factor $A + C$ measured after the Chernobyl accident ranged from 4 to 50 ($\times 10^{-9}$) m^{-1} (Garland et al., 1992). The loss rate B of the resuspension factor found after the Chernobyl accident corresponds to half-times of between 200 and 800 days, with a median of approximately one year (Garland et al., 1992). Resuspension is higher under arid conditions; in the PATHWAY model (Whicker and Kirchner, 1987) developed for the southwestern United States, a resuspension factor of $10^{-5}\ m^{-1}$ is applied.

Doses due to inhalation or to ingestion of resuspended material subsequent to re-deposition on crops are, in general, very small, but will be affected by human behaviour. For example, doses to traditional owners in areas affected by weapons testing at the Maralinga site in South Australia were enhanced by sleeping on or near the ground, by children actively playing on dusty areas, by food collection involving digging, and by sitting in open vehicles over dirt tracks amongst other dust-raising activities (Giles et al., 1990; TAG, 1990; Haywood and Smith, 1997). Nonetheless, generally only for actinides is the inhalation of resuspended soil particles relevant as a long-term pathway.

7.2.2.7 Transfer to Animal Products

The transfer of radionuclides from feed to milk, meat, and other animal products is most often calculated using element-dependent TFs, defined as the ratio of the activity in milk or meat to the daily activity intake under equilibrium conditions. Alternative approaches as described in Howard et al. (2009) and in IAEA (2009) simply use the ratio of the activity concentration in the animal product and mean activity concentration in the daily diet. In a strict sense, such TFs are applicable only for a constant long-term rate of activity intake by adult animals.

To account for time-dependent (dynamic) intakes, one or more biological half-lives are introduced into models to simulate the buildup and loss of activity in milk, meat, or other animal products.

The processes controlling the transfer of radionuclides to animals are indirectly dependent on climate. The type of animal and the basal metabolic rate of domestic animals may be controlled by climate. The water intake is typically higher in warmer climates. In general, its impact on the transfer to milk and meat is considered to be less important.

7.2.2.8 Food Processing and Culinary Preparation

The concentration of radionuclides in foods and feeds are affected by food processing and culinary preparation. During processing of feed or foodstuffs (milling, grinding, rinsing, peeling, boiling, etc.) the concentration of activity in the final product can increase or decrease as compared to the raw product from which it is processed.

The processing factor is defined as the ratio of the activity concentrations in the processed and the raw product; for example, during the processing of flour, the outer parts of the grains are removed. Since the distribution of the minerals in the grain is inhomogeneous, this usually leads to lower activity concentration in the flour, compared to the raw product. The production of cheese may also cause considerable changes of the activity concentrations. Cesium tends to stay in the liquid phase. The strontium concentration in hard cheese is a factor of 5–10 higher than in milk, whereas in the whey, the activity is considerably reduced.

Processing factors are very specific to the product consumed; for example, some crops can be consumed fresh and nearly completely, such as for many fruit and vegetables. In such cases, only a little part of the activity is removed. Other products undergo a long and specific procedure, which may be accompanied by relevant activity losses. Food processing is not influenced by climate, but it is very dependent upon site-specific habits and traditions. A detailed compilation of food processing factors is given in IAEA (2009).

7.2.2.9 External Exposure

The external exposure from radionuclides deposited on the ground depends on the activity per unit area and its depth profile, the radionuclide involved, and the exposure geometry. The exposure is estimated by the following equation with shielding at different locations being taken into consideration using a correction factor:

$$D_{g,i}(T) = A_d \cdot r_{bi} \cdot \int_0^T s(t) \cdot g_{b,i}(t) \cdot \exp(-\lambda_r t)dt \tag{7.5}$$

where $D_{g,i}(t)$ = dose from gamma radiation of deposited radionuclide i from time of deposition to time T (Sv), A_d = deposited activity per unit area (Bq/m^2), r_{bi} = reduction factor for staying at different locations, $s(t)$ = correction function for shielding effect due to migration of the radionuclides into deeper soil layers, and $g_{b,i}(t)$ = age-dependent dose factor for exposure from ground (Sv m^2 Bq^{-1} s^{-1}).

Dose factors $g_{b,i}$ for external exposure from radionuclides deposited on the ground are calculated by Jacob et al. (1990). The dose factors are derived assuming that the radionuclides are on the surface of an infinite smooth area. For this purpose, lawn has been used as a reference.

However, in real situations, there are shielding effects due to adjacent buildings, vegetation, or other objects. To account for this, the reduction factor $r_{g,i}$ for occupancy at different locations is needed. It is defined as the ratio of the γ-dose rate at a specific location and the γ-dose rate over infinite lawn. Reduction factors are calculated for a range of exposure geometries (e.g., urban, suburban, and rural areas) by Meckbach and Jacob (1988).

Additional to shielding, the γ-dose rate over a contaminated area decreases due to the following:

- Reduction in activity due to radioactive decay
- Removal processes such as runoff from the surface, street cleaning, and removal of material
- Radionuclide migration to deeper soil layers, which causes shielding due to the increasing coverage of activity with soil

To account for the time dependence of shielding, the following approach was developed by Jacob (1989). The exponential function $s(t)$ has two components:

$$s(t) = a_1 \cdot \exp(-\lambda_1 \cdot t) + a_2 \cdot \exp(-\lambda_2 \cdot t) \tag{7.6}$$

where λ_1, λ_2 = migration rates (λ_1 = 1.5E-3 d^{-1}, λ_2 = 3.9E-5 d^{-1}), and a_1, a_2 = contribution fractions of the migration rates (a_1 = 0.36, a_2 = 0.64). These migration rates have been derived from measurements of the external exposure from cesium deposited after the Chernobyl accident. They can be considered as representative for a mixed (wet and dry) deposition in temperate environments. For other elements, no such data are available, but long-term external exposure is of minor importance for most of them. Therefore, the above data for cesium are applied for all elements.

Little information is available about the shielding function in urban environments for tropical areas. Most tropical countries use light building materials that provide less shielding against radiation than the more dense materials used in temperate countries. Also, large windows, verandas, and balconies with little shielding are typical for tropical areas, so the attenuation of the γ-dose rate in buildings is lower than in temperate climates.

Studies on γ-exposure and shielding in tropical areas have demonstrated significant differences to temperate areas. In German studies (Meckbach and Jacob, 1988; Meckbach et al., 1988), indoor exposures due to outdoor contamination were simulated for typical German urban environments with brick houses, ($\rho = 1.8\ g\ cm^{-3}$). Shielding properties of building materials that are typical for tropical countries were compared by Salinas et al. (2006a, b). In Figure 7.3, the attenuation of mono-energetic photons as a function of energy for bricks, concrete, asbestos, and wood is shown. The thickness of walls is relatively reduced in warm countries due to the low requirements for insulation. From the figure, it is apparent that the shielding effect, for example, for ^{137}Cs, is relatively low using lighter materials; it is less than one third for asbestos and wood buildings compared with brick.

Along with the properties of building materials, high temperatures in the tropics also impact human habits. To optimise ventilation, windows are kept open. People stay outdoors much longer; outdoor sports, such as football, are played all year. Conversely, in more developed urban areas, people may stay indoors longer to enjoy air-conditioning systems. In general, with regard to external exposure, the seasonality of population habits in temperate-climate countries is more pronounced than in tropical areas.

Snow cover, typical of temperate countries, may also act as shielding for deposited radioactivity, while heavy rain showers, typical of tropical areas,

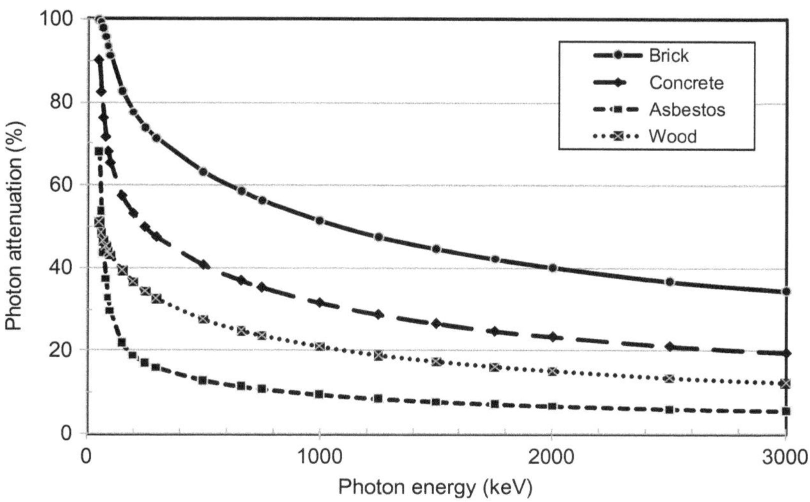

FIGURE 7.3 Attenuation of photons by different building materials: Bricks ($\rho = 1.15\ g\ cm^{-3}$, $d = 11.4$ cm), concrete ($\rho = 2.4\ g\ cm^{-3}$, $d = 6.0$ cm), asbestos ($\rho = 1.93\ g\ cm^{-3}$, $d = 1.6$ cm), and wood ($\rho = 0.79\ g\ cm^{-3}$, $d = 3.5$ cm). *Data from Salinas et al. (2006a).*

may remove some material. During melting, snow also takes away part of the material deposited on surfaces. Aspects, such as the contact time of snow with contaminated surfaces and temperature effects on the solubility of material retained on urban surfaces, may also play a role in the history of a contamination event. However, at this time, the lack of regional data does not preclude the use of factors describing weathering in temperate areas for tropical environments.

Some differences between temperate and tropical areas also may be related to the stage of social and economic lifestyle of the population. Long-term followup of external dose rates performed in Goiânia (see Chapters 1 and 5) with thermoluminescent dosimeters (TLDs) shows that weathering loss rates from the urban environment are nearly half of those observed in European urban areas after the Chernobyl accident (Rochedo et al., 2001). This finding was not expected, since the high rainfall rate observed in Goiânia (1700 mm a^{-1}), including heavy showers during summer, should cause faster long-term weathering losses. This result may be due to differences related to the fraction of paved surfaces (which facilitate runoff) and the maintenance of urban surfaces by activities such as street cleaning. Such activities contribute to a faster weathering, and the frequency of such actions is probably higher in European urban areas than in developing countries. This is not a typical difference related to climate but because of the type of urban environment and maintenance activities.

7.2.3 Comparison of Modelled Exposures for a Temperate and a Subtropical Environment

As a reaction to the Chernobyl accident, the dynamic radioecological model ECOSYS (Müller and Pröhl, 1993) was developed in Germany for estimating time-dependent activities in animal feeds and foods and to assess exposures to reference population groups subsequent to single depositions of radionuclides. The model takes into account internal exposure via inhalation and ingestion; furthermore, it calculates external exposure from the passing cloud and from radioactivity deposited on the ground. The site-specific parameter values of the model are typical for southern German agricultural conditions. The ingestion dose is calculated as a function of time considering 18 plant species, 11 animal food products, and 18 processed products. The ingestion and inhalation exposure is estimated for six age groups using age-dependent consumption and inhalation rates and age-dependent dose factors. The model has been integrated into various national and European real-time decision-support systems, such as ARGOS (Hoe et al., 2002) and RODOS (Raskob and Ehrhardt, 2007), to be applied subsequent to large-scale radioactive contamination events. Europe, as part of the temperate climatic zone, is subject to a pronounced seasonality causing large variations in, for example, plant growth, feeding regimes, food intake, and the time of outdoor activities.

Therefore, a key purpose of the model is the simulation of such seasonal effects. Since, even within Europe, environmental and agricultural conditions vary considerably, the model has been designed to facilitate adaption to other situations.

In one case, the ECOSYS model was adapted to the conditions of the Hong Kong area (Poon et al., 1997), which is in the subtropical climatic zone. It required a number of major modifications to meet these specific climatic environmental and agricultural conditions.

Modifications of the model were required specifically in the two areas:

1. The spectrum of crops and domestic animals important in Hong Kong differed significantly from that in central Europe. Important crops for Germany could be neglected for Hong Kong, whereas other foodstuffs not considered in the model had to be introduced. Furthermore, the growth periods of plants had to be modified to meet the situation in Hong Kong.
2. In the German version, ECOSYS estimates the ingestion dose assuming that all foodstuffs consumed are produced locally. This assumption may be acceptable in central Europe, but it is inappropriate for an urban area such as Hong Kong, which imports quite a large fraction of its food from remote sources and overseas countries.

The climate in Hong Kong is characterised by two pronounced seasons: the wet (monsoon) and the dry season. The mean annual precipitation in Hong Kong is above 2000 mm with about 90% of the rain falling from April to September. The average monthly temperature varies from 15.3°C in January to 28.8°C in July. In such climatic conditions, farming is possible during the whole year, whereas in temperate climates there is a period in winter without farming or with dormant vegetation. To account for the all-season growth, two growth periods are simulated.

Furthermore, the model parameters were adapted to meet the conditions in an East Asian subtropical environment and to apply appropriate parameters for deposition and interception, weathering, uptake from soil, migration in soil, resuspension, cattle feeding, and human food intake. A detailed description of the model is given in Poon et al. (1997).

To illustrate the potential differences in radiological consequences in a temperate and a subtropical environment, model results are compared in Figure 7.4 for ^{137}Cs, ^{90}Sr, and ^{131}I using the German and the Hong Kong version of the ECOSYS model, respectively. For this purpose, ingestion and total doses (ingestion, inhalation, and external exposure from the cloud and from the ground) for adults were calculated. To illustrate seasonality, short-term activities in air (occurring on the first day of each month) of 10^6 Bq m^{-3} for ^{137}Cs, ^{90}Sr, and ^{131}I respectively, were compared. Furthermore, it was assumed that all the foods consumed were produced locally on an area that was exposed to the radionuclide activity in air, specified above. This may have been a pessimistic assumption in many areas; however, it was made to

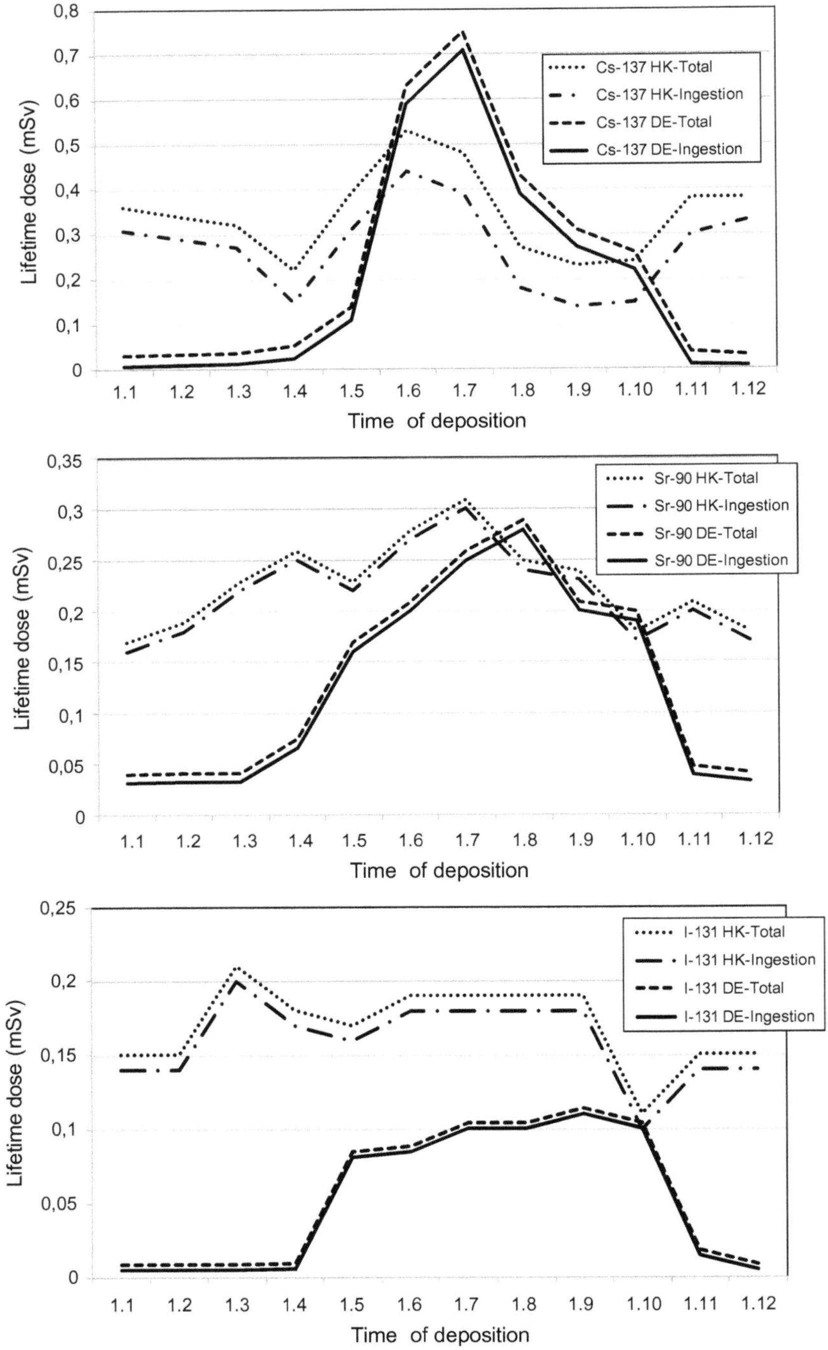

FIGURE 7.4 Comparison of the dependence of ingestion and total doses (integrated over 50 years) on the time of deposition for sites in a temperate (Germany, DE) and a subtropical (Hong Kong, HK) area. Single dry depositions are assumed to occur on the first day of each month. A time-integrated activity concentration of 10^6 Bq s m^{-3} has been assumed for ^{137}Cs (upper part), ^{90}Sr (middle), and ^{131}I (lower part), respectively.

ensure the comparability of the results. For ^{137}Cs and ^{90}Sr, a time-integrated activity concentration in air causes a deposition in the range of about 500–2000 Bq m^{-2}. The deposition varies due to the dependence of the deposition velocity on the stage of plant development. For iodine, the deposition depends very much on the chemical form, which has a strong impact on the deposition velocity. Iodine occurs in three chemical forms: (1) elemental (I_2), (2) bound to particles, and (3) as methyliodide (CH_3I). Deposition velocities of these three forms are in the order of 10^{-2} m s^{-1}, 10^{-3} m s^{-1}, and 10^{-4} m s^{-1}, respectively. In these calculations, contributions of 30%, 50%, and 20% for elemental iodine, particle-bound iodine, and methyliodide were assumed, respectively. Ratios in this order were observed after the Chernobyl accident in Europe (see Hötzl et al., 1987).

The results are compared in Figure 7.4. In all cases, the total dose is dominated by ingestion. In the case of ^{137}Cs and ^{131}I, a small contribution to the total dose is due to external exposure from the ground. For ^{90}Sr, a small part of the exposure is caused by inhalation.

A pronounced seasonal dependence of the ingestion dose on the time of occurrence of deposition is observed for the temperate case, reflecting the differences in plant growth between winter and summer. The largest variation is observed for ^{137}Cs due to effective translocation of the nuclide in many plants. The seasonal variability of ^{90}Sr is less than that for ^{137}Cs due to the higher root uptake of strontium, which causes higher doses for a deposition in winter. Furthermore, translocation is lower than for cesium, which causes crops to be relatively less affected by a summer deposition. For ^{131}I, the seasonal dependence is less pronounced in Hong Kong; there is only little variation between depositions during the vegetation time period.

Running the model for the subtropical environment, it is apparent that the seasonal variations are much less than those in a temperate environment because there is plant growth all year round. The ingestion dose for ^{137}Cs, ^{90}Sr, and ^{131}I for deposition in winter in Hong Kong is about a factor of 2 lower than in summer; whereas in Germany the difference is a factor of about 90, 9, and 20 for ^{137}Cs, ^{90}Sr, and ^{131}I, respectively.

A little dip in dose can be seen for all three radionuclides in tropical autumn at the transition from wet (summer) to dry (winter) season. This is the period when the summer vegetables are harvested and the winter vegetables start to grow. However, this dip should not be overestimated, since in reality the transition from summer to winter vegetables is probably smoother than modelled.

It is interesting to note that the maximum for ^{137}Cs in the temperate climate is higher by a factor of 1.6 than the maximum for the subtropical area. For ^{90}Sr, there is no real difference, but the maximum for ^{131}I is higher by a factor of 2 in the subtropical area. These findings are the result of the interaction of different growth periods and consumption rates. In the temperate climate, the crops comprising the most important contributors to dose are harvested within a relatively small period; whereas, in subtropical areas this period is longer, which results in a "smearing" effect. For ^{90}Sr, this effect is less pronounced due to the lower translocation (see Section 7.2.2.3). In the case of ^{131}I, the dominating contributor is the intake of leafy vegetables that are consumed in higher amounts in Hong Kong.

The comparison shows that in both temperate and subtropical environments, seasonal factors need to be taken into account. In temperate environments, the seasonality is especially due to the differences in temperatures and solar radiation between winter and summer. In subtropical environments, these two factors are not limiting, but the distribution of rainfall during the year is the controlling factor for farming. In tropical areas, seasonality is even less pronounced.

To demonstrate the influence of the deposition mode, Figure 7.5 compares total dose and the ingestion dose due to wet depositions of 1000 Bq m^{-2} ^{137}Cs that occurs on the first day of each month during rainfall events of 1 and 10 mm, respectively. It is obvious that the ratio of total and ingestion dose differs considerably from the dry deposition case (Figure 7.4).

The ingestion dose largely depends on the amount of rainfall; it is lower if the same activity is deposited during a rainfall event with a higher amount of rainfall, since the interception fraction is then smaller. In the dry deposition case, the ingestion dose was by far the more dominant. For wet deposition, the relative importance of external dose is higher for wet depositions with high amounts of rain, since then the ingestion dose is less due to the low

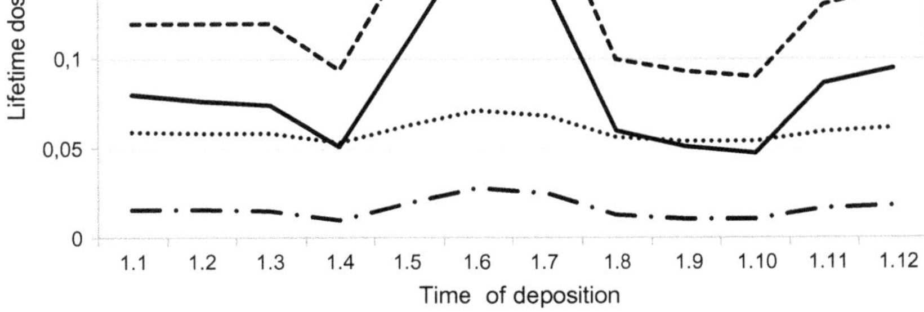

FIGURE 7.5 Comparison of the dependence of ingestion and total doses (integrated over 50 years) on the time of deposition for a subtropical site (Hong Kong area). Single wet depositions of 1000 Bq m^{-2} ^{137}Cs during rainfalls of 1 and 10 mm are assumed to occur on the first day of each month.

interception of radionuclides by plants. Therefore, the seasonality of dose is less pronounced for depositions with high rainfall due to the higher, season-independent contribution of external exposure.

The calculations in Figures 7.4 and 7.5 predict general differences in the seasonality of exposures subsequent to single depositions between sites in the temperate and in the subtropical zone. They provide trends, but the real doses always depend on the site-specific circumstances.

7.3. A RADIOLOGICAL CONSEQUENCE MODEL SUITABLE FOR TROPICAL SYSTEMS

As noted previously, a number of computer models have been implemented to simulate the transfer of radionuclides through air, water, and terrestrial ecosystems. However, the majority of models have focussed mainly on the temperate and cold regions of the world and almost exclusively in the Northern hemisphere, for example, NUREG 1.109 (Till and Meyer, 1983), FARMLAND (Brown and Simmonds, 1995), ECOSYS (Müller and Pröhl, 1993), and PATHWAY (Whicker and Kirchner, 1987). Whilst these can be adapted to suit tropical and subtropical systems, as shown in Section 7.2.3, they still tend to emulate temperate systems and apply default temperate data for the input parameters. This is not unreasonable under current circumstances given that the basic modelled processes are consistent irrespective of region and the lack of tropical data.

Nonetheless, tropical areas cover a great variety of environmental conditions. Reliable radioecological models require the integration of a wide range of site-specific conditions into the dose estimation. An early implementation of a methodology to handle spatial diversity and a variety of potential exposure scenarios, as required for a broader range of regional applications implicit in the involvement of tropical climates and lifestyles, is presented in the RadCon model (Crawford and Domel, 2000; Crawford et al., 2000). RadCon was implemented to calculate the dose over a two-dimensional region with parameter values at any location reflecting the underlying characteristics at that location.

RadCon was included in one of the International Atomic Energy Agency (IAEA) BIOMASS programmes, which compared the performance of a range of dose assessment models using intercomparison exercises. The exercises involved the estimation of ^{137}Cs concentrations in plants, animals, and humans and dose to humans in the Bryansk Region of Russia following the Chernobyl accident. RadCon performed well in these studies. Reproduced in Figure 7.6 are the estimates by RadCon of the concentrations of ^{137}Cs in leafy vegetables, potatoes, cereal, hay, beef, and milk plotted with the measured concentrations for the scenario. For these exercises, site-specific parameter values (such as TFs) were used. The dotted lines represent the 95% confidence intervals for the estimates. These are high due to the fact that large uncertainties were given

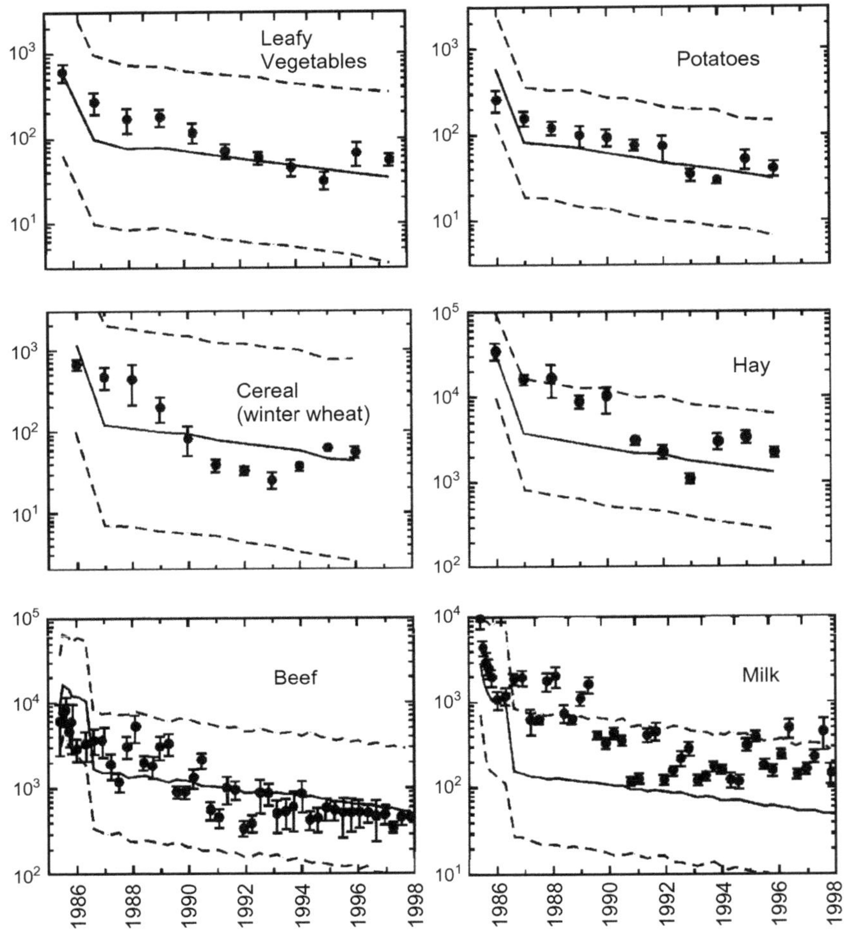

FIGURE 7.6 ^{137}Cs concentrations in leafy vegetables, potatoes, cereal, hay, beef, and milk from the Bryansk Region of Russia given as concentration (Bq/kg fresh weight) against time, from the Chernobyl accident in years (1986–1998). Measured data are represented by the filled circles. RadCon's predictions are given by the solid line and the two dotted lines give the 95% confidence interval estimated by RadCon. *Adapted from IAEA-BIOMASS-4 (2003).*

on each parameter and these were propagated to the solution. The results show reasonable agreement with the measured data. For a full description of the scenario, as well as details of the results and analysis generated by the participants, the reader is referred to IAEA-BIOMASS-4 (2003).

7.3.1 The RadCon Model

The main motivation behind the implementation of RadCon was to assist in the identification of the most relevant parameters in dose assessment estimation at some distance from the source. In RadCon, atmospheric dispersion has been decoupled from the dose calculation and a separate atmospheric transport code is required. The dispersion model can be any code that is suitable for the conditions under study and that generates time-varying air concentration and ground deposition over a two-dimensional region of simulation. RadCon calculates the dose to individuals from the exposure pathways presented in Figure 7.1.

The mathematical models for the calculation of dose in RadCon have been adapted from the internationally used and recognised models, ECOSYS-87 (Müller and Pröhl, 1993) and CLRP (Krajewski, 1994). For the calculation of contamination from crops, uptake by plant roots from contaminated soil is considered, as well as direct deposition onto the foliage. In estimating uptake from the soil, the effective soil contamination is adjusted for radioactive decay, leaching, and fixation of the radionuclides to the soil – factors reducing the effective availability for root uptake. For the deposition onto foliage, two crop types are considered: those that are consumed totally (e.g., leafy vegetables) and those that are used only partially (e.g., cereals, potatoes, fruit). The concentration in the crops from direct deposition is determined by the amount of activity deposited followed by an adjustment for activity loss due to weathering, radioactive decay, and dilution due to growth. For crops that are partly consumed, translocation from the deposited material to the edible part is modelled. A factor is also included to account for the removal of contamination by food preparation (e.g., washing, cooking), and an additional factor accounts for the importation of uncontaminated food. However, no food storage is considered, because it is assumed that the crop is harvested at time of consumption.

7.3.1.1 RadCon Implementation

RadCon was designed and implemented with an initial focus on its potential use in the Australian and Southeast Asian areas. Given the large variability in environmental conditions within that range, emphasis was placed on the ability of the model to handle the diversity of parameters such as diets of individuals, soil types, and the variety of food crops consumed across the region. To achieve this, the RadCon user interface employs a Geographical Information System (GIS), where different conditions are specified over grids of the computational domain. The calculation then applies parameters based on the specified conditions when estimating dose for each grid cell. This approach has made the model widely applicable beyond the initial regions of interest and, hence, makes it generally useful across the tropical and subtropical climatic regimes and temperate zones as well.

To communicate with the user, RadCon provides a graphical user interface and is portable across computer platforms. To achieve portability, RadCon has been implemented in the Java (Flanagan, 1997) programming language. To use RadCon, the following implementations are required:

- Files containing the time-varying air and ground concentrations (from an atmospheric dispersion model or by direct measurement)
- An image of the region (e.g., a map), required for visualization purposes
- Files containing the required tables of parameter values and their dependencies on radionuclide, soil type, individual lifestyle, and so forth. The required information includes human diet, animal diet, transfer rates (both from soil to plants and plants to animals), dose conversion rates, and so on. A detailed description of the information in these files is given in Crawford and Domel (2000; also, see Chapter 5).

This utility enables the results to conform much more closely to site-specific conditions.

Instead of requesting the user to specify required parameter values over the region, a system was designed to enable the user to select from a predefined set of alternatives for each of the characteristics. For example, the soil types have been grouped into clay, sand, loam, or coral. (This last soil type recognises the significantly different soil-to-plant transfer that occurs in coral and that these soils are common in tropical islands.) Lifestyles of individuals have been grouped into rural, urban, suburban, and so forth, and the substantial cultural differences known to exist between ethnic groups have been taken into account to a limited degree. The subset of characteristics affecting each model parameter was identified, and the actual parameter values are stored against the different characteristics in the data files used by RadCon.

For the ingestion pathway, alternatives can be chosen for the following:

- Soil type: sand, loam, clay, or coral
- Diet of the target group: a race default diet or one of the preset diets (the alternative diets are predefined and the composition and quantities of each of the food items consumed by each of the age groups are set)
- Diets of each of the animals (they have been predefined in a manner similar to the human diets)
- Ethnic characteristics of the target group
- Age of the target group
- Lifestyle, indoor, and outdoor occupancy factors as indicated by the actual lifestyle of the chosen target group

7.3.1.2 Parameter Sensitivity Analysis in RadCon

One of the reasons for supporting the development of RadCon was to identify those parameters that contribute most to the final dose to humans, such that any future research could be directed toward better scientific understanding of those factors. To assist with this, two parameter sensitivity analysis techniques have been implemented in RadCon:

One-at-a-time sensitivity measures. Ranking of the importance of the parameters can be obtained quickly using this technique. The ranking is obtained by calculating the dose with each parameter set to a base value, for example, the mean, followed by calculation of the dose by perturbing each parameter by a user-selected percentage (e.g., 1%) while leaving all other parameters constant. The change in dose gives the rank of the corresponding parameter.

Extended one-at-a-time. This technique (Downing et al., 1985) examines the change in output as each parameter is individually increased from its mean by a factor of its standard deviation. This sensitivity measure takes into account the parameter's variability and the associated influence on model output.

A parameter sensitivity study was implemented for an Australian case (Hess et al., 1998). The results demonstrated a distinct difference in parameter ranking between the short-term and the long-term studies (1 and 5 years upward). Deposition and translocation onto crops obtained the top parameter ranking at 1 year and soil type and related effects obtained the top parameter ranking at 5 years and later.

7.4. AQUATIC PATHWAY DOSE ASSESSMENT

The aquatic dosimetry pathway to humans is, in principle, the same as any other dosimetric pathway. In general, one needs to determine the quantities and types of radiation, their distributions in the environment, the critically exposed groups and individuals, and the range of potential exposure routes to those people. Particularly since the Chernobyl accident, a number of models of the environmental behaviour of radionuclides in river and lake systems have been developed and tested in the international programmes Biosphere Model Validation Study (BIOMOVS) and Validation of Model Predictions (VAMP). More recently, the model testing was expanded to include other aquatic systems, including coastal waters, within the IAEA Environmental Modelling for Radiation Safety (EMRAS) programme. These models have been evaluated and compared using case studies derived for temperate systems, but the models are generically applicable to tropical systems as well. A summary of those comparisons has been prepared by Monte et al. (2008).

However, it needs to be clearly stated at the outset that the proportional contribution of the aquatic pathway to overall radiological dose can vary considerably for a number of reasons, but generally tends to be relatively minor following actions resulting in wide dispersions. For example, Aarkrog (1997) stated that doses from consumption of marine foods are two to three orders of magnitude less than the dose from the terrestrial food chain to any exposed individuals. Similarly, a subtropical study of ^{210}Po and ^{210}Pb in Syrian fish (Al-Masri et al., 2000) reported that those radionuclides contributed only a small dose. The major residual dose arising from the Marshall Islands nuclear testing is from ^{137}Cs via terrestrial pathways, and within the lagoons of Mururoa and

Fangataufa, sites of French nuclear weapons testing in the tropical South Pacific, the radiological dose was estimated to be less than 0.01 mSv/year (Povinec et al., 1999).

The actual implications of significant environmental effects due to a major accident at a coastal nuclear facility were tested recently by the occurrences at the Fukushima Daiichi nuclear power plant in Japan, beginning in early March 2011. The accidents were consequences of an earthquake and, more significantly, the subsequent 14 m tsunami that disabled the plant's backup power supply and cooling facilities. Very little in the way of substantial impacts related to radiological effects in the aquatic environment have been noted, although some localised contamination was observed. That is not to say that the situation was insignificant; it did raise a high level of public concern and alarm, reaction, and resulting community angst. The commercial implications have certainly been substantial. If anything, the incident has increased the need for site-specific radioecological studies to be undertaken in preparation for any future such occurrences to improve understanding of any likely outcomes.

Despite its perceived seriousness, the radiological dose from Fukushima via the aquatic pathway has been very minor so far. In a modelling study undertaken to predict the ^{137}Cs ingestion dose from marine fish consumption in Hong Kong after an accidental release from a nuclear power plant, Poon and Au (2003) assumed that an amount of 30,000 TBq of ^{137}Cs could be released into the atmosphere. This was based on a worst-case scenario developed earlier by Cook et al. (1990) and was an amount of radioactivity far in excess of that known to have been released at Fukushima. Further, Poon and Au (2003) assumed that all the released radionuclides would be deposited into the local bay, a highly conservative and unlikely outcome that would maximise dose estimates to the local population. As expected, the dose to the critical group, the local fish farmers, was highest at 1.6 mSv in the first year after the accident, accumulating to 3.6 mSv in total over 50 years and would require some monitoring. The population at large would receive an annual dose of only 175 μSv. The fact that the doses were estimated to be so low, even to the most exposed individuals, reinforces the relative insignificance of the aquatic pathway to overall dose assessment.

The relatively lower dose to humans from the aquatic pathway is primarily because typical aquatic systems are continually flushed or are large with respect to the added radioactivity and, hence, tend to dilute and disperse the material. However, a consequence of this phenomenon in freshwater systems is that streams and rivers become vectors of contamination to population centres and agricultural areas downstream of radiological incidents. Kiev, downstream of Chernobyl, provides a non-tropical example of where aquatic pathways were important in contributing to overall and ongoing radiological dose. Similarly, in the southern United States, modelled dose following a forest fire (RAC, 2002) identified the aquatic pathway as giving the highest dose to humans via ^{137}Cs washed from the Los Alamos site into the Rio Grande and subsequently accumulated into fish. This pathway was to a hypothetical full-time resident living downstream at the confluence of the Rio Grande with the stream draining the Los Alamos site, exposed to deposited sediments, drinking river water, and eating a substantial proportion of fish in his/her diet. Another particular exception to the relative insignificance of aquatic pathways to radiological dose is shallow, closed drainage systems that will be discussed in the next Section (7.4.1). Releases specifically into aquatic systems (e.g., from Sellafield into the Irish Sea) have in the past also resulted in significant exposures. As previously noted in this chapter, very few examples of aquatic dose assessment modelling exist for tropical systems, but the general principles are valid irrespective of the environment.

7.4.1 Overall Exposure Pathways

There are two major aquatic classifications: marine and freshwater. Dilution and dispersion modelling is beyond the scope of this chapter; however, basic mixing models for marine and freshwater systems are included in Kathren (1984) and additional useful information is available from Hakanson (1997). Inputs to these model ecosystems can include accidental and/or routine releases of radioactivity from mines (particularly of U, but also for other naturally occuring radioactive material (NORM)-associated minerals), reactors (power or research), nuclear reprocessing and/or storage facilities, medical and industrial facilities (including hospitals), and military (including nuclear actions) and terrorist activities.

Marine systems can be broken down into estuarine, coastal, and open waters. For radiological assessment purposes, unless they are the release point for nuclear contaminated effluents, marine systems tend to be less significant than freshwater systems for a number of reasons (see previous discussion on examples from Fukushima and Hong Kong). First, they do not provide drinking, sanitation, or irrigation water. Second, there are typically much higher dilution and dispersion rates in marine systems than in rivers or lakes; in turn, this leads to lower concentrations in water and hence fish and other aquatic foods. (Note that, because of tidal movement in estuaries and the mixing of high and low salinity waters, dilution-dispersion models in these systems should be carefully applied to account for poor mixing). Third, high salt concentrations tend to reduce bioaccumulation by competitive inhibition of biological uptake mechanisms (e.g., high K and Ca concentrations reduce Cs and Sr bioaccumulation). Further, any particle-reactive radionuclides will eventually be transported into deep sedimentary sinks that tend to be insignificant contributors to human dose. Despite these factors, it is always important to evaluate what the concentrations of radionuclides within these zones are, or are likely to be, and make an assessment of their relative contribution to overall dose to the critical groups living in or exploiting these environments.

In addition, whilst sinks in aquatic systems may reduce human doses, comprehensive dose risk analyses should consider the exposure of non-human species and communities present in the locations where the removed radioactivity eventually concentrates (see further discussion in Section 7.5).

Freshwater systems can be categorised into streams (rivers) and lakes (both closed and open). These habitats are more critical for dose assessment than marine systems for the reasons outlined in the previous paragraph. In addition, sediments in freshwater systems can become deposited on alluvial floodplains and, with irrigation waters, contribute to terrestrial pathways via radionuclide transfers from soil to crop or forage and eventually to human diet, as described earlier in this chapter. A generic freshwater exposure pathway is given in Figure 7.7.

Closed lakes or lakes with a high water residence time are of particular concern in that they provide a circumstance where the radioactivity does not become diluted over time. Under such conditions, the levels of radionuclides, particularly Cs, can build up in predatory species by a process of biomagnification via their food web (see the following discussion and Chapter 6). Local fishermen and their families are at greater risk in this case.

For either system, marine or freshwater, it is important to determine the catchment area of the released radioactivity and the aerial deposition across that zone following atmospheric dispersal of radioactivity. This is particularly true for freshwater systems, as the size of affected catchment will have a significant bearing on the concentrations of radioactivity in the water. The soil type in the catchment will have a substantial effect due to the relative abilities of various types of soil to retard migration of radionuclides deposited across the region into local waterways (Smith, 2003). Other factors, such as rain

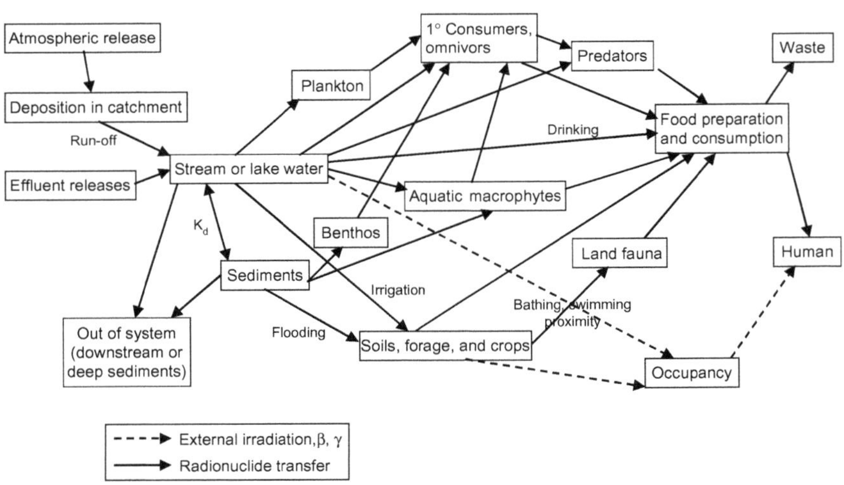

FIGURE 7.7 Generic exposure pathways for a freshwater system.

intensity, will also have a direct bearing on catchment retention (Garcia-Sanchez and Konoplev, 2009).

Groundwaters provide a special case as they may have enhanced radioactivity from human activity or from natural radionuclides, particularly from the U- and Th-series. The groundwater contamination may not be apparent at the surface, so these conditions need to be considered on a case-by-case basis.

7.4.2 Critical Groups

Critical groups or individuals are defined as those people who, by their location and lifestyle, are most at risk of being exposed to radiological dose from any particular pathway. It is important to realise that men, women, and infants will have different exposures due to factors including (but not limited to) body mass, metabolism, consumption rates, diet, and lifestyle. Some of these factors are accounted for in the derived limits applied to different ages and genders by the International Commission on Radiological Protection (ICRP, 1991, 2002). Similarly, different dose limits are applied for workers as distinct from members of the public. Given the exposure pathways depicted in Figure 7.7, it is apparent that the critical groups from freshwater systems will be those who may be in one or more of the following categories:

- Those relying on river, lake, or ground water for drinking, cooking/food preparation, irrigation, bathing, or recreation
- Those harvesting aquatic foods for personal consumption or trade
  - Fish, and other fauna (farming and fishing)
  - Aquatic plants
- People living and farming (livestock and crops) on alluvial floodplains
- Those otherwise exposed to water as a consequence of their normal activities
  - Miners, river traders, dam workers, etc.

The 2006 State of World Fisheries and Aquaculture (SOFIA) report (FAO, 2007) provided clear evidence for a dramatically greater reliance on fish catching and farming in tropical countries as compared to more temperate regions of the planet. In terms of inland capture by continent, Asia (65% of world tonnage), Africa (25%), and South America (5%) outstripped the rest of the world by far. The top 10 countries capturing inland fish by order of tonnage rank were China, India, Bangladesh, Myanmar, Uganda, Indonesia, Tanzania, Egypt, Cambodia, and Brazil. Similarly for aquaculture, the most productive areas were China (70% by itself) and the rest of Asia and the Pacific contributing a further 22%. Europe and North America combined to produce just over 5%. Hence, this exposure pathway is much stronger in tropical and subtropical regions of the planet and has the potential to become even stronger. In Africa, freshwater fish farming is in its infancy, but the potential growth and possible extent is estimated as being vast (Kapetsky, 1994).

In many freshwater tropical systems, productive irrigation fields for paddy rice are also used for fish production (e.g., Smith et al., 2002). This provides a closed system for potential biomagnification in addition to expanding the dose exposure pathways of families relying on these systems for sustenance and barter.

For marine systems, the most likely exposure is via consumption of radionuclide-contaminated foods (fish, shellfish, and algae), but exposure from swimming or occupation is also possible. Hence, harvesters of fish and shellfish, their families, and clients are the most likely critical groups (e.g., Poon and Au, 2003), followed by those exposed recreationally, either as anglers or swimmers.

Once critical groups have been identified, the transfer of radionuclides and radiation to those people via the various exposure pathways needs to be measured or calculated. This includes determining the quantities of each food type consumed and the proportion of those foods acquired from within the contaminated region as well as the source of any drinking water. For occupancy, exposures to external radiation, duration, and type of activity are important. In addition, mitigating factors such as cooking methods and food storage need to be considered (see the next section). Age effects also need to be considered. In Thailand, one study (Porntepkasemsan and Srisuksawad, 2008) assessed the dose derived from ^{226}Ra in canal and ground waters consumed by humans across a range of ages, assuming a consumption rate of 2 L.d^{-1}. Whilst the values estimated (<15 μSv.a^{-1}) were all below the World Health Organization (WHO) recommended reference dose level of 100 μSv.a^{-1}, the doses to children and teens were significantly higher than those to adults. A number of other important considerations are expanded upon in the following section.

For the purposes of this section, the focus is on radiological dose to humans. However, it is important to note that, as a consequence of the recognition of humanity as simply part of a broader connected ecology of the planet, there is increasing consideration placed on potential impacts on other species. In Section 7.5, there will be some discussion of radioecological risk assessment. This will cover the application of quantitative probabilistic modelling together with environmental dose assessment and using data on radiological dose-response relationships for non-human species. The concepts covered will relate to preserving biodiversity by appropriate application of the precautionary principle.

7.4.3 Critical Radionuclides

Critical radionuclides are those most likely to contribute substantially to radiological dose. The factors that determine the critical radionuclides often include the source of the radioactivity (e.g., fallout, reactor effluent, U processing facility, medical effluent, natural enhancement, etc.); the fission yield; half-life; the type, energy, and abundance of radioactive emissions; geochemical behaviour; and bioaccumulation or concentration ratios (discussed next).

The most critical radionuclides from fallout or reactor accidents include isotopes of Cs, Sr, I, and ^{60}Co and some short-lived species in addition to the fuel constituents. For routine releases, the same radionuclides are present, except for Co plus ^{3}H. From other sectors of the nuclear fuel cycle and naturally enhanced radioactivity, the U-series radionuclides of concern include ^{226}Ra and ^{210}Po. Nuclear waste processing will also produce a number of potentially harmful nuclides. Some critical nuclides and associated information are included in Table 7.7. A more complete list of potentially harmful fission and activation product radionuclides is given in IAEA (1989).

7.4.4 Concentration Ratios

Concentration ratios (C_rs), sometimes also referred to as bioaccumulation or TFs (ICRU, 2001), indicate the most likely radioactivity concentration in an organism as a consequence of its exposure to radioactivity in the water. Similar to that previously described for soil-to-plant transfer, the aquatic C_r is defined using the following equation:

$$C_r = \text{radionuclide in organism (Bq/kg)/radionuclide in water (Bq/L)} \quad (7.7)$$

The concentration in the organism should be referred to using the mass in the form that is normally consumed and this form should be stated. Most often the fresh weight C_r is used, but for dried fish, for example, the dry weight C_r would be more appropriate. You will note that the units for a C_rs are in L/kg, which are dimensionless if the volume of water is expressed as a mass. Values > 1 indicate active uptake and a potential for biomagnification. Some examples of C_rs are given in Table 7.7.

TABLE 7.7 Some Potentially Important Radionuclides, Their Annual Limits on Intake for Workers, Half-lives, and Selected C_r Values

Nuclide	Decay	$T_{½}$ (d, days; y, years)	ALI Ingestion by Workers (Bq)	Recommended Bioaccumulation or C_rs (C_r, L/kg)	Biota Affected (m, marine)
^{60}Co	β, γ	5.3 y	4.0×10^{6}	2000	Seals[m]
^{90}Sr	β	30 y	7.1×10^{5}	60	Freshwater fish
^{131}I	β, γ	8 d	9.1×10^{5}	10,000	Macroalgae[m]
^{137}Cs	β, γ	30 y	1.5×10^{6}	2000	Freshwater fish
^{210}Po	α	138 d	8.3×10^{4}	30,000	Zooplankton[m]
^{226}Ra	α, γ	1600 y	7.4×10^{4}	2000	Phytoplankton[m]
^{239}Pu	α, γ	24,000 y	8.0×10^{4}	3000	Mollusc[m]

From Delacroix et al. (1998) and IAEA (1982, 2004).

It should be remembered that all processes, including C_rs, discussed hereafter are time dependent and, for the purposes of simplicity, most of the discussions have assumed that equilibrium has been achieved. Nonetheless, nonequilibrium will be more typical in dynamic systems and should be considered when deriving accurate dose estimates. The AQUASCOPE model (Smith, 2003) is an example of a model that includes time dependency in aquatic dose estimation.

Assuming that the concentrations of radioactivity in the water and organism have reached a steady state, then knowing any two values in the equation above, the third value can be estimated. For example, knowing the radionuclide concentration in fish flesh and the C_r allows estimation of the average radionuclide concentration in the water in the period prior to catching the fish.

Values of C_r can vary widely both between species for the same radionuclide and between radionuclides. Even when these two parameters are stipulated, the range of C_rs can be extensive for a number of reasons, primarily as a consequence of natural variability between individuals and because of diverse biogeochemical conditions at different times and locations (including lack of equilibrium). These will be expanded upon in the next section.

An important issue to note for dose assessment purposes is that if the estimates of radiological dose based on the recommended C_r values are approaching restrictive trigger or action levels, then it would be reasonable to evaluate the actual, site-specific values for the circumstances in question. As well as giving more accurate and precise measures for the dose assessment at hand, the data derived from site-specific studies can be used to improve the generic C_r values supplied by organisations such as the IAEA (2004). There is relatively recent evidence (Twining et al., 1996) to suggest that the generic C_r values for Cs and Sr transfer into freshwater fish, which were mostly generated using studies on temperate Northern Hemisphere species, are not the same as those for tropical species (see Chapter 6). This finding supports the acquisition of site- and region-specific data.

One problem with this approach is that it takes time to scientifically derive such data. In such cases, baseline studies are recommended if predictive modelling indicates the potential for significant exposure. A separate empirical alternative is to back up the C_r modelling approach by undertaking biomonitoring of the local foodstuffs to provide real-time radionuclide concentrations for dose estimation, as well as for evaluation or validation of the applied C_r and predictive models.

Until the site-specific data are better evaluated, generic models are available for estimation of the concentrations of significant radiological contaminants in important aquatic environmental compartments. One useful model for estimating the concentrations of $^{89,90}Sr$, ^{131}I, and $^{134,137}Cs$ in surface water and freshwater fish is AQUASCOPE (Smith, 2003). Again, this was developed from a non-tropical perspective, but it does provide a reasonable means for making a first-pass estimation in any system as it assumes mechanisms and controlling factors that are widely applicable, such as geochemical effects on bioavailability, as discussed next.

7.4.5 Geochemistry and Bioavailability

There are several environmental variables that can have direct and indirect influence on the fate of radionuclides entering aquatic systems and on the degree of bioaccumulation leading to ingested dose. Volumes, flow rates, dilution, and dispersion are key issues, but they will not be discussed further as the details are beyond the scope of this chapter. Some basic dilution and dispersion modelling for radioactivity in aquatic systems can be found in Kathren (1984). Most of the other factors affect the retention of radionuclides in the water column and bioavailability.

One of the first factors influencing the fate of radionuclides entering aquatic systems is their transfer from water into sediment. This can occur as a result of several mechanisms including: adsorption to, or inclusion within, biogenic particles (e.g., faecal pellets) and gravitational removal; resuspension of sediment particles due to physical disruption; or contact between the water and bed sediments. A simplified, generic approach to determine the effect of such contact is to apply the sediment-water distribution coefficient, K_d (ICRU, 2001). This is estimated using the following equation:

$$K_d = \text{concentration in sediment}\,(\text{Bq/kg})/\text{concentration in water}\,(\text{Bq/L}) \quad (7.8)$$

A table of sediment-water K_d values for a range of radionuclides in marine and freshwater systems is given in IAEA (2004). The value of K_d depends on the types of sediments (e.g., gravels, sands, silts, and clays) and their mineralogies (e.g., carbonates vs. Fe/Mn oxides). In addition, the quantity of radioactivity removed from the water to the sediments will depend not only on the K_d and the time available, but also on the 'active' mass of sediment involved compared with the volume of water. It should also be noted that the quality of the sediment can change within and between locations.

Hence, sediments can act as a sink for radioactivity from water and tend to reduce radiological dose to humans by limiting exposure via ingestion of drinking water or bioaccumulated radionuclides. However, they can also increase dose by two mechanisms. The first is that, as a sink, sediments may retain radioactivity that could otherwise have been flushed from the system and then release it over long periods of time (albeit at low rates). The second is that sediments provide a mechanism for transferring radioactivity onto floodplains and intertidal areas by resuspension and deposition, consequently increasing radiological dose via the terrestrial pathway. This includes direct radiation as well as via soil-to-crop or forage transfer and subsequent ingestion. It is important to evaluate the potential for floodplain, tidal flat, and beach deposition in any critical exposure scenario.

Of the radioactivity remaining in the water column, there are two other factors that control the transfer into human foods (aquatic plants and animals): (1) the chemical makeup of the water, which influences the ability of the radioactive elements to cross biological membranes (bioavailability) and (2) the biokinetics of the radioactivity within and between organisms in the food web. Both of these factors are discussed in more detail in Chapter 6. The following is a general overview.

In relation to water quality, it is well known that K and Ca actively compete for biological absorption sites with Cs and Sr; hence, when concentrations in water of K or Ca are high the degree of bioaccumulation of Cs or Sr is reduced. There are enough data available for this effect to be modeled in freshwaters by the following equations (IAEA, 1994):

$$\text{Log}\, C_{rCs} = 4.3 - 0.72 \times \log K_w + 0.29T - 0.23 \log (SS) \tag{7.9}$$

where K_w is the concentration of K in water (μM/L), T is the trophic level (1 for piscivorous fish and 0 otherwise), and SS is the suspended sediment load in mg/L, and:

$$\text{Log}\, C_{rSr} = \exp (5.18 - 1.21 \ln Ca_w) \tag{7.10}$$

where Ca_w is the concentration of Ca in water (μM/L).

Whilst these generic relationships exist for Cs and Sr, it is reasonably understood that competitive uptake will occur between any radioisotope and its chemically analogous nutrient element. Potassium is a monovalent nutrient and competes against the monovalent Cs ion (as does divalent Ca with Sr). Hence, Ca will also compete with Ra, and so on. In general, waters with greater hardness, alkalinity, or salinity will tend to have reduced bioavailability.

The higher values of Cs in predatory fish are due to biomagnification. From Figure 7.7, it is apparent that organisms at higher trophic levels not only take up radionuclides directly from the water, but also accumulate radioactivity from other plants or prey within the food web. If these intermediate organisms also concentrate radioactivity above the levels in the water (i.e., C_rs > 1), then the effect is enhanced when these organisms are consumed.

7.4.6 Biodistribution

Due to the many physiological mechanisms controlling uptake and distribution of elements within organisms, the acquired radionuclides are typically distributed unevenly amongst the various tissues (e.g., Twining et al., 1996). Hence, bioaccumulation of radioactivity into organisms does not necessarily equate to increased dose to the consumer.

An example where high accumulation levels may be of greater concern is Cs, which tends to accumulate in muscle flesh. Strontium and Ra are predominantly associated with skin, bones, and scales so they may be of less concern depending upon how the fish are prepared and consumed. In tropical freshwater mussels, high accumulation of Ra has been observed (Jeffree, 1990; Martin et al., 1998) along with other radionuclides. Some of the Ra activity is undoubtedly associated with the shell of the organism, since Ra is an analogue of the macronutrient Ca. However, much of it is also associated with soft tissues and intracellular granules within those tissues and will normally be ingested. Sauerland et al. (2005) provided one tropical example of radiological dose assessment to humans and biotas from ^{226}Ra in Magela Creek in tropical northern Australia, an area closely associated with uranium mining. They estimated a locally derived ^{226}Ra concentration ratio of 19×10^{-3} L/kg in freshwater mussels. From that estimation they derived a limit of no more than a 10 mBq/L increase in ^{226}Ra above background in creek water concentrations. This limit was dose constrained to an increment of 0.3 mSv per year to a 10-year-old child.

7.4.7 Food Preparation

As indicated in the previous section, excluding parts of the organism from the diet can substantially reduce radiological exposure to some radionuclides accumulated in aquatic foods. This may be the normal situation, for example, when fish are filleted and scaled prior to consumption. However, the exposures may be much higher if the animals are consumed whole or where viscera, skin, scales, and bones are included in dishes such as soups or stews.

Rinsing and cooking with uncontaminated water may also reduce radiological exposure. The IAEA (1994) has identified food-processing retention factors for Sr and Cs in fish and for Ra, Pb, and Po in crustaceans and molluscs. These are shown in Table 7.8. These factors represent the fraction of radionuclides present in the raw food that are retained in cooked food after processing. However, care should be taken to include some of the washed-out activity if the fluids are retained, such as in soups or stocks.

TABLE 7.8 Food Processing Retention Factors

Organism	Processing Method	Sr	Cs	Ra	Pb	Po
Fish	Boiling meat	0.9	0.2–0.9			
	Frying meat		0.8–0.9			
Crustacean	Cooking			0.04–0.5	0.0–0.4	0.04–0.8
Mollusc	Washing and removal of shell			0.01	0.5	0.02

From IAEA (1994).

Other options that may be considered for reducing exposures, particularly to short-lived radionuclides (e.g., ^{131}I; $T_{½}$ 8 days), are methods that allow the foods to be stored until the radioactivity can decay. These methods include drying, salting, smoking, and canning. Some of these processes may also remove radioactivity in addition to the benefits of storage. However, long-term storage may be less of an option in underdeveloped economies, particularly in areas of subsistence agriculture.

7.4.8 Countermeasures

Countermeasures are actions taken to reduce and minimise radiological dose following environmental releases of radioactivity. A critical review of aquatic countermeasures for freshwater systems is provided in Smith et al. (2001) and for information on countermeasures in general (datasheets originally developed in the STRATEGY project (Howard et al., 2004, 2005) and revised in the Euranos project". You can access: http://www.euranos.fzk.de/index.php?action=euranos&title=products. Generally, apart from drinking water treatment actions, countermeasures to reduce radioactivity in aquatic environments are inadequate, and they have not been applied to marine systems to date. Countermeasures that have been attempted include addition of KCl to lakes affected by fallout from Chernobyl to try to reduce Cs bioaccumulation. Similarly, lime was added to reduce Sr uptake. Both were ineffective, primarily because the amount of salt required to significantly alter the chemistry of the lake was beyond practical limits.

Alternate countermeasures that may be more effective include treatment of drinking water supplies, behavioural change, and avoidance of exposure pathways until the radioactivity levels reduce to acceptable levels. Hence, preventing fishing, swimming, and drinking in and from contaminated streams until radioactivity levels have dropped will substantially reduce radiological dose. For remote or subsistence fishermen and villagers, this may be impractical. Government intervention may be required to provide alternate food and drinking water supplies in these cases.

Education and consultative programmes for the affected groups or individuals are valuable. They are also more effective when the local communities have input into the discussion and decision-making processes.

7.4.9 Summary of Aquatic Dose Pathways

Aquatic pathways typically contribute less radiological dose than terrestrial exposures. Nonetheless, they do provide a means of exposure that should be assessed in some circumstances, particularly as lifestyles closely associated with aquatic environments are common in tropical regions. Similarly, seafood can comprise a significant, and in some circumstances, a major component of the diet in tropical countries, particularly those comprising island states common in the Pacific Ocean. The FAO (2007) noted the higher dependence of developing nations on fishing. Other key habitats include lakes that are in a closed catchment or have a long water retention time.

Following atmospheric dispersion, the overall exposure from freshwater pathways is strongly correlated with the amount of fallout within the catchment area. Dilution and dispersion, together with geochemical factors that control the partitioning of radioactivity into sediments and the bioavailability of the various radionuclides, have a marked bearing on the radiological dose. Biomagnification through food webs can significantly increase dose. Understanding the exposure pathways will help to define the critical groups and individuals.

External exposures can be occupational or as a result of lifestyle and may be enhanced in association with alluvial deposition areas (floodplains) or irrigation from contaminated streams. Internal exposures can be from direct ingestion of contaminated drinking water or as a result of eating contaminated foods. Physicochemical countermeasures are less likely to be effective than behavioural change in reducing exposure and dose.

7.5. ENVIRONMENTAL DOSE ASSESSMENT AND RADIOECOLOGICAL RISK ASSESSMENT

7.5.1 Development of the Concept of Environmental Dose Assessment

Previously, any consideration of the impacts of radioactive substances on the broader environment hinged on the assumption that by protecting humans we conferred protection on populations of non-human species (e.g., ICRP, 1991). Developing scientific and sociopolitical perceptions of the value of the environment have questioned the basis for this assumption (e.g., OECD/NEA, 2003), particularly in light of the requirement to maintain and enhance biodiversity. There is also the need to appreciate that some species may be more radiosensitive than humans (Daniel et al., 2003) or exist in niches more exposed to higher levels of radioactivity (e.g., in or about deep-sea waste disposal sites) or both. Consequently, there is an ongoing effort to reach consensus on an achievable philosophical and technical approach to assessing the effects of ionising radiation in the environment.

Countries such as Canada (e.g., Bird et al., 2003) and the United States (DOE, 2002) have developed approaches to assess radiological dose (and to some extent risk) to non-human species, but as yet there is no international agreement or recommendation (e.g., via the IAEA) on a practical approach to this challenge. The ICRP has developed a framework (ICRP, 2008) based on the concept of reference plants and animals. In Europe, the EC 6th Framework Environmental Risk from Ionising Contaminants: Assessment and Management (ERICA) initiative resulted in the development of the 'ERICA integrated approach to assessment and management of environmental risks from ionising radiation, using practical tools' (Howard and Larsson, 2008).

The ERICA programme and its closely related precursors, Framework for Assessment of Environmental Impact (FASSET) and Environmental Protection from Ionising Contaminants (EPIC), have been followed by the Protection of the Environment from Ionizing Radiation in a Regulatory Context (PROTECT) programme (http://www.ceh.ac.uk/protect/). The IAEA has been aware of these developments and is primarily responsible for the related EMRAS programme. Within some working groups of these studies, several aspects of the evaluation of exposures to non-human species as a consequence of release and movement of radioactivity in the environment are discussed in detail. Most of these programmes have evaluated international opinion and approaches toward achieving some international consensus, but there is still some way to go in this regard.

As part of the FASSET Programme, Jones and Gilek (2004) produced a good summary of the approaches that had been developed at that time for the assessment of risks to the environment from both ionising radiation and hazardous chemicals. These approaches comprised site-specific or stylised scenarios (the latter consisting of simplified or reference-type biospheres) and both deterministic and probabilistic methods. In all cases, there was found to be a necessary compromise between data availability and practical limits to management of the assessed problem. The balance between these factors has always been less well achieved for tropical areas due to the paucity of data.

On a similar issue, Beresford et al. (2008a) provided an international comparison of models and approaches for the estimation of the radiological exposure of non-human biota as an output from the Biota Working Group of the IAEA EMRAS programme. The group aimed at approaching an international consensus on the best methodology to apply. The outcomes of the comparison show that each of the models derives somewhat different results under the same scenarios, reflecting mostly the developmental nature of all the models. However, the variability of C_r values has been identified as one of the parameters to which model outputs were most sensitive. A recent overview of ecological risk assessments for nuclear sites (SENES, 2007) provides a good reference point for a variety of studies carried out in this field covering fuel cycle activities and NORMS in terrestrial, marine, and freshwater environments. They provided a number of conclusions, but the key findings were that, with the evidence provided, the environmental impacts of enhanced radioactivity in the environment were generally low and of little concern. This is good news; however, their evaluations were partially based on reference criteria of acceptability that are currently not well established and the studies used for the analyses were conducted for purposes other than dose assessment.

7.5.2 Key Recommendations for Future Approaches

Incorporating the broadest perspective on information and opinions as possible, the PROTECT group established some key recommendations for any approach to assessing radioecological risks, initially reported by Hingston et al. (2007) and later updated by Howard et al. (2008), as follows:

- Protection should focus on population level effects rather than on impacts to individuals. The goals should be translated into measurable targets with advice provided on tolerable risks associated with those endpoints. This will be more easily interpretable when considering effects of chronic doses to higher levels of organisation, such as communities and ecosystems.
- Linking radiological protection methodologies with approaches used for chemical toxicological assessments was strongly advocated. This was done to make the best use of a well-developed field of risk analysis with very similar environmental objectives.
- Internationally agreed upon approaches for setting environmental thresholds (namely, the Species Sensitivity Distribution and Application Factor methods) should be applied to derive numeric criteria (as dose rates). The use of purely expert judgement should be avoided where possible.
- Currently applied or suggested numeric criteria (e.g., screening values, 'regulatory action levels') should be evaluated with a view to PROTECT recommending criteria to be applied within a tiered assessment process. Outcomes should be supported with clearly documented decisions and identified data strengths and limitations. This is a quality assurance step working toward international consistency as well as identifying areas of specific research need.

In relation to the first recommendation from the PROTECT group, some recent work (e.g., Alonzo et al., 2008a, b) has shown that sublethal effects resulting in lower energy transfer to successive generations (such as reduced feeding rates, reduced egg number or weight, and/or delay in reproduction), as a consequence of chronic low dose-rate exposures to individuals, can result in substantial population effects if extended to multigenerational assessments. These effects were, in many cases, more significant to the long-term viability of the population than was individual survival.

7.5.3 Probabilistic Ecological Risk Assessment Models and Species Sensitivity Distributions

Taking the second and third key points further (Section 7.5.2), quantitative and probabilistic Ecological Risk Assessment (ERA), based on comparison of the probabilities of exposure and response, has an increasingly accepted place in management of the nonradiological ecological impact of human activity (e.g., ANZECC/ARMCANZ, 2000; Posthuma et al., 2001). This form of ERA is premised on the existence of many, usually laboratory-based, studies of the responses of various biotas when exposed to environmental stressors. The advantage of this approach is that as much relevant information as possible is used to make the assessment (as distinct from relying on a single datum from the most sensitive organism); hence, the result is more robust. It also enables a range of endpoints, effects, species diversity, and sensitivities to be included, which ensures that the risk analysis is relevant to the functional and structural capacity of the ecosystem as a whole. The data on biotic responses to absorbed dose or dose-rate are used to generate Species Sensitivity Distributions (SSD) from which criteria may be derived corresponding to an agreed level of acceptable effect on species within the exposed ecosystem. The approach is precautionary in that the selected endpoints are generally protective of most of the exposed populations and encompass sensitive effects. If particular organisms or communities are crucial, or perceived to be at greater risk, the assessment can be more closely refined to suit them. In addition, the probabilistic nature of the assessment takes into account the variability in the data and enables extrapolations to account for, or protect against, that uncertainty.

There are a number of useful generic references and critiques of the overall application of probabilistic ecological risk assessment and the specific approaches used to address the issue. Useful starting points for those interested in the field could include Calabrese and Baldwin (1993), Forbes and Calow (2002), Posthuma et al. (2001), Kefford et al. (2005), and Suter (2007).

Publication of the FASSET Radiation Effects Database (FRED; Woodhead and Zinger, 2003), and subsequently FREDERICA (Copplestone et al., 2008), brings together large amounts of published information on the effects of ionising radiation on different biotas and raises the prospect of applying quantitative and probabilistic ERA to radioactive pollutants in ecosystems, effectively Radioecological Risk Assessment (RRA). The database is available online at www.frederica-online.org, with tutorials on how to apply the data provided at http://www.ceh.ac.uk/PROTECT/ERICAdeliverables.html.

Garnier-Laplace et al. (2006) were the first to use this approach for radiological impacts in freshwater and terrestrial systems. For chronic gamma radiation, they derived a value of 10 μGy/hour for freshwater environments (using an assessment factor method due to the paucity of data and subjective expert opinion on the degree of uncertainty in the assessment balanced with the need for a conservatively safe determination) and 67 μGy/hour for terrestrial ecosystems using a SSD approach. Beresford et al. (2008b) evaluated background dose rates for selected terrestrial organisms to be in the range of 0.07 to 0.6 μGy/hour (although this did not include the exposure of soil organisms to radon). Hence, the derived criteria values are a factor of 10 to 1000 times typical background dose rates.

7.5.4 Data Quality and Quantity

Despite the existence of the FREDERICA database, the last key point, that of data quality and applicability, from the list (Section 7.5.2) provided by the PROTECT group needs to be emphasised. Similar views were expressed by

Garnier-Laplace et al. (2004) when looking at data and extrapolation interpretations. This was especially in reference to the lack of data on chronic low-dose exposures, which is likely to be the most common type experienced in environmental situations. Of primary importance is that different species express a wide range of radiosensitivities (see Chapter 1) and bioaccumulation rates (see Chapters 5 and 6). Hence, extrapolation from one species to another already brings in a high degree of uncertainty. This variability is reduced within classes of similar organisms, but the differences may still vary over orders of magnitude. Beresford et al. (2004) reviewed the problem of estimating C_r values for wild species using a number of approaches. These included using TF data for organisms occupying similar environmental niches (e.g., mussels for oysters, as both are filter-feeding molluscs) or for radionuclides with similar environmental characteristics to those studied (e.g., Ra for Sr, as both are divalent cations), as well as a number of other generalising extrapolations. Whilst they determined that the approaches had merit, there was some lack of confidence in their application and very few had been validated effectively, although activities within the IAEA EMRAS programme are addressing this matter. It was with these reservations in mind that the final PROTECT dot point stressed the need for clear documentation of the rationale used for any extrapolations made in the process of undertaking an environmental dose assessment.

Further, in relation to the FREDERICA database, much of the data are from acute (short-term in relation to the lifespan of the test organism) exposures and conversion to equivalent chronic effects is not straightforward. Factors such as the Relative Biological Effectiveness (RBE) of any absorbed dose will vary depending upon the type of radiation (alpha, beta, and/or gamma) and its linear energy transfer (LET; the rate at which its energy is transferred to the tissue by ionisation, see Chapter 1) and the dose-rate will affect the ability of the affected cells to respond and repair. The tissues that absorb the dose will also have different sensitivities (accounted for, to some degree, in human dose assessment by tissue weighting factors). Synergistic and antagonistic effects of multiple stressors also need to be considered. Very few of these factors have yet been addressed to any substantial extent when trying to convert the existing data for use in environmental impact modelling.

As discussed earlier, site-specific data are always better than generic modelling. A few examples of such data collection within tropical systems include: Santos et al. (2002) in Brazil; Jibiri et al. (2007) in Nigeria; Radhakrishna et al. (1996) and Shahul Hameed et al. (1997) in India; Rahman and Voigt (2004) and Rahman et al. (2005) in Bangladesh; arid zone studies near the nuclear test site at Maralinga in central Australia (Giles et al., 1990); many studies undertaken in the vicinity of the Ranger uranium mine in northern Australia (e.g., Martin et al., 1998; Twining, 1989); and the large body of work done at the Savannah River Ecology Laboratory in the southern United States (e.g., Brisbin et al., 1989; Whicker et al., 1990a; Kelly and

Pinder, 1996). Whenever critical areas of concern have been identified for more detailed investigation as part of any environmental risk assessment process, site-specific data should be acquired for a detailed assessment of potential effects.

7.6. CONCLUSIONS

In the last decades, the assessment of radiological consequences due to discharges of radionuclides to the environment mainly focussed on sites in the temperate zone. Due to the increasing application of nuclear techniques and the increasing production of nuclear energy in tropical and subtropical countries, the question arises whether radioecology in those countries needs special consideration. From the discussion, it can be concluded that there is no special, specific tropical radioecology. All mechanisms leading to transport and migration from one environmental compartment to another can be found in all parts of the world. However, it is also clear that the degree of such transport processes is dependent on site-specific climatic and environmental conditions, the land use and habits of the people, and the endemic biota. To address the specific features of environmental transport of radionuclides in tropical areas appropriately, the following items need to be taken into consideration:

- Transfer processes for radionuclides are generic. However, specific circumstances may modify the transfer of radionuclides and these can be more marked in tropical and subtropical regions. For example, during the rainy season the heavy rainfall that occurs daily in the early afternoon causes lower interception fractions of radionuclides that are deposited during such events. Similarly, tropical regions are characterised by an enormous variation of crops and fruit, which are cultivated throughout the year, providing the possibility for several harvests per year.
- Seasonality of the transfer of radionuclides in tropical countries is mainly controlled by the sequence of dry and wet periods instead of the annual variations of temperature and light, as is seen in temperate zones. The phenological data and time dependency of leaf-area indices are very important for the reliability of model results. These data describe the morphological and physiological development of plants, which can vary considerably from region to region. The stage of the plant's development controls the dry deposition, the interception of wet-deposited nuclides, and the translocation of nuclides within the plants.
- Contamination processes, such as dry deposition, interception of wet-deposited activity, and translocation from the leaves to the edible parts of the plant, are dependent on the development of the plant. The maxima of these processes are, in general, observed during the peak growing season. In the tropics, there are no pronounced peak seasons, since plants can grow all year. Therefore, there are no considerable seasonal variations of the radiological consequences of an accident.

- Tropical countries have soil characteristics peculiar to their area (Kellman and Tackaberry, 1997; Scheffer and Schachtschabel, 2002; Chapter 5). In the tropics, with high temperatures and high precipitation, weathering of the parent rock is much more advanced than in other parts of the world. Soils are depleted of calcium, magnesium, potassium, and silicon; the main soil constituents are iron and aluminium oxides. Plant productivity in the tropics is very high. However, sustainable soil cultivation is difficult due to structural problems, vulnerability to erosion, rapid degradation of organic matter, and leaching of nutrients. Due to advanced clay decomposition, the cation exchange capacity is often very low; in other words, ^{137}Cs is only weakly bound. This is an important difference to soils in temperate climates, where cesium is strongly bound to clay, leading to a low uptake by roots. Large variability can be found for soil-plant TFs. Moreover, the soil-plant TFs are correlated with leaching rates to deep soil layers; high radionuclide mobility implies high TFs as well as high leaching rates.
- For urban environments, the shielding provided by construction materials in the tropical and subtropical regions is generally less than in colder climates. Due to higher temperatures, requirements for insulation are less in the tropics; therefore, walls are thinner and lightweight construction is favoured. However, increasing use of air-conditioning and cyclone-proof construction methods will lead to an evolution of these parameter values in developing tropical nations.
- Weathering loss of radionuclides from tropical and subtropical urban surfaces may be slower than in temperate climates. This has been observed during studies performed in Goiânia (Brazil). The loss was slower in urban areas compared to observations made in Europe after the Chernobyl accident, although the rainfall is much higher. This may be due to the different road surfaces and lower frequencies of street cleaning and maintenance activities.
- Most tropical areas in the world comprise developing economies where land-use may change faster than in most temperate climate industrialised countries. Eventual remediation activities should also take into account possible future land-use changes. Remediation strategies that do not physically remove the contamination from long-term accessible environments are to be used with care.
- Human diets in tropical countries are generally dominated by plant food products. Site-specific feeding regimes should be analysed. Storage and processing times and methods should be representative of the agricultural and economic conditions of the considered region. Consideration of the use of the RADCON model should be made.
- Weather conditions affect the lifestyle of the population and the time of outdoor activities, which are performed all year round. However, in developed urban areas, people may also stay longer in indoor environments due to air-conditioning systems.

- Aquatic pathways for radiological dose are typically, but not always, less important than the terrestrial pathways. However, there is a range of communities that are culturally more connected to them or that include significant or major components of aquatic foods. This can be particularly true in tropical regions associated with floodplains, coastal environments, or island nations. Tropical countries dominate global aquaculture production. Hence, under some exposure scenarios, aquatic pathways should be considered more closely. Generic models are available to estimate the concentrations of radionuclides in various biotic and abiotic compartments associated with marine and freshwater systems. However, at present these models still tend to reflect temperate systems. They have the advantage of reflecting environmental processes and mechanisms that should not be unduly affected by climatic regime, but additional site-specific, or at least regional-specific data on local species with respect to TFs will enhance their utility.
- The recently increased awareness of the need to protect the broader environment from the adverse impacts of radiological releases is also pertinent to tropical and subtropical regions of the planet. This is particularly true in light of the perceived requirement for additional nuclear energy in the coming decades. At present, the emphasis for radiological dose assessment modelling continues to reflect the pre-eminent need to protect human life. Hence measurements of broader impacts from environmental dose are more focussed at the community level, rather than on the individual. In any case, the trend in environmental risk management will continue to concentrate on making better estimations of radiological exposures to both humans and the environment. This will be in addition to reducing exposures and mitigating any potentially adverse outcomes irrespective of the degree of refinement of the assessments used. To those ends, the current developments are taking the traditional dose assessment methodologies and combining them with the approaches being applied in more traditional environmental assessment methodologies, such as in toxicology. RRAs, using SSDs or similar approaches, will become more prevalent in coming years. For that to be achieved, much more data, particularly on the effects of chronic low dose, are required. This is all the more pertinent for tropical systems given the relative lack of such data in those environments and the fact that the current data for temperate species can only be poorly extrapolated to tropical and subtropical species and environments. In any case, where critical environmental assessments are required, site- and species-specific data will be preferred.
- Tropical and subtropical areas cover a wide range of conditions. Plant production may occur on small, subsidence farms or systems on large plantations. Animal husbandry varies from nomadism to industrialised production of meat, milk, and eggs. All this is done in a wide range of environmental conditions. Therefore, one should be aware that the aim of this chapter is to analyse and discuss the impact of climatic factors on

radioecological processes. This means, in any specific assessment, an analysis of the site-specific conditions is required to ensure that local peculiarities do not cause unexpected exposures.

ACKNOWLEDGMENTS

The authors would like to acknowledge the discussions with Dr. Elaine Rochedo (Instituto de Radioproteção e Dosimetria/Comissão Nacional de Energia Nuclear, Rio de Janeiro, Brazil) regarding the assessment of radiological consequences of accidental releases in tropical areas. Constructive critical reviews of the draft document were provided by Dr. M. Johansen, Dr. K. Thiessen, and Dr. N. Beresford. Their comments were most helpful.

REFERENCES

Aarkrog, A., 1997. Radioactive contamination of the marine environment. Ciba Foundation Symposium 203. Health Impacts of Large Releases of Radionuclides. John Wiley & Sons, Chichester pp. 68–73.

Al-Masri, M.S., Mamish, S., Budeir, Y., Nashwati, A., 2000. ^{210}Po and ^{210}Pb concentrations in fish consumed in Syria. J. Environ. Radioact. 49, 345–352.

Alonzo, F., Gilbin, R., Zeman, F.A., Garnier-Laplace, J., 2008a. Increased effects of internal alpha irradiation in *Daphnia magna* after chronic exposure over three successive generations. Aquat. Toxicol. 87, 146–156 2008.

Alonzo, F., Hertel-Aas, T., Gilek, M., Gilbin, R., Oughton, D.H., Garnier-Laplace, J., 2008b. Modelling the propagation of effects of chronic exposure to ionising radiation from individuals to populations. J. Environ. Radioact. 99, 1464–1473.

ANZECC/ARMCANZ, 2000. Australian National Guidelines for Fresh and Marine Water Quality. Australia and New Zealand Environment and Conservation Council/Agricultural and Resource Management Council of Australia and New Zealand, Canberra, ACT, Australia.

Beresford, N.A., Broadley, M.R., Howard, B.J., Barnett, C.L., White, P.J., 2004. Estimating radionuclide transfer to wild species—data requirements and availability for terrestrial ecosystems. J. Radiol. Prot. 24, A89–A110.

Beresford, N.A., et al., 2008a. An international comparison of models and approaches for the estimation of the radiological exposure of non-human biota. Appl. Radiat. Isot. 66 (11), 1745–1749.

Beresford, N.A., Barnett, C.L., Jones, D.G., Wood, M.D., Appleton, J.D., Breward, N., et al., 2008b. Background exposure rates of terrestrial wildlife in England and Wales. In: Howard, B., Larsson, C., (Eds.), The ERICA Project, September. J. Environ. Radioact. 99 (9), 1430–1439.

Bird, G.A., Thompson, P.A., Macdonald, C.R., Shepherd, S.C., 2003. Assessment of the impact of radionuclide releases from Canadian nuclear facilities on non-human biota. Darwin, Australia, July 2002 (unedited papers, IAEA-CSP-17). In: Proceedings of the third international symposium on the protection of the environment from ionising radiation. International Atomic Energy Agency, Vienna, Austria, pp. 241–247.

Birkeland, P., 1999. Soils and geomorphology. Oxford University Press, Oxford.

Brisbin, I.L., Breshear, D.D., Brown, K.L., Ladd, M., Smith, M.H., Smith, M.W., et al., 1989. Relationships between levels of radiocaesium in components of terrestrial and aquatic food webs of a contaminated streambed and floodplain community. J. Appl. Ecology 26, 173–182.

Brown, J., Simmonds, J.R., 1995. FARMLAND: A Dynamic Model for the Transfer of Radionuclides Through Terrestrial Foodchains NRPB-R273. NRPB, Chilton.

Bunzl, K., Schimmack, W., Kreutzer, K., Schierl, R., 1989. Interception and retention of Chernobyl derived ^{134}Cs, ^{137}Cs and ^{106}Ru in a spruce stand. Sci. Total Environ. 78, 77–87.

Bunzl, K., Kracke, W., Schimmack, W., 1992. Vertical migration of $^{239/240}$Pu, ^{241}Am and ^{137}Cs fallout in a forest soil under spruce. Analyst 117.

Bunzl, K., Förster, H., Kracke, W., Schimmack, W., 1994. Residence times of fallout $^{239+240}$Pu, ^{239}Pu, ^{241}Am, and ^{137}Cs in the upper horizons of an undisturbed grassland soil. J. Environ. Radioact. 22, 11–27.

Bunzl, K., Kofuji, H., Schimmack, W., Tsumura, A., Ueno, K., Yamamoto, M., 1995a. Residence times of global weapons testing fallout Np-237 in a grassland soil compared to Pu-239/240, Am-241 and Cs-137. Health Phys. 68, 89–93.

Bunzl, K., Kracke, W., Schimmack, W., Auerswald, K., 1995b. Migration of fallout $^{239+240}$Pu, ^{241}Am and ^{137}Cs in the various horizons of a forest soil under pine. J. Environ. Radioact. 28, 17–34.

Bunzl, K., Jacob, P., Schimmack, W., Alexakhin, R.M., Arkhipov, N.P., Ivanov, Y., et al., 1997. ^{137}Cs mobility in soils and its long-term effect on the external radiation exposure. Radiat. Environ. Biophys. 36, 31–37.

Calabrese, E.J., Baldwin, L.A., 1993. Performing Ecological Risk Assessments. Lewis Publishers, Boca Raton.

Carini, F., 1999. Radionuclides in plants bearing fruit: An overview. J. Environ. Radioact. 46, 77–97.

Carini, F., 2001. Radionuclide transfer from soil to fruit. J. Environ. Radioact. 52, 237–279.

Chamberlain, A.C., 1970. Interception and retention of radioactive aerosols by vegetation. Atmos. Environ. 4, 57–58.

Choi, Y.H., Lim, K.M., Yu, D., Park, H.G., Choi, Y.G., Lee, C.M., 2002. Transfer pathways of ^{54}Mn, ^{57}Co, ^{85}Sr, ^{103}Ru, and in rice and radish plants directly contaminated at different growth stages. Ann. Nucl. Energy 29, 429–446.

Colle, C., Madoz-Escande, C., Leclerc, E., 2009. Foliar transfer into the biosphere: Review of translocation factors to cereal grains. J. Environ. Radioact. 100 (9), 683–689.

Cook, I., Holloway, N.J., Nixon, W., Philips, D.W., 1990. Consultancy on the Environmental Aspects of the Daya Bay Nuclear Power Station for the Government of Hong Kong: Risk Assessment. Government Printer, Hong Kong.

Copplestone, D., Hingston, J., Real, A., 2008. The development and purpose of the FREDERICA radiation effects database. In: Howard, B., Larsson, C., (Eds.), The ERICA Project, September, 2008. J. Environ. Radioact. 99 (9), 1456–1463.

Crawford, J., Domel, R.U., 2000. RadCon: A Radiological Consequences Model. User Guide, ANSTO M-128, ISBN 0-642-59983-1, 2000.

Crawford, J., Domel, R.U., Harris, F.F., Twining, J.T., 2000. RadCon: A Radiological Consequences Model, Technical Guide, Version 2.0. ANSTO E-744, ISBN 0-642-59982-3, May, 2000.

Daniel, D.P., Garnier-Laplace, J., Gilek, M., Kautsky, U., Larsson, C.M., Pentreath, J., et al., 2003. Radiation effects on plants and animals. Deliverable 4 FASSET Project, ed. D.S. Woodhead and I. Zinger p 196 (Contract FIGE-CT-2000-00102).

Delacroix, D., Guerre, J.P., Leblanc, P., Hickman, C., 1998. Radionuclide and Radiation Protection Data Handbook. Radiat. Prot. Dosimetry 76 (1–2).

Dixon, J., Gulliver, A., Gibbon, D., 2001. Global challenges, potentials and priorities. In: Hall, M. (Ed.), Farming Systems and Poverty: Improving Farmers' Livelihoods in a Changing World. FAO and World Bank, Rome and Washington D.C, pp. 307–357 (Chapter 8).

DOE, 2002. A graded approach for evaluating radiation doses to aquatic and terrestrial biota. Technical Standard DOE-STD-1153-2002US Department of Energy, Washington DC.http://web.ead.anl.gov/resrad/home2/biota.cfm.

Downing, D.J., Gardner, R.H., Hoffman, F.O., 1985. An examination of response-surface methodologies for uncertainty analysis in assessment models. Technometrics 27, 151–163.

Eaton, R., 2002. Fallout from a re-entering satellite carrying a small reactor. In: Restoration of environments with radioactive residues. IAEA, Vienna, pp. 385–394.

Ertel, J., Voigt, G., Paretzke, H.G., 1989. Weathering of $^{134/137}$Cs following leaf contamination of grass cultures in an outdoor experiment. Radiat. Environ. Biophys. 28, 319–326.

Espinosa, A., Aragon, A., Hogdson, A., Stradling, N., Birchall, A., 1998. Assessment of doses to members of the public in Palomares from inhalation of plutonium and americium. Radiat. Prot. Dosimetry 79, 1–4.

Espinosa, A., Aragón, A., Martínez, J., et al., 2000. Experience with environmental behaviour of Pu at Palomares, Spain. In: Site Characterization Techniques Used in Environmental Restoration Activities. IAEA, Vienna, pp. 173–186 IAEA-TECDOC-1148.

FAO, 2007. The State of World Fisheries and Aquaculture – 2006. FAO Fisheries and Aquaculture Department, Food and Agriculture Organization of the United Nations, Rome. http://www.fao.org/docrep/009/A0699e/A0699E04.htm#4.1.2.

Flanagan, D., 1997. Java in a Nutshell, 2nd ed, O'Reilly 7 Associates, Inc, Sebastopol ISBN 1-56592-304-9, 1997.

Forbes, V.E., Calow, P., 2002. Species sensitivity distributions revisited: A critical appraisal. Hum. Ecol. Risk Assess. 8, 473–492.

Franke, G., 1975. Nutzpflanzen der Tropen und Subtropen. Verlag S, Hirzel, Leipzig.

Frissel, M., Deb, D., Fathony, M., Lin, Y., Mollah, A., Ngo, N., et al., 2002. Generic values for soil-to-plant transfer factors of radiocesium. J. Environ. Radioact. 58, 113–128.

Garcia-Sanchez, L., Konoplev, A.V., 2009. Watershed wash-off of atmospherically deposited radionuclides: a review of normalized entrainment coefficients. J. Environ. Radioact. 100, 774–778.

Garland, J.A., Pattenden, N.J., Playford, K., 1992. Resuspension following Chernobyl. In: International Atomic Energy Agency, (Ed.), Modelling of Resuspension, Seasonality, and Losses During Food Processing. IAEA, Vienna IAEA-TECDOC-647.

Giles, M.S., Twining, J.R., Williams, A.K., Jeffree, R.A., Domel, R.U., 1990. Study No. 2 (Radioecology). Report to the Technical Assessment Group for the Maralinga Rehabilitation Project. In: Technical Assessment Group Study Program Reports, vol. 1. Aust. Gov. Pub. Serv, Canberra, p. 62.

Garnier-Laplace, J., Gilek, M., Sundell-Bergman, S., Larsson, C.-M., 2004. Assessing ecological effects of radionuclides: data gaps and extrapolation issues. J. Radiol. Prot. 24, A139–A155.

Garnier-Laplace, J., Della-Vedova, C., Gilbin, R., Copplestone, D., Hingston, J., Ciffroy, P., 2006. First derivation of predicted-no-effect values for freshwater and terrestrial ecosystems exposed to radioactive substances. Environ. Sci. Technol. 40 (20), 6498–6505.

Hakanson, L., 1997. Transport and processes in freshwater ecosystems. In: Health Impacts of Large Releases of Radionuclides. John Wiley & Sons, Chichester, pp. 46–67, Ciba Foundation Symposium 203.

Haywood, S.M., Smith, J., 1997. Assessment of the potential radiological impact of residual contamination in the Maralinga and Emu areas. National Radiological Protection Board report NRPB-R237, Oxford, 1997.

Heinemann, K., Vogt, K.J., 1980. Measurements of the deposition of iodine onto vegetation and the biological half-life of iodine on vegetation. Health Phys. 39, 463–474.

Hess, G.D., McDonald, N.R., Manins, P.C., Cameron, R.F., Clark, G., Crawford, J., et al., 1998. A Study of the Environmental Impact on Australia of a Nuclear Accident in Indonesia. Department of Foreign Affairs, Australia.

Hingston, J., Andersson, P., Beresford, N., Brown, J., Copplestone, D., Garnier-Laplace, J., et al., 2007. A review of approaches to protection of the environment from chemicals and ionising radiation: requirements and recommendations for a common framework. PROTECT Deliverable 3European Commission. (Contract number 036425 (F16R))www.ceh.ac.uk/PROTECT/.

Hoe, S., Müller, H., Gering, F., Thykier-Nielsen, S., Sørensen, J.H., 2002. ARGOS 2001: A decision support system for nuclear emergencies. In: Proceedings of the Radiation Protection and Shielding Division Topical Meeting April 14–17, 2002, Santa Fe, New Mexico, USA. (2002).

Hoffman, F.O., Thiessen, K.M., Frank, M.L., Blaylock, B.G., 1992. Quantification of the interception and initial retention of radioactive contaminants deposited on pasture grass by simulated rain. Atmos. Environ. 26 A, 3313–3321.

Hötzl, H., Rosner, G., Winkler, R., 1987. Ground depositions and air concentrations of Chernobyl fallout radionuclides at Munich-Neuherberg. Radiochimica Acta 41, 181–190.

Howard, B.J., et al., 2004. A Critical Evaluation of the Strategy Project. Rad. Prot. Dosim. 109, 63–67.

Howard, B.J., et al., 2005. The strategy project: decision tools to aid sustainable restoration and long−term management of contaminated agricultural ecosystems, J. Env. Radioact. 83 275–295.

Howard, B.J., Larsson, C.-M., 2008. The ERICA integrated approach and its contribution to protection of the environment from ionising radiation. In: Howard, B., Larsson, C., (Eds.), The ERICA Project, September, 2008. J. Environ. Radioact. 99 (9), 1361–1363.

Howard, B.J., Beresford, N.A., Copplestone, D., Andersson, P., Garnier-Laplace, J., Brown, J.E., et al., 2008. Final Activity report for the PROTECT project, EC Contract Number: 036425 (FI6R). CEH-Lancaster, Lancaster.

Howard, B.J., Beresford, N.A., Barnett, C.L., Fesenko, S., 2009. Quantifying the transfer of radionuclides to food products from domestic farm animals. J. Environ. Radioact. 100, 737–773.

IAEA, 1982. Generic models and parameters for assessing the environmental transfer of radionuclides from routine releases. IAEA Safety Series No. 57 Procedures and DataExposures of Critical Groups, Vienna 1982.

IAEA, 1989. Measurement of radionuclides in food and the environment. A guidebook. IAEA Technical Report Series No. 295. Vienna, 1989.

IAEA, 1994. Handbook of parameter values for the prediction of radionuclide transfer in temperate environments. IAEA Technical Report Series No. 364. Vienna, 1994.

IAEA, 2001. Generic models for use in assessing the impact of discharges of radioactive substances to the environment. International Atomic Energy Agency, Vienna Safety Report Series No. 19.

IAEA, 2004. Sediment distribution coefficients and concentration factors for radionuclides in the marine environment. IAEA Technical Report Series No. 422. Vienna, 2004. (Replaces IAEA Technical Report Series No 247).

IAEA-BIOMASS-4, 2003. Testing of environmental transfer models using Chernobyl fallout data from the Iput River catchment area. April Bryansk Region, Russian Federation.

IAEA, 2009. Quantification of radionuclide transfer in terrestrial and freshwater environments for radiological assessments. IAEA TECDOC-1616, Vienna 2009.

ICRP, 1991. 1990 Recommendations of the International Commission on Radiological Protection. Ann. ICRP ICRP Publication 60. Pergamon Press, Oxford.

ICRP, 2002. Basic anatomical and physiological data for use in radiological protection: Reference values. Ann. ICRP ICRP Publication 89. Pergamon Press, Oxford.

ICRP, 2008. Environmental Protection - the Concept and Use of Reference Animals and Plants. Ann. ICRP 38 (4–6) 2008. ICRP Publication 108. Pergamon Press, Oxford.

ICRU, 2001. International Commission on Radiation Units and Measurements Report 65. Quantities, units and terms in radioecology. Journal of the ICRU 1 (2), 48.

Jacob, P., 1989. External exposures from radionuclides deposited in rural and urban environments. In: Proceedings of the 15th Regional Congress of IRPA. 10.-14.9. Visby (Schweden).

Jacob, P., Rosenbaum, H., Petoussi, N., Zankl, M., 1990. Calculation of organ doses from environmental gamma rays using human phantoms and Monte Carlo methods. Part II: Radionuclides distributed in the air or deposited on the ground. Gesellschaft für Strahlen- und Umweltforschung (GSF), Neuherberg GSF-Bericht 12/90.

Jeffree, R.A., 1990. Radium uptake by freshwater invertebrates. Part 5.4. The Environmental Behaviour of Radium. IAEA Technical Report Series No. 310, vol. 1. Vienna, pp. 509–528.

Jibiri, N.N., Farai, I.P., Alausa, S.K., 2007. Estimation of annual effective dose due to natural radioactive elements in ingestion of foodstuffs in tin mining area of Jos-Plateau, Nigeria. J. Environ. Radioact. 94, 31–40.

Jonas, R., 1984. Ablagerung und Bindung von Luftverunreinigungen an Vegetation und anderen atmosphärischen Grenzflächen. Bericht Jül-1949. Kernforschungsanlage Jülich.

Jones, C., Gilek, M., 2004. Overview of programmes for the assessment of risks to the environment from ionising radiation and hazardous chemicals. J. Radiol. Prot. 24, A157–A177.

Kapetsky, J.M., 1994. A strategic assessment of warm water fish farming potential in Africa. CIFA Technical Paper. No. 27FAO, Rome 67 p.

Kathren, R.L., 1984. Radiooactivity in the Environment. Sources, Distribution, and Surveillance. Harwood Academic Publishers, New York.

Kefford, B.J., Palmer, C.G., Jooste, S., Warne, M.St.J., Nugegoda, D., 2005. What is meant by "95% of Species"? An argument for the inclusion of rapid tolerance testing. Hum. Ecol. Risk Assess. 11 (5), 1025–1046.

Kellman, M., Tackaberry, R., 1997. Tropical Environments. Routledge, London.

Kelly, M.S., Pinder III, J.E., 1996. Foliar uptake of ^{137}Cs from the water column by aquatic macrophytes. J. Environ. Radioact. 30 (3), 271–280.

Kinnersley, R.P., Goddard, A.J.H., Minski, M.J., Shaw, G., 1997. Interception of caesium-contaminated rain by vegetation. Atmos. Environ. 31, 1137–1145.

Kirchner, G., 1994. Transport of iodine and caesium via the grass-cow-milk pathway after the Chernobyl accident. Health Phys. 66, 653–665.

Krajewski, P., 1994. Individual evaluation of model performance for Scenario S. 1993/1994 IAEA Exercise in Validation of Model Predictions (VAMP), Central Laboratory for Radiological Protection, Poland (CLRP).

Martin, P., Hancock, G.J., Johnston, A., Murray, A.S., 1998. Natural-series radionuclides in traditional north Australian Aboriginal foods. J. Environ. Radioact. 40 (1), 37–58.

Meckbach, R., Jacob, P., 1988. Gamma exposures due to radionuclides deposited in urban environments. Part II: Location factors for different deposition patterns. Radiat. Prot. Dosimetry 25, 181–190.

Meckbach, R., Jacob, P., Paretzke, H.G., 1988. Gamma exposures due to radionuclides deposited in urban environments. Part I: Kerma rates from contaminated urban surfaces. Radiat. Prot. Dosimetry 25, 167–180.

Miller, C.W., Hoffman, F.O., 1983. An examination of environmental half-time for radionuclides deposited on vegetation. Health Phys. 24, 731–744.

Monte, L., Boyer, P., Brittain, J.E., Goutal, N., Heling, R., Kryshev, A., et al., 2008. Testing models for predicting the behaviour of radionuclides in aquatic systems. Appl. Radiat. Isot. 66, 1736–1740.

Müller, H., Pröhl, G., 1993. ECOSYS-87: A dynamic model for assessing radiological consequences of nuclear accidents. Health Phys. 64, 232–252.

OECD/NEA, 2003. Radiological Protection of the Environment: Summary Report of the Issues. Summary Report of the NEA Forum on "Radiological Protection of the Environment: The Path Forward to a New Policy?" held in collaboration with the International Commission on Radiological Protection (ICRP) at Taormina, Sicily, Italy 12-14 February 2002Nuclear Energy Agency / Organisation for Economic Co-Operation and Development, Paris.

Poon, C.B., Au, S.M., 2003. Predicting the ^{137}Cs ingestion dose from marine fish consumption in Hong Kong after an accidental release. J. Radiol. Prot. 23, 97–104.

Poon, C.B., Au, S.M., Pröhl, G., Müller, H., 1997. Adaptation of Ecosys-87 to Hong Kong environmental conditions. Health Phys. 72, 856–864.

Porntepkasemsan, B., Srisuksawad, K., 2008. Assessment of ^{226}Ra age-dependent dose from water intake. Appl. Radiat. Isot. 66 (11), 1654–1656 November, 2008.

Posthuma, L., Suter, G.W., Traas, T.P., 2001. Species Sensitivity Distributions in Ecotoxicology. Lewis Publishers, Boca Raton, FL, USA.

Povinec, P., Woodhead, D., Blowers, P., Bonfield, R., Cooper, M., Chen, Q., et al., 1999. Marine radioactivity assessment of Mururoa and Fangataufa atolls. Sci. Total Environ. 237–238, 249–267.

Pröhl, G., 2009. Interception of dry and wet deposited radionuclides by vegetation. J. Environ. Radioact. 100, 675–682.

RAC, 2002. Analysis of exposure and risks to the public from radionuclides and chemicals released by the Cerro Grande fire at Los Alamos. Summary report submitted by the Risk Assessment Corporation to the New Mexico Environment Department. 12 June 2002, 16 p.

Radhakrishna, A.P., Somashekarappa, H.M., Narayana, Y., Siddappa, K., 1996. Distribution of some natural and artificial radionuclides in Mangalore environment of south India. J. Environ. Radioact. 30, 31–54.

Rahman, M.M., Voigt, G., 2004. Radiocaesium soil-to-plant transfer in tropical environments. J. Environ. Radioact. 71, 127–138.

Rahman, M.M., Koddus, A., Ahmad, G.U., Voigt, G., 2005. Soil-to-plant transfer of radiocaesium for selected tropical plant species in Bangladesh. J. Environ. Radioact. 83, 199–211.

Raskob, W., Ehrhardt, J., 2007. Status of the RODOS system for off-site emergency management after nuclear and radiological accidents and its enhancement under the EURANOS project. International Journal of Nuclear Governance. Economy and Ecology 1, 363–371.

Robison, W.L., Conrado, C.L., Hamilton, T.F., 1997. A comparative study on the ^{137}Cs transfer from soil to vegetation in the Marshall Islands. Lawrence Livermore National Laboratory Report UCRL-JC-128490, 1997.

Rochedo, E.R.R., 2005. The role of climate and other site specific aspects in environmental impact assessments after a nuclear accident. Int. Conf. on: Monitoring, Assessments, and Uncertainties for Nuclear and Radiological Emergency Response. Rio de Janeiro, 21–25 November 2005.

Rochedo, E.R.R., Wasserman, M.A., Pires Do Rio, M.A., 2001. The relevance of site specific data on environmental impact assessment. Proceedings of the V Regional Congress on Radiation Protection and Safety. Recife, PE, Maio - em CD-ROM, 2001.

Roed, J., 1987. Dry deposition in rural and urban areas in Denmark. Radiat. Prot. Dosimetry 21, 33–36.

Salinas, I.C.P., Conti, C.C., Lopes, R.T., 2006a. Effective density and mass attenuation coefficient for building material in Brazil. Appl. Radiat. Isot. 64, 13–18.

Salinas, I.C.P., Conti, C.C., Rochedo, E.E.R., Lopes, R.T., 2006b. Gamma shielding factor for typical houses in Brazil. Radiat. Prot. Dosimetry 121, 420–424.

Santos, E.E., Lauria, D.C., Amaral, E.C.S., Rochedo, E.R., 2002. Daily ingestion of ^{232}Th, ^{238}U, ^{226}Ra, ^{228}Ra and ^{210}Pb in vegetables by inhabitants of Rio de Janeiro City. J. Environ. Radioact. 62, 75–86.

Sauerland, C., Martin, P., Humphrey, C., 2005. Radium-226 in Magela Creek, northern Australia: Application of protection limits from radiation for humans and biota. Radioprotection 40 (Suppl. 1), S451–S456.

Scheffer, F., Schachtschabel, P., 2002. Lehrbuch der Bodenkunde. Spektrum Akademischer Verlag, Stuttgart.

SENES, 2007. Overview of representative ecological risk assessments conducted for sites with enhanced radioactivity. Prepared for the World Nuclear Association, SENES Consultants Limited, Richmond Hill, Ontario.

Shahul Hameed, P., Shaheed, K., Somasundaram, S.S.N., Iyengar, M.A.R., 1997. Bioaccumulation of ^{210}Pb in the Kaveri River ecosystem. India. J. Environ. Radioact. 37 (1), 17–27.

Shinonaga, T., Pröhl, H., Müller, H., Ambe, S., 1999. Experimentally determined mobility of trace elements in soybean plants. Sci. Total Environ. 225, 241–248.

Shinonaga, T., Schimmack, W., Gerzabek, M., 2005. Vertical migration of ^{60}Co, ^{137}Cs and ^{226}Ra in agricultural soils as observed in lysimeters under crop rotation. J. Environ. Radioact. 79, 93–106.

Smith, J., 2003. AQUASCOPE Version 1.0. Documentation. Simplified models for predicting $^{89,90}Sr$, ^{131}I, $^{134,137}Cs$ in surface water and freshwater fish. Centre of Ecology & Hydrology, Dorset.

Smith, K.E.C., Jones, K.C., 2000. Particles and vegetation: implications for the transfer of particle-bound organic contaminants to vegetation. Sci. Total Environ. 246, 207–236.

Smith, E.G., Leung, L., Phaloeun, C., Sopheap, U., 2002. Rice field fish farming integrated with rodent pest management in Cambodia. Cambodian Journal of Agriculture 6, 9–14.

Smith, J.T., Voitsekhovitch, O.V., Håkanson, L., Hilton, J., 2001. A critical review of measures to reduce radioactive doses from drinking water and consumption of freshwater foodstuffs. J. Environ. Radioact. 56, 11–32.

Suter, G.W., 2007. Ecological risk assessment, second ed. CRC Press, Boca Raton.

TAG (Technical Assessment Group), 1990. Rehabilitation of former nuclear test sites in Australia. Report by the Technical Assessment Group, Australian Department of Primary Industries and EnergyAustralian Govt. Pub. Service, Canberra 1990.

Till, J.E., Meyer, H.R., 1983. Radiological Assessment: A textbook on Environmental Dose Analysis. NUREG/CR-3332US Nuclear Regulatory Commission, Washington, D.C September, 1983.

Tsukada, H., Nakamura, Y., 1998. Transfer factors of 31 elements in several agricultural plants collected from 150 farm fields in Aomori, Japan. J. Radioanal. Nucl. Chem. 236, 123–132.

Twining, J.R., 1989. Principal coordinate analysis of the distribution of radium-226 between water, sediment and the waterlily *Nymphaea violacea* (Lehm), in the vicinity of a uranium mine in the Northern Territory, Australia. J. Environ. Radioact. 10, 99–113.

Twining, J.R., Ferris, J.M., Markich, S.J., 1996. Bioaccumulation of ^{85}Sr and ^{137}Cs by an Australian subtropical freshwater teleost (*Bidyanus bidyanus*). Sci. Total Environ. 192, 245–257.

Twining, J., Payne, T., Itakura, T., 2004. Soil–water distribution coefficients and plant transfer factors for ^{134}Cs, ^{85}Sr, and ^{65}Zn under field conditions in tropical Australia. J. Environ. Radioact. 71, 71–87.

US NRC, NUREG, 1977. Regulatory Guide 1.109. Calculation of Annual Doses to Man from Routine Releases of Reactor Effluents for the Purpose of Evaluating Compliance with 10 CFR Part 50, Appendix I.

Walker, R., Gessel, S., Held, E., 1997. The ecosystem study on Rongelap Atoll. Health Phys. 73, 223–233.

Wang, C.-J., Lai, S.-Y., Wang, J.-J., et al., 1997. Transfer of radionuclides from soil to grass in Northern Taiwan. Appl. Radiat. Isot. 48, 301–303.

Wasserman, M.A., Bartoly, F., Viana, A.G., Silva, M.M., Rochedo, E.R.R., Perez, D.V., et al., 2008. Soil to plant transfer of ^{137}Cs and ^{60}Co in Ferralsol, Nitisol, and Acrisol. J. Environ. Radioact. 99, 546–553.

Wasserman, M.A., Viana, A.G., Bartoly, F., Vidal Perez, D., Rochedo, E.R., Wasserman, J.C., et al., 2005. Bio-geochemical behaviour of ^{90}Sr and ^{137}Cs in tropical soil. Radioprotection 40 (Suppl. 1), S135–S142.

Whicker, R., Ibrahim, S., 2006. Vertical migration of ^{134}Cs bearing soil particles in arid soils: Implications for plutonium redistribution. J. Environ. Radioact. 88, 171–188.

Whicker, W., Kirchner, T.B., 1987. Pathway: A Dynamic food-chain model to predict radionuclide ingestion after fallout deposition. Health Phys. 52, 717–737.

Whicker, F.W., Pinder III, J.E., Bowling, J.W., Alberts, J.J., Brisbin Jnr, I.L., 1990a. The distribution of long-lived radionuclides in an abandoned reactor cooling reservoir. Ecol. Monographs 60, 471–496.

Whicker, W., Kirchner, T.B., Breshears, D.D., Otis, M., 1990b. Estimation of radionuclide ingestion: the "Pathway" food-chain model. Health Phys. 59, 645–657.

Woodhead, D.S., Zinger, I., 2003. Radiation effects on plants and animals. In: Woodhead, D.S., Zinger, I. (Eds.), Framework for Assessment of Environmental Impact (FASSET) deliverable 4, 196. http://www.erica-project.org/.

Index

Note: Page numbers followed by "*f*" indicate figures, and "*t*" indicate tables.

D

E

H

I

J

K

L

M

Q

R

V

W

X

Y

www.ingramcontent.com/pod-product-compliance
Ingram Content Group UK Ltd.
Pitfield, Milton Keynes, MK11 3LW, UK
UKHW020127290926
15471UKWH00003B/336

9 780080 975108